AF389615

Industrial Chemistry Reaction: Kinetics, Mass Transfer and Industrial Reactor Design (II)

Industrial Chemistry Reaction: Kinetics, Mass Transfer and Industrial Reactor Design (II)

Editors

Elio Santacesaria
Riccardo Tesser
Vincenzo Russo

MDPI • Basel • Beijing • Wuhan • Barcelona • Belgrade • Manchester • Tokyo • Cluj • Tianjin

Editors
Elio Santacesaria
Eurochem Engineering LtD
Milan
Italy

Riccardo Tesser
University of Naples
Federico II
Napoli
Italy

Vincenzo Russo
University of Naples
Federico II
Napoli
Italy

Editorial Office
MDPI
St. Alban-Anlage 66
4052 Basel, Switzerland

This is a reprint of articles from the Special Issue published online in the open access journal *Processes* (ISSN 2227-9717) (available at: https://www.mdpi.com/journal/processes/special_issues/Kinetics_Mass_ransfer_Industrial_Reactor).

For citation purposes, cite each article independently as indicated on the article page online and as indicated below:

LastName, A.A.; LastName, B.B.; LastName, C.C. Article Title. *Journal Name* **Year**, *Volume Number*, Page Range.

ISBN 978-3-0365-8194-1 (Hbk)
ISBN 978-3-0365-8195-8 (PDF)

Contents

About the Editors

Elio Santacesaria

Elio Santacesaria obtained a chemistry degree at the Pavia University in 1966, with a thesis developed under the guidance of the Nobel prize winner, Giulio Natta. He worked as researcher at the Polytechnic of Milano from 1969 to 1986, conducting research in the fields of: chemistry of propellants, catalysis, kinetics, reactors design and simulation, and separation science. In 1986, he became the chair of Industrial Chemistry at the University of Naples "FEDERICO II". He continued research in almost all the aforementioned areas. Santacesaria had been a National Coordinator of research on "Colloid, Interphase and Surfactants" and founded the Italian Group of Colloid and Interphase Science (GICI) of the Italian Chemical Society. He has been the Coordinator of this group for more than ten years (1985–1995). He has been the organizer of many national and international congresses, workshops, and post-doctoral schools. He was elected President of the Industrial Chemistry Division of the Italian Chemical Society for 2004–2009. He has been the Dean of the Industrial Chemistry Degree Course from 1994 until 2009. He has collaborated with many Italian and foreign chemical industries. He has published more than 260 papers on qualified journals and 30 Patents. In 2011, he was awarded the "Emanuele Paternò" Gold Medal by the Italian Chemical Society. In November 2012, he retired from the university and became responsible for the Company Eurochem Engineering Ltd located in Milano. In September 2016, he was awarded the Giacomo Fauser Silver Plate entitled to by the Italian Group of Catalysis (GIC) of Italian Chemical Society (SCI) for a related conference.

Riccardo Tesser

Riccardo Tesser received an Industrial Chemistry master's degree from the University of Napoli in 1989. He has ten years of industrial experience working in the Montefibre Research Centre (NA), conducting research activities and developing technological improvements in dimethyl terephthalate production and batch and continuous PET polymerization. He achieved an Associate Professor position in the Chemical Sciences Department of the University of Naples "Federico II" in Industrial Chemistry. His main fields of research in industrial chemistry are industrial catalysis and kinetic studies, reaction and reactor modeling of multiphase systems, kinetic investigation of catalytic reactions, optimization algorithms in chemical engineering and the design and realization of lab-scale pilot plants. More recently, his main field of activity has been in biofuel production and the use of renewable resources as raw materials for the chemical industry. The results of these scientific activities have been published in 164 papers in a wide range of international, peer-reviewed chemistry journals and three books, and he has 11 international patents.

Vincenzo Russo

Vincenzo Russo obtained a PhD in Chemical Sciences at the University of Napoli Federico II (Italy) in 2014 and a PhD in Chemical Engineering at Åbo Akademi (Finland). He has been an Associate Professor in the Department of Chemical Sciences of the University of Naples Federico II since November 2019. His research activity can be summarized as "Bridging the Fundamentals of Molecular Phenomena with Modern Concepts of Chemical Reactor Engineering". He approaches his research topics with sound methodology, starting with the investigation of the chemical and physical phenomena occurring in the reaction network. He carries out detailed kinetic investigations, quantitatively describing the experimental data in lab-scale reactors, both continuous and discontinuous. He bases kinetic models on the investigation of both the physical properties of the catalysts tested (homogeneous or heterogeneous) and the mass and heat transfer phenomena involved in the reaction network. Studying advanced reactor modeling activity has led him to the design and optimization of pilot-scale chemical plants, allowing the scale-up of the overall process. Finally, he tests and designs novel reactor concepts with the aim of optimizing and intensifying the chemical plant (i.e., reactive chromatography and microreactors). The results of these scientific activities have been published in 130 papers in a wide range of international, peer-reviewed chemistry journals, three books and four book chapters.

Editorial

Industrial Chemistry Reactions: Kinetics, Mass Transfer and Industrial Reactor Design (II)

Elio Santacesaria [1,*], **Riccardo Tesser** [2] and **Vincenzo Russo** [2]

[1] Eurochem Engineering Ltd., 20139 Milano, Italy
[2] NICL—Department of Chemical Science, University of Naples Federico II, 80126 Naples, Italy; riccardo.tesser@unina.it (R.T.); v.russo@unina.it (V.R.)
* Correspondence: elio.santacesaria@eurochemengineering.com

1. Introduction

Due to the success of the first edition of the Special Issue "Industrial Chemistry Reactions: Kinetics, Mass Transfer and Industrial Reactor Design" in terms of both the quantity and quality of the published papers, we thought it would be prudent to announce a second edition.

Recently conceived, powerful calculation methods have enabled the development of more sophisticated approaches to catalysis, kinetics, reactor design, and simulation. It is well known that many chemical reactions of a large scale and of great interest for industrial processes require information related to topics ranging from thermodynamics and kinetics to transport phenomena related to mass, energy, and momentum. For reliable industrial-scale reactor design, all these pieces of information must be incorporated into appropriate equations and mathematical models that facilitate accurate and reliable simulations for scale-up purposes. One challenge in this regard is to collect, in a memorable volume, the main advances and trends in the field of industrial chemistry that have been brought about by the contributions of various champions of scientific and technological progress by reviewing their past activity in the field or providing, through original manuscripts, examples of modern approaches to the investigation of industrial chemical reactions.

Therefore, the aim of this Special Issue is to collect worldwide contributions from experts in the field of industrial reactor design pertaining to kinetic and mass-transfer studies. The call for reviews and original papers solicited research in the following areas:

- Kinetic studies of complex reaction schemes (multiphase systems);
- Kinetics and mass transfer in multifunctional reactors;
- Reactions in mass-transfer-dominated regimes (e.g., fluid–solid and intraparticle diffusive limitations);
- Kinetics and mass transfer modeling with alternative approaches (e.g., stochastic modeling);
- Pilot plant and industrial-size reactor simulation and scale-ups based on kinetic studies (e.g., the lab-to-plant approach).

This Special Issue collected eleven papers that could be framed within different macro areas and whose distribution is depicted in Figure 1.

As depicted, the distribution is well-balanced among the five different macro areas, namely,

(i)   kinetics for complex reaction schemes;
(ii) multifunctional reactors;
(iii) reactions in mass-transfer-dominated regimes;
(iv) modeling with alternative approaches;
(v)   scale-ups.

Citation: Santacesaria, E.; Tesser, R.; Russo, V. Industrial Chemistry Reactions: Kinetics, Mass Transfer and Industrial Reactor Design (II). *Processes* **2023**, *11*, 1880. https://doi.org/10.3390/pr11071880

Received: 20 June 2023
Accepted: 21 June 2023
Published: 22 June 2023

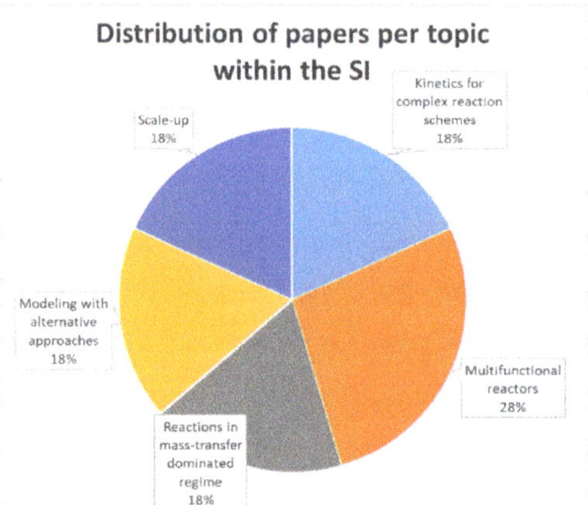

Figure 1. Statistical information related to the Special Issue, "Industrial Chemistry Reactions: Kinetics, Mass Transfer and Industrial Reactor Design (II)": distribution of papers per macro area.

In the following sections, the main achievements published in this SI will be reviewed and summarized, highlighting the points of novelty with respect to each macro area.

2. Kinetics for Complex Reaction Schemes

Two articles were published in the field of kinetics for complex reaction schemes. Santacesaria et al. [1] reviewed and analyzed the possibility of using ethanol as a liquid organic hydrogen carrier (LOHC), analyzing in detail both the main and side reactions (see Figure 2A for the latter), finding that the main rate expressions can be strongly dependent on the adsorption phenomena with regard to the surface reaction mechanisms.

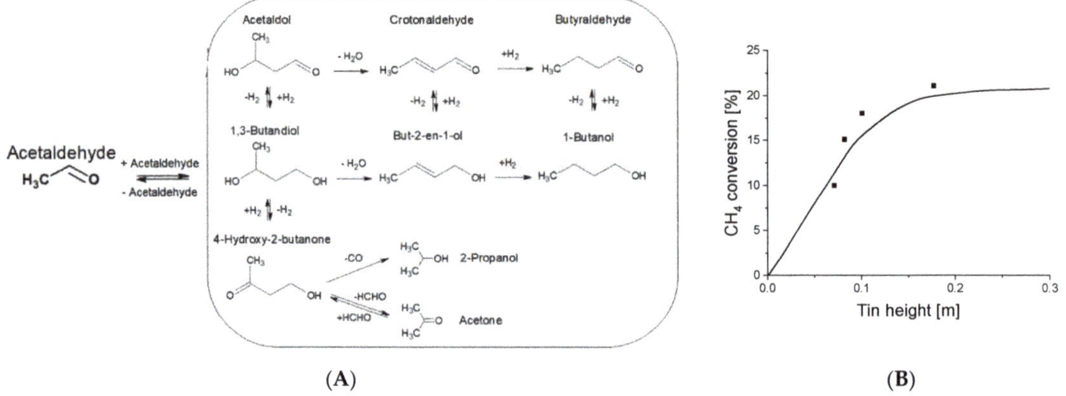

(A)   **(B)**

Figure 2. (**A**) Detailed reaction scheme of acetaldehyde condensation [1]. (**B**) CH_4 conversion versus reactors' axial coordinates for 6.0 cm ID reactor [2].

Palo et al. proposed a kinetic model of thermal methane cracking on molten metal [2] that correctly described the experimental data collected using a pilot plant even when using simplified ideal packed bed reactor models (see Figure 2B).

3. Multifunctional Reactors

Three papers were published in the field of multifunctional reactor technologies. Alqahtani et al. [3] explored the possibility of using a membrane reactor for hydrogen production, specifically with respect to its modeling and simulation. Figure 3A shows the hydrogen concentration profiles along the residence time, illustrating that the model can also predict instability regions for the operation, making it a valuable tool in reactor design.

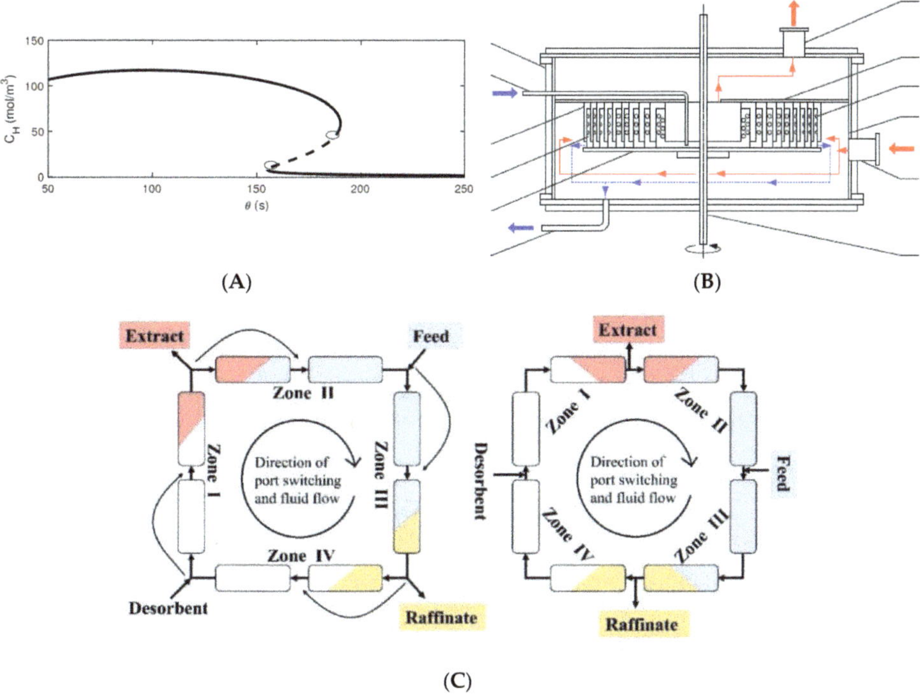

(A)

(B)

(C)

Figure 3. (**A**) Variations in hydrogen concentration with residence time in a membrane reactor [3]. (**B**) Diagram of a rotating zigzag bed [4]. (**C**) Simulated moving bed reactor [5].

Liu et al. proposed a rotating zigzag bed reactor for a modeling and experimental study on CO_2 absorption using NaOH solution [4], adopting the experimental apparatus sketched in Figure 3B. The authors managed to measure and simulate the gas–liquid mass transfer interfacial parameters with high precision.

Finally, Zhang et al. reviewed both the classical and the most recent technologies in relation to simulated moving bed reactors (see Figure 3C) [5], reviewing the possibility of using such technologies in specific applications.

4. Reactions in Mass-Transfer-Dominated Regimes

Two articles were published in the field of chemical reactions in the presence of mass-transfer limitations. Hua et al. published a paper on a Multiphysics numerical simulation model related to an argon-stirred ladle [6]. As revealed in Figure 4A, the model accounts for the mass transfer aspects involved in the overall reaction network, paying particular attention to the diffusion of the chemical components between the involved phases. The results were promising and in line with the collected experimental data.

Tao et al. explored the hydrodynamic performance of a countercurrent total spray tray under sloshing conditions [7]. The influence of the main experimental conditions, including gas velocity and rolling-induced sweeping, on pressure drops, entrainment, and plate fluctuations was determined, thereby retrieving information for a better physical understanding of the system under investigation (see scheme reported in Figure 4B).

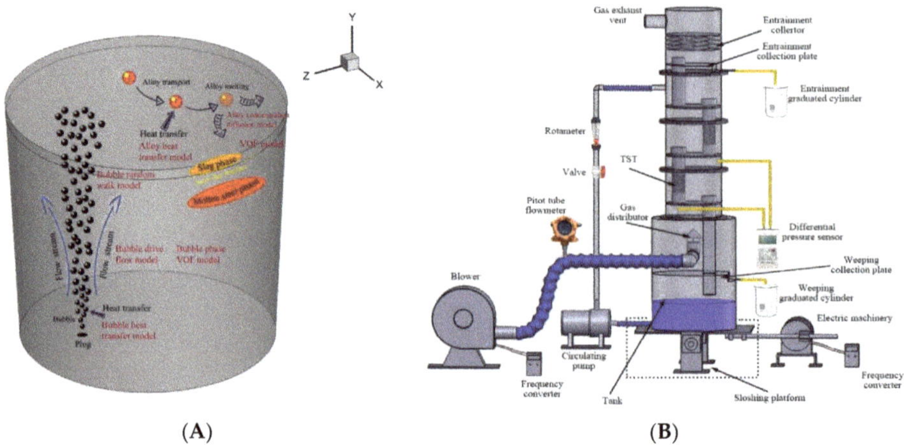

(A)   (B)

Figure 4. (**A**) Diagram of the simulation model applied to the argon-stirred ladle [6]. (**B**) Experimental setup used by Tao et al. to investigate the hydrodynamic performance of a countercurrent total spray tray under sloshing conditions [7].

5. Modeling Using Alternative Approaches

The research groups of Huang and Lan published two articles on the topic of modeling using alternative approaches. Huang et al. conducted a model-based analysis of ethylene carbonate hydrogenation in industrial tubular reactors [8], comparing the performances of different reactors operating in adiabatic or heat-exchanged modalities (see Figure 5A). They identified the optimal operation-condition-related windows within which to optimize both the conversion and selectivity of the process.

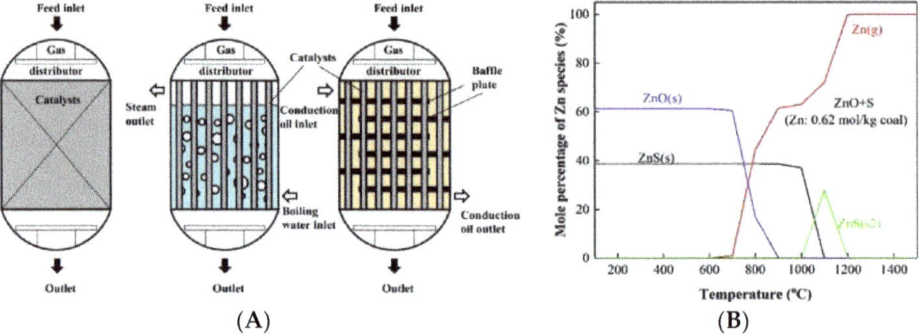

(A)   (B)

Figure 5. (**A**) Different industrial tubular reactors modeled in the work conducted by Huang et al. [8]. (**B**) Thermodynamic equilibrium distribution distributions of Zn when S and Zn coexist in the coal [9].

Lan et al. investigated the effects of S and mineral elements (Ca, Al, Si, and Fe) on the thermochemical behavior of Zn during the co-pyrolysis of coal and waste tires [9]. The authors conducted an in-depth experimental and thermodynamic analysis to model the mentioned system, allowing them to predict the chemical composition of the mixtures obtained at different temperatures after thermal treatment (Figure 5B). Undoubtedly, this study yielded information required for the further scale-up of the operation.

6. Scale-Up

Finally, two articles were devoted to the scale-up process. Wang et al. studied the scaling-up of the catalytic decomposition of chlorinated hydrocarbons [10]. A dedicated experiment (see device in Figure 6A) was conducted to prove the concept, analyzing the quality of the products obtained for the future scaling-up of the process.

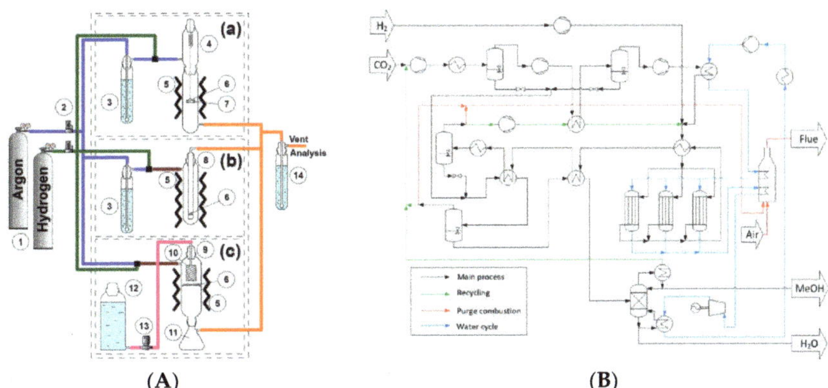

Figure 6. (**A**) Experimental setup of the catalytic decomposition of chlorinated hydrocarbons: (**a–c**), reactors (**b**); 1 gas cylinders; 2 flow mass controllers; 3 saturator; 4 quartz reactor; 5 heating zone; 6 catalyst sample; 7 foamed quartz basket; 8 lab flowthrough quartz reactor; 9 scaled-up quartz reactor; 10 evaporator; 11 carbon collector; 12 vessel; 13 pump; 14 alkali trap. [10]. (**B**) One-step process flowsheet [11].

Lacerda de Oliveira Campos et al. proposed a process and techno-economic analysis of methanol synthesis from hydrogen and carbon dioxide with intermediate condensation steps [11]. The authors proposed a detailed flowsheet (see Figure 6B) that was designed to optimize the operation conditions of the whole plant.

7. Conclusions

Due to the success of the first edition, the second edition of this Special Issue on Industrial Chemistry Reactions: Kinetics, Mass Transfer, and Industrial Reactor Design (II) was launched and received significant attention. Part of the success of this second edition was due to the great variety of options that can be obtained when chemical reactions, catalysis, kinetics, transport phenomena, and multiphase systems are considered. The high quality of the collected papers, together with the highly interdisciplinary approach, allowed for the production of relevant papers in this sector, which will surely serve as a reference for future research.

Funding: This research received no external funding.

Conflicts of Interest: The authors declare no conflict of interest.

References

1. Santacesaria, E.; Tesser, R.; Fulignati, S.; Galletti, A.M.R. The Perspective of Using the System Ethanol-Ethyl Acetate in a Liquid Organic Hydrogen Carrier (LOHC) Cycle. *Processes* **2023**, *11*, 785. [CrossRef]
2. Palo, E.; Cosentino, V.; Iaquaniello, G.; Piemonte, V.; Busillo, E. Thermal Methane Cracking on Molten Metal: Kinetics Modeling for Pilot Reactor Design. *Processes* **2023**, *11*, 1537. [CrossRef]
3. Alqahtani, R.T.; Ajbar, A.; Bhowmik, S.K.; Alghamdi, R.A. Study of Static and Dynamic Behavior of a Membrane Reactor for Hydrogen Production. *Processes* **2021**, *9*, 2275. [CrossRef]
4. Liu, Z.; Esmaeili, A.; Zhang, H.; Wang, D.; Lu, Y.; Shao, L. Modeling and Experimental Studies on Carbon Dioxide Absorption with Sodium Hydroxide Solution in a Rotating Zigzag Bed. *Processes* **2022**, *10*, 614. [CrossRef]
5. Zhang, X.; Liu, J.; Ray, A.K.; Li, Y.; Tesser, R.; Russo, V.; Zhang, X.; Liu, J.; Ray, A.K.; Li, Y. Research Progress on the Typical Variants of Simulated Moving Bed: From the Established Processes to the Advanced Technologies. *Processes* **2023**, *11*, 508. [CrossRef]

6. Hua, C.; Bao, Y.; Wang, M. Multiphysics Numerical Simulation Model and Hydraulic Model Experiments in the Argon-Stirred Ladle. *Processes* **2022**, *10*, 1563. [CrossRef]
7. Tao, J.; Zhang, G.; Yao, J.; Wang, L.; Wei, F.; Tesser, R.; Russo, V.; Tao, J.; Zhang, G.; Yao, J.; et al. Study on the Hydrodynamic Performance of a Countercurrent Total Spray Tray under Sloshing Conditions. *Processes* **2023**, *11*, 355. [CrossRef]
8. Huang, H.; Cao, C.; Wang, Y.; Yang, Y.; Lv, J.; Xu, J. Model-Based Analysis for Ethylene Carbonate Hydrogenation Operation in Industrial-Type Tubular Reactors. *Processes* **2022**, *10*, 688. [CrossRef]
9. Lan, Y.; Jin, S.; Wang, J.; Wang, X.; Zhang, R.; Ling, L.; Jin, M. Effects of S and Mineral Elements (Ca, Al, Si and Fe) on Thermochemical Behaviors of Zn during Co-Pyrolysis of Coal and Waste Tire: A Combined Experimental and Thermodynamic Simulation Study. *Processes* **2022**, *10*, 1635. [CrossRef]
10. Wang, C.; Bauman, Y.I.; Mishakov, I.V.; Stoyanovskii, V.O.; Shelepova, E.V.; Vedyagin, A.A. Scaling up the Process of Catalytic Decomposition of Chlorinated Hydrocarbons with the Formation of Carbon Nanostructures. *Processes* **2022**, *10*, 506. [CrossRef]
11. Lacerda de Oliveira Campos, B.; John, K.; Beeskow, P.; Herrera Delgado, K.; Pitter, S.; Dahmen, N.; Sauer, J. A Detailed Process and Techno-Economic Analysis of Methanol Synthesis from H_2 and CO_2 with Intermediate Condensation Steps. *Processes* **2022**, *10*, 1535. [CrossRef]

 *processes* 

Article

Thermal Methane Cracking on Molten Metal: Kinetics Modeling for Pilot Reactor Design

Emma Palo [1], Vittoria Cosentino [1,*], Gaetano Iaquaniello [1], Vincenzo Piemonte [2] and Emmanuel Busillo [3]

[1] NextChem Spa, Via di Vannina 88/94, 00156 Rome, Italy
[2] Department of Science and Technology for the Sustainable Development and One Health, University Campus Bio-Medico di Roma, Via Álvaro del Portillo 21, 00128 Rome, Italy
[3] Department of Chemical Engineering, University La Sapienza di Roma, Via Eudossiana 18, 00184 Rome, Italy
* Correspondence: v.cosentino@nextchem.it

Abstract: Up to 80% of hydrogen production is currently carried out through CO_2 emission-intensive natural gas reforming and coal gasification. Water-splitting electrolysis using renewable energy (green H_2) is the only process that does not emit greenhouses gases, but it is a quite energy-demanding process. To significantly contribute to the clean energy transition, it is critical that low-carbon hydrogen production routes that can replace current production methods and can expand production capacity to meet new demands are developed. A new path, alternative to steam reforming coupled with CCS (blue H_2) that is based on methane cracking, in which H_2 production is associated with solid carbon instead of CO_2 (turquoise H_2), has received increasing attention recent years. The reaction takes place inside the liquid bath, a molten metal reactor. The aim of this article is to model the main kinetic mechanisms involved in the methane cracking reaction with molten metals. The model developed was validated using experimental data produced by the University of La Sapienza. Finally, such a model was used to scale up the reactor architecture.

Keywords: hydrogen; methane cracking; molten metal process; modeling; CO_2 free process

Citation: Palo, E.; Cosentino, V.; Iaquaniello, G.; Piemonte, V.; Busillo, E. Thermal Methane Cracking on Molten Metal: Kinetics Modeling for Pilot Reactor Design. *Processes* **2023**, *11*, 1537. https://doi.org/10.3390/pr11051537

Academic Editors: Elio Santacesaria, Riccardo Tesser and Vincenzo Russo

Received: 30 March 2023
Revised: 8 May 2023
Accepted: 15 May 2023
Published: 17 May 2023

1. Introduction

The European Union aims to reach climate neutrality by the 2050, which means that Europe needs to be carbon neutral years before. Clean energy technology deployment must accelerate rapidly to meet climate goals [1]. Today, hydrogen is mainly produced from fossil fuels, resulting in close to 900 Mt of CO_2 emissions per year. Approximately 48% of H_2 comes from natural gas through the steam reforming process, 30% comes from naphtha/oil reforming in the chemical industry and 18% comes from coal gasification. Only the remaining 4% is generated by water electrolysis, allowing a CO_2-free process, only if the electricity comes from renewable sources [2,3].

Therefore, in order to significantly contribute to the clean energy transition, it is critical to develop low-carbon hydrogen production routes. Natural gas reforming with carbon capture and storage (CCS) (Blue H_2) and electrolysis (Green H_2) are today the two main alternatives [1]. However, a new path based on methane pyrolysis (Turquoise H_2) has garnered increasing interest in recent years and may represent an interesting option during the transition to a long-term sustainable society [4].

Methane cracking is based on the splitting of methane into hydrogen and carbon, without any associated CO_2 emissions [5]. The main reaction is endothermic and characterized by strong kinetic limitations due to a high activation energy, which is between 356 and 452 kJ/mol [2].

$$CH_{4\,(g)} = 2H_{2\,(g)} + C_{(s)} \quad \Delta H^0_{298K} = -74.8 \text{ kJ/mol} \tag{1}$$

The reaction can occur in a thermal, catalytic and combined process. Thermal cracking occurs theoretically above 300 °C; however, only around 1000 °C can a reasonable high conversion be reached [2]. Figure 1 shows the CH_4 conversion calculated at equilibrium at different pressures and temperatures using Aspen Plus software v11 by Aspen Tech, which presents the same trend reported in the literature [6].

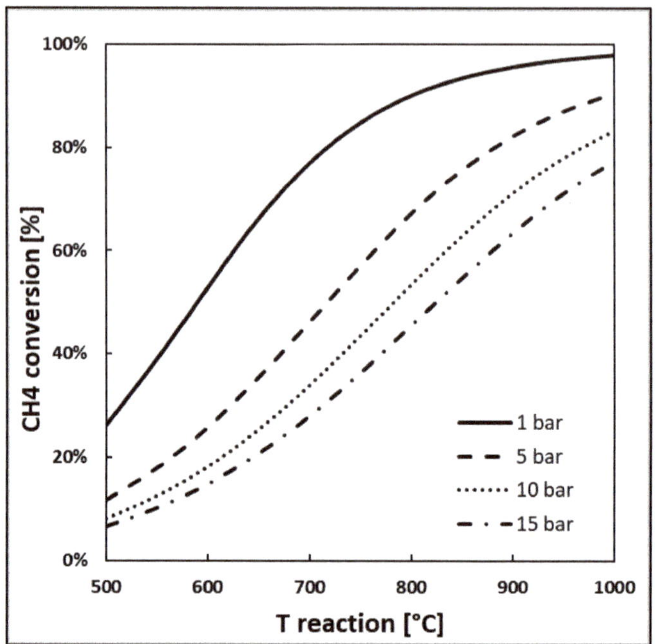

Figure 1. CH_4 equilibrium conversion at different temperatures [°C] and pressures [bar] calculated on Aspen Plus v11.

Catalyst addition enables a reduction in the activation energy to the range of 96.1–236 kJ/mol, according to the catalyst type. Different kinds of catalysts are studied in the literature for this application. Solid catalysts have been used to lower the activation energy of the reaction and to increase the reaction rate at moderate temperatures (below 1000 °C). Such catalysts are either metallic catalysts (Ni, Pd, Pt, Fe, Ni-Pd, etc.) usually supported on metal oxides (Al_2O_3, MgO, etc.), or carbonaceous catalysts (activated char, carbon black, etc.) [2]. The active metals used in conventional methane cracking can be based on transition metals (Ni > Co > Fe) and noble metals (Pt, Pd), with catalytic activity, however, impacted by coking in time on stream tests, due to the blockage of active sites [7]. Ni-based catalysts are the most active and stable at temperatures between 500 and 700 °C, but they rapidly deactivate with increasing temperatures. Moreover, Ni is toxic, therefore any downstream use of a carbon product is enabled if purification is carried out. Fe-based catalysts are cheaper and use a nontoxic metal, but require higher temperatures (700–900 °C) [8]. Carbonaceous catalysts, such as activated char/biochar/coal char and carbon black, have also been reported for methane cracking. Their main benefits are a lower cost, no contamination of by-products carbon and more resistance to sulfur [9].

One of the main concerns regarding catalytic cracking is the clogging of the catalyst formed by the carbon, which usually therefore needs to be regenerated in steam or air. Regeneration is not always simple and for some metals, such as Ni, it is almost impossible to recover the initial activity of the catalyst. Furthermore, even if carbonaceous catalysts do not theoretically necessitate regeneration, their catalytic activity also drops after a few hours of operation [2].

To conclude, regardless of whether methane cracking is catalytic or not, the main problem is the cumulative deposition of coke, which can plug not only the catalyst, but the reactor itself after a few hours of operation [10].

A possible solution is to carry out the reaction in a molten media, such as in liquid bubble column reactors [11,12]. The feed gas is then injected at the bottom of the reactor, which rises as bubbles through the molten metal. In this way, the carbon can be easily separated in continuous operation since the low-density solid carbon floats on the surface of the liquid, preventing carbon accumulation and the blockage of the reactor [2,12,13]. This solution still represents a challenge regarding its applicability on an industrial scale [14].

Molten media present other advantages, such as their ability to improve heat transfer and thermal inertia due to their heat capacity, to increase the residence time due to the liquid viscosity, to enable efficient heating using renewable electricity and thus replace fossil fuel combustion, and the liquid phase may catalyze the reaction [2,12].

Molten media can be both molten salts and molten metals. Molten metals show a higher performance in terms of catalytic activity than molten salts. In particular, Ni, Pd, Pt, Co and Fe have a high catalytic behavior, while In, Ga, Sn, Pb and Bi are considered inert metals with low catalytic activity, even if their combination could highly modify the alloy activity and bring unexpected results, exceeding the performance of active metal alloys [2,13]. Furthermore, molten metals should offer isothermal conditions during cracking, and generally have high thermal conductivities. Therefore, they can homogenize the temperature, enhancing methane decomposition [2,12].

Starting from all the above considerations, it was decided that the thermal methane cracking using molten tin in a liquid bubble reactor would be analyzed. The first patent on molten methane cracking was proposed by Tyrer in 1931, although this solution was not considered until the 1990s [15]. Among the studies of the last twenty years, Steinberg was one of the first to propose a bubble reactor configuration with a tin/copper molten metal bath to promote heat transfer and carbon separation [16]. Serban et al. conducted experiments using a stainless-steel feed tube or a porous metal sparger, injecting the methane gas from the top into liquid tin or lead, with or without a packed bed of SiC; they achieved a conversion of 57% at 750 °C with a packed bed/tin combination and a porous metal sparger [17]. Geißler et al. studied the thermo-chemical modeling of hydrogen production in a liquid metal bubble column reactor filled with tin, using a packed bed and a methane volume flow rate of 50 mL/min, obtaining a maximum hydrogen yield of 30% at 1000 °C [18]. Catalan et al. proposed a model that takes into account the coupling between kinetics and hydrodynamics for thermal and catalytic cracking. The model combines a drift flux correlation for gas hold-up, one of the most critical parameters in this type of application, and a thermodynamically consistent kinetic model of methane decomposition [13]. Subsequently, this first model was improved so that the interfacial area and gas hold-up can be calculated as a function of the operating conditions in the reactor, with no assumptions about the bubble size [19].

The purpose of this work is to analyze in particular the kinetic mechanisms involved in the reaction in the molten media. The main factors to be considered are the high temperature, high residence time in the liquid phase and the diameter of the bubbles along the reactor. A simplified mathematical model to predict methane conversion is developed and solved for both lab-scale and industrial-scale reactors. The results of the first case are compared with experimental data produced by the Department of Chemical Engineering University of La Sapienza in Rome, in order to carry out a first validation of the approach used.

This work is connected to the evaluations of the process scheme and reactor design described in two applied patent applications, one of which will be published by 2023 [12].

2. Materials and Methods

2.1. Experimental Set-Up

Experimental tests were carried out at the Industrial Chemistry laboratory, Department of Chemical Engineering University of La Sapienza in Rome.

Three tubular fixed-bed quartz reactors with different diameters were tested: 1.5 cm, 3.3 cm, 6.0 cm ID; in all three cases, a height of up to 20 cm was filled.

All the reactors were heated by an electric furnace, with the temperature varying between 950 °C and 1070 °C. The reactors were placed vertically inside the furnace. The temperature was controlled using a K-type thermocouple. The selected temperature range complies with the literature: looking at thermodynamics, the reaction is favored at high temperatures; however, almost zero methane conversions are recorded below 900 °C [13].

The flow rate of the methane could be varied in a range of 30–140 mL/min. The methane flow rate was controlled by a flowmeter, which was injected into the reactor from a capillary centered at the bottom of the reactor and bubbling inside the molten tin.

A scheme of the experimental set up and the reactor configuration schemes are reported in Figures 2 and 3.

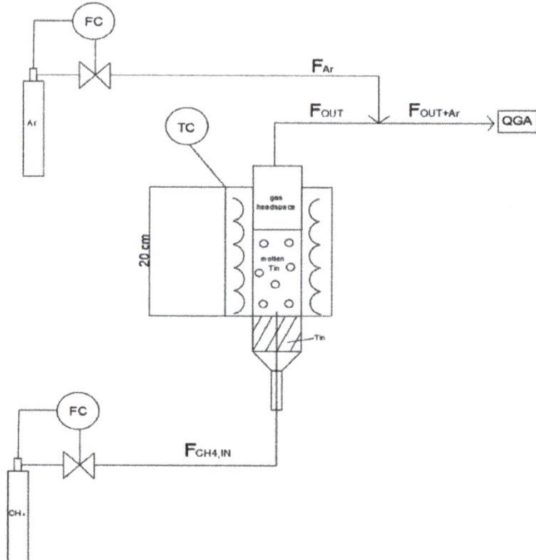

Figure 2. Experimental set up scheme.

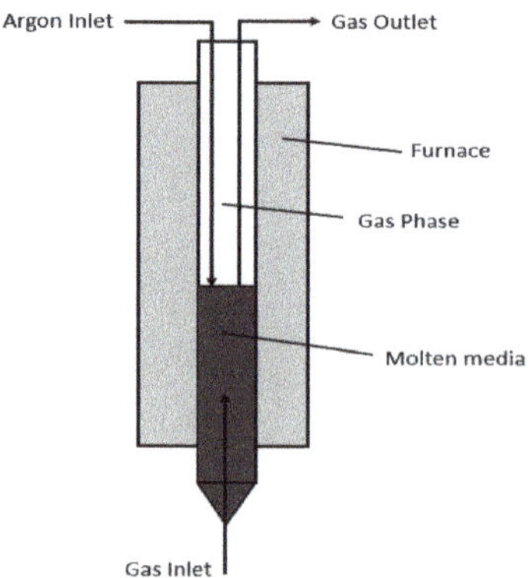

Figure 3. Experimental lab reactor: reactor scheme.

The possibility of cooling the tin surface with Argon was evaluated for the 6 mm ID reactor. In this way, it was possible to slow down the cracking kinetics and avoid a reaction in the gas phase.

The carbon produced was separated by a trap at the reactor exit and the gas was sent to a mass spectrometer analyzer QGA-Hiden.

The experimental CH_4 conversion for each reactor diameter is reported in Table 1.

Table 1. Experimental data of methane conversion.

Reactor ID Diameter [cm]	Tin Height [cm]	Exp. CH_4 Conversion [%]
1.5	7.1	22.0
1.5	9.9	31.0
1.5	15.0	40.0
1.5	20.0	72.0
3.3	7.0	18.0
3.3	15.0	37.0
3.3	20.0	75.0
6.0	7.0	10.0
6.0	8.0	15.0
6.0	10.0	18.0
6.0	18.0	21.0

2.2. Simplified Kinetics Model

The methane cracking mechanism in a liquid bubble reactor is determined by both the reaction kinetics and the hydrodynamics of gas rising in the bath. In this study, only the kinetics aspects are analyzed inside the molten media phase by assuming that there are bubbles with a constant diameter along the entire reactor height. Therefore, the mass transfer phenomena effects are not currently considered. This simplified hypothesis will be removed in an updated model based on a dedicated experimental campaign.

In general, in a molten metal, as the methane bubbles rise, two types of reactions occur: one at the center of the bubbles where the methane is not in contact with the liquid, and one at the bubble interface where the gas is in direct contact with the molten media [2]. In this study, it is assumed that the prevalent contribution to methane conversion is due to the heterogeneous surface reaction at the gas/liquid–metal interface; the reaction rate is therefore proportional to the specific contact surface area, which depends on the size distribution of the bubbles [13].

The metal bath, due to its high thermal conductivity, is assumed to have a uniform temperature, independent of the heat required by the reaction [6,20].

The mathematic model developed is based on the methane mass balance inside the molten tin. The methane mass balance is defined in space, along the axial axis of the reactor, and in time. The equations reported from here on were solved using the Gproms software.

The methane mass balance ($A = CH_4$) can be written as follows:

$$Vg * \frac{dCa(z,t)}{dt} = Q\,(z,t) * Ca(z,t)\Big|_z - Q(z,t) * Ca(z,t)\Big|_{z+\Delta z} - (-ra) * S \qquad (2)$$

A first-order linear reaction kinetics with respect to methane is considered [21]:

$$(-ra) = k(T)\,Ca(z,t) \qquad (3)$$

where $k(T)$ is the rate coefficient of the catalytic reaction (Equation (17)) and $Ca(z,t)$ is the concentration of methane as a function of height and time.

The decomposition of methane is a reversible reaction. In the model developed, only the forward reaction is considered, since it can be assumed that the contribution of the backward reaction tends to be zero at the high temperatures analyzed [22,23].

The methane flow rate, as a function of height and time, can be written as follows:

$$Q(z,t) = Q^0(1 + \varepsilon A * Xa) \tag{4}$$

Xa represents methane conversion, defined as follows:

$$Xa = \frac{Q^0 * Ca^0 - Q(z,t)Ca(z,t)}{Q^0 * Ca^0} \tag{5}$$

where Q^0 represents the methane inlet flow rate, Ca^0 represents the methane initial concentration and εA is derived from stoichiometry and defined as follows:

$$\varepsilon A = \sum \beta i * yi^0 \tag{6}$$

where $\sum \beta i$ represents the sum of the stoichiometric coefficients and yi^0 represents the molar fraction of methane.

The final balance obtained by combining all the equations above is as follows:

$$\frac{dCa}{dt} = -\frac{1}{\pi * R^2}\left[Q\,(z,t)\frac{dCa(z,t)}{dz} + Ca(z,t)\frac{dQ(z,t)}{dz}\right] - k(T) * Ca(z,t) * a \tag{7}$$

$$Q(z,t) = Q^0\left[1 + \varepsilon A * \left(\frac{Q^0 * Ca^0 - QCa}{Q^0 * Ca^0}\right)\right] \tag{8}$$

with the following initial conditions:

$$\begin{cases} Z = 0, & Ca = Ca^\circ \\ Z = 0, & Q = Q^\circ \end{cases} \tag{9}$$

The specific contact surface area between the methane bubble and the liquid metal, a, is an important process parameter, since by increasing its value, a higher methane conversion can be reached. It is calculated as follows [24]:

$$a = \frac{6\varepsilon}{d_b * (1 - \varepsilon)} \tag{10}$$

where d_b is the bubble diameter and ε is the gas hold up.

As reported in Equation (10), the specific contact surface is also inversely proportional to the diameter of the bubble. This confirms that small bubbles lead to high methane conversions, as also reported in the literature [2].

As explained at the beginning of this section, the bubble diameter is assumed to be constant along the reactor height. The bubble diameter is estimated using Tate's law, assuming that there are no viscous effects, no interactions between the bubbles, that there is a vertical rising pathway, and that there are no variations in the molten surface tension [25]:

$$d_b = \left(\frac{12R_0\sigma}{(\rho_{Tin} - \rho_g)g}\right)^{\frac{1}{3}}, \tag{11}$$

where

- $R_0 =$ radius of the sparger orifice $[\text{m}]$
- $\sigma =$ surface tension of the tin $\left[\frac{\text{N}}{\text{m}}\right]$
- $\rho_{Sn} =$ density of tin $\left[\frac{\text{kg}}{\text{m}^3}\right]$
- $\rho_g =$ density of methane $\left[\frac{\text{kg}}{\text{m}^3}\right]$

Gas hold-up can be defined as the volume fraction of gas in the total volume of the gas–liquid phase in the bubble column. A smaller gas hold-up requires a larger reactor volume given the same hydrogen production rate, methane conversion and reaction temperature. It can be estimated as the ratio of the gas volume over the total volume of the molten bed.

$$\varepsilon = \frac{V_g}{V_{Sn}} \tag{12}$$

where

$$V_g = Q * \tau_m \tag{13}$$

$$\tau_m = \frac{H_{Sn}}{v_b} = \text{rising time} \tag{14}$$

$$V_{Sn} = \pi * R^2 * H_{Sn} \tag{15}$$

$$v_b = 29.69 * d_b^{0.316} = \text{rising velocity} \left[\frac{\text{cm}}{\text{s}}\right] \tag{16}$$

with

- $H_{Sn} = $ height of the tin bath$[\text{m}]$
- $R = $ reactor radius$[\text{m}]$

As the bubble rising velocity is a function of the bubble diameter only, it is assumed constant along the entire reactor height.

The kinetic constant is defined by Arrhenius' law:

$$k(T) = k_0 * exp^{\left(-\frac{Ea}{R*T}\right)} \tag{17}$$

Various studies in the literature were analyzed in order to find the values of the pre-exponential factor k_0 and the activation energy Ea for thermal cracking in molten tin. The values from Rodat et al. [26] were selected since they better fitted the experimental data.

In Table 2, all the parameters fixed to solve the methane balance are reported.

Table 2. Parameters fixed to solve methane material balance.

Parameter	Value
R_0 [mm]	0.125
σ [N/m]	0.493
ρ_{Sn} [kg/m^3]	6486.1
ρ_g [kg/m^3]	0.145
g [m/s^2]	9.81
k^0 [^{1}/s]	6.6×10^{13} [26]
Ea [kJ/mol]	370 [26]

3. Results

The mathematical model described above is solved using Gprom by entering different tin heights into the software. For each tin height, it is possible to calculate the rising time τ_m (Equation (14)), the volume of the molten bed V_{Sn} (Equation (15)), the gas volume V_g (Equation (13)) and the gas hold up ε (Equation (12)).

By combining these parameters to solve the mass balance, it is possible to obtain the methane conversion as a function of the tin height.

To validate this approach, the model results should be compared with the experimental data reported in Table 1. Therefore, the model was solved for 1.5, 3.3 and 6.0 cm ID reactors, considering an initial methane flowrate of 30 mL/min. For all these three cases analyzed, the bubble diameter (Equation (11)) was 2 mm and, accordingly, the rising velocity (Equation (16)) was 178.5 mm/s.

3.1. Case A: 1.5 cm and 3.3 cm ID Reactors

The first two cases analyzed use a 1.5 and 3.3 cm ID reactor. In Figure 4, the conversion profile vs. the tin height is reported. It is evident that the experimental data for the first two reactors are not consistent with the mathematical prediction since the experimental conversions results are much higher than the modelling ones, which tend towards an asymptote for relatively low values.

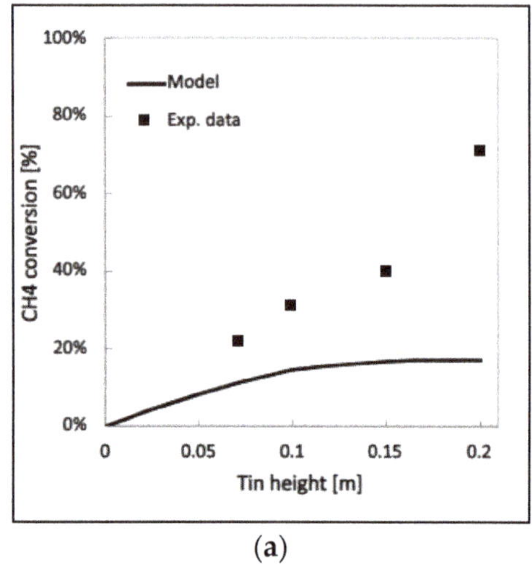

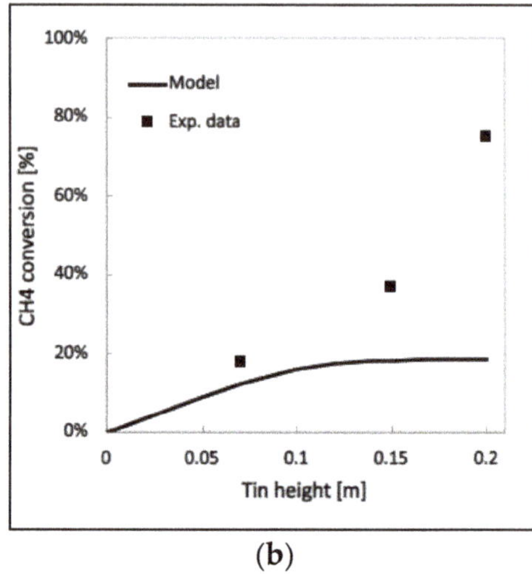

(a) **(b)**

Figure 4. Model and experimental CH_4 conversion [%] vs. tin height [m]. (**a**) Case for 1.5 cm ID reactor. (**b**) Case for 3.3 cm ID reactor.

This behavior can be attributed to the small size of the reactors. The wettability of the quartz tube with tin is very low and therefore the bubbles tend to flow along the walls of the tube where the pressure drops are lower. This effect is called the wall effect. Bubble velocity increases on the quartz wall and therefore bubbles leave the tin bed quickly.

This phenomenon is inversely proportional to the diameter of the reactor and is accentuated when the tin height, and therefore the pressure drop, increases. The main effect is that the residence time of methane in the tin bath becomes very limited, leading to a methane conversion that occurs predominantly in the gas phase above the molten metal. The latter must be avoided since the solid carbon, formed outside the bath in the headspace above the molten metal, can deposit in an uncontrolled manner on the reactor surfaces, causing the consequent problems of fouling and/or clogging.

The experimental data in Figure 4 are therefore much higher than the prediction of the model since they consider both the conversion contribution in the liquid phase and in the gaseous phase. As the height of the tin increases, the values diverge more and more, confirming that the conversion occurs mostly in the headspace above the bath.

3.2. Case B: 6.0 cm ID Reactor

By following the modeling approach already described, it is possible to calculate the methane conversion at different tin heights also for the 6.0 cm ID reactor. The model profile is then compared with the experimental data from Table 1, and the results are shown in Figure 5.

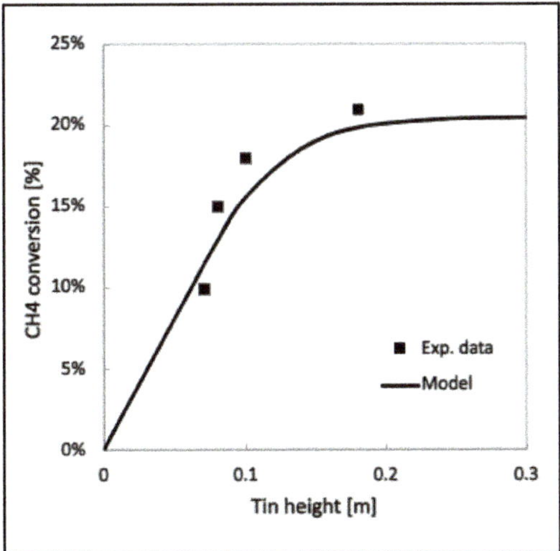

Figure 5. Model and experimental CH_4 conversion [%] vs. tin height [m] for 6.0 cm ID reactor.

The model conversion still presents an asymptotic behavior. However, unlike the previous cases, the experimental data are more congruent with the model prediction. This result can be attributed on the one hand to an increase in the diameter, which discourages the wall effect with the consequent flow of the bubbles. On the other hand, in the experimental tests of the 6.0 ID reactor, the conversion was measured only in the molten tin, removing the contribution in the gaseous phase. This was possible thanks to a flow of cold Argon, with the same flow rate as the methane feed, which is sent to the surface of the tin bath; in this way, the temperature in the headspace decreases with a slowdown of the kinetics, preventing conversion into the gaseous phase.

The described procedure is only applicable for the 6.0 cm ID reactor due to geometrical constraints.

The main result is that, despite the fact that the mass of the tin is greater than in the previous cases (the volume is greater since at the same relative height, the diameter is bigger), the experimental points in Figure 5 have shifted downwards with respect to those in Figure 4. The values of conversion for the 1.5 cm and 3.3 cm ID reactors should then be corrected by subtracting the conversion in the gaseous phase:

$$X_{a\_tin} = X_{a\_exp} - X_{a\_gas} \tag{18}$$

The conversion into the gaseous phase can be estimated using experimental tests on the empty reactor, which will be the object of a future experimentation campaign.

4. Discussion

By analyzing the results obtained, it may be concluded that the simplified model describes the experimental data quite accurately. Figure 5 also reflects the methane conversion trend reported in the literature for a similar reactor configuration [2,13]. The curve flattening is due to the small size of the reactor, which leads to a maximum conversion of 20%. This value corresponds to a tin height of about 20 cm, above which the molten metal seems to have no more influence.

It can be concluded that the approach developed is validated.

After this validation, the model is used to predict the methane conversion for an industrial liquid bubble reactor, which is schematized in Figure 6.

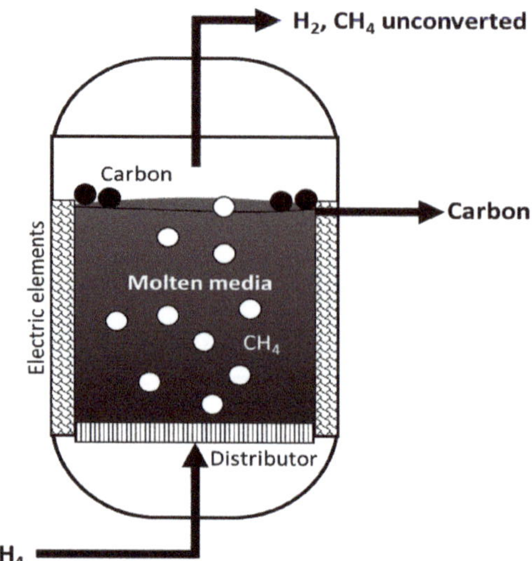

Figure 6. Liquid bubble reactor scheme for CH_4 cracking.

The reactor details are described in the patent applied, which will be published in 2023 [12]. It consists of a metallic reactor internally coated with refractory material and filled with molten tin. The metal is maintained at a constant temperature of 1000 °C by electric elements immersed in the molten bath. The methane enters inside the molten tin as bubbles through a gas distributor placed at the bottom of the reactor. The geometry and the dimensions of the reactor have been optimized in the same patents mentioned above.

The main parameters fixed are reported in Table 3.

Table 3. Parameters fixed for an industrial reactor.

Parameter	Value
Reactor diameter [m]	1
Temperature [°C]	1000
Pressure [bar]	15
Methane inlet flowrate [kg/h]	80.0

As already stated, the conversion is strongly influenced by the diameter of the gas bubble since it increases as the bubble diameter decreases. For this reason, three different bubble diameters are analyzed: 0.2 cm, 2.0 cm and 5.0 cm. For each of them, rising velocities (Equation (16)) of 17.85 cm/s, 36.96 cm/s and 49.37 cm/s were calculated, respectively.

Like the lab cases, the model is solved on Gproms, by entering different tin heights into the software: for each of them, the rising time τ_m (Equation (14)), the volume of the molten bed V_{Sn} (Equation (15)), the gas volume V_g (Equation (13)), the gas hold up ε (Equation (12)) and the corresponding methane conversion are calculated.

In Figure 7, the methane conversion profile vs. the tin height is reported for each bubble diameter. Figure 7 shows the equilibrium conversion as well, evaluated in the same operating conditions. In addition, in this case, the conversion presents the same trend as shown in Figures 4 and 5; what changes is the value at which the curve flattens out and the corresponding tin height. The latter is independent of the diameter of the bubble and is equal to 1.5 m in all cases. Instead, the methane conversion, as expected and also reported in the literature [6,26], is strongly influenced by the bubble diameter: with big bubbles, a maximum methane conversion of about 37% is obtained; this value can become double only by decreasing the diameter of the bubbles and keeping all other operating conditions equal. This demonstrates that the bubble diameter is one of the key factors for this process architecture.

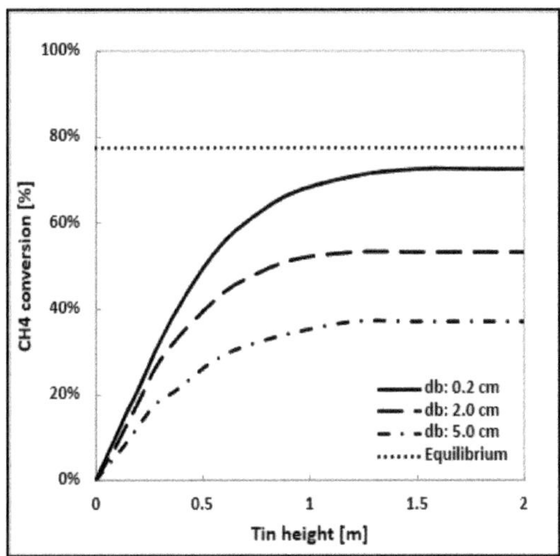

Figure 7. Model CH_4 conversion [%] vs. tin height [m] for industrial reactor at different bubble diameter db [cm].

The bubble diameter depends both on the characteristics of the fluid and on the geometry of the sparger. Conversions close to equilibrium can be obtained by suitably designing the sparger to guarantee a bubble diameter in the order of mm or even less.

5. Conclusions

Methane cracking on molten media represents a promising process for low-emission hydrogen production. It becomes important to understand the thermodynamic and kinetic mechanisms on which the cracking reaction occurs, in order to predict the methane conversion and optimize the reactor design.

For this purpose, a mathematical model was developed in this work. The model is based on the methane mass balance inside the molten tin. The differential equations are solved using Gproms software.

The model results were compared with the experimental data, showing an almost equal trend in the methane conversion during the molten phase.

Afterwards, the mathematical model developed is used to estimate the methane conversion in an industrial reactor at different bubble diameters. The results obtained are promising since quite high conversions can be achieved with small bubbles.

This proves that the bubble diameter is one of the main factors that plays a role in methane cracking. Finding the right compromise between the size of the sparger orifice and the tin bath height is a key parameter in the development of the technology.

Author Contributions: Conceptualization, V.C. and E.P.; methodology, V.P., G.I.; validation, E.P., G.I. and V.P.; data curation, E.B.; writing—original draft preparation, V.C.; writing—review and editing, V.C.; supervision, E.P., G.I. and V.P. All authors have read and agreed to the published version of the manuscript.

Funding: This research received no external funding.

Institutional Review Board Statement: Not applicable.

Informed Consent Statement: Not applicable.

Data Availability Statement: Not applicable.

Conflicts of Interest: The authors declare no conflict of interest.

References

1. IEA-International Energy Agency. *Energy Technology Perspectives 2023*; IEA-International Energy Agency: Paris, France, 2023.
2. Msheik, M.; Rodat, S.; Abanades, S. Methane Cracking for Hydrogen Production: A Review of Catalytic and Molten Media Pyrolysis. *Energies* **2021**, *14*, 3107. [CrossRef]
3. IEA-International Energy Agency. *Hydrogen*; IEA-International Energy Agency: Paris, France, 2021.
4. Sanchez-Bastardo, N.; Schlogl, R.; Ruland, H. Methane Pyrolysis for CO_2-Free H_2 Production: A Green Process to Overcome Renewable Energies Unsteadiness. *Chem. Ing. Tech.* **2020**, *92*, 1596–1609. [CrossRef]
5. Harrison, S.B. Turquoise Hydrogen Production by Methane Pyrolysis. Digital Refining PTQ Q4. 2021. Available online: https://www.sbh4.de/assets/turquoise-hydrogen-production-by-methane-pyrolysis%2C-petroleum-technology-quarterly-october-2021.pdf (accessed on 29 March 2023).
6. Leal Pérez, B.J.; Jiménez, J.A.M.; Bhardwaj, R.; Goetheer, E.; Annaland, M.v.S.; Gallucci, F. Methane pyrolysis in a molten gallium bubble column reactor for sustainable hydrogen production: Proof of concept & techno-economic assessment. *Int. J. Hydrog. Energy* **2021**, *46*, 4917–4935.
7. Bartholomew, C.H. Mechanism of catalyst deactivation. *Appl. Catal. A Gen.* **2001**, *212*, 17–60. [CrossRef]
8. Silva, J.A.; Santos, J.B.O.; Torres, D.; Pinilla, J.L.; Suelves, I. Natural Fe-based catalysts for the production of hydrogen and carbon nanomaterials via methane decomposition. *Int. J. Hydrog. Energy* **2021**, *46*, 35137–35148. [CrossRef]
9. Suelves, I.; Lázaro, M.J.; Moliner, R.; Pinilla, J.L.; Cubero, H. Hydrogen production by methane decarbonization: Carbonaceous catalysts. *Int. J. Hydrog. Energy* **2007**, *32*, 3320–3326. [CrossRef]
10. Parfenov, V.; Nikitchenko, N.V.; Pimenov, A.A.; Kuz'min, A.E.; Kulikova, M.V.; Chupichev, O.B.; Maksimov, A.L. Methane Pyrolysis for Hydrogen Production: Specific Features of Using Molten Metals. *Russ. J. Appl. Chem.* **2020**, *93*, 625–632. [CrossRef]
11. Rahimi, N.; Kang, D.; Gelinas, J.; Menon, A.; Gordon, M.J.; Metiu, H.; McFarland, E.W. Solid carbon production and recovery from high temperature methane pyrolysis in bubble columns containing molten metals and molten salts. *Carbon* **2019**, *151*, 181–191. [CrossRef]
12. Epstein, M.; Iaquaniello, G.; Salladini, A.; Borgogna, A.; Palo, E. Processo e Apparato per la Produzione di Idrogeno Mediante Cracking di Metano e di Idrocarburi a Bassa Emissione di CO_2. Patent PTC/IB2022/060932, 14 November 2022.
13. Catalan, L.J.; Rezaei, E. Coupled hydrodynamic and kinetic model of liquid metal bubble reactor for hydrogen production by noncatalytic thermal decomposition of methane. *Int. J. Hydrog. Energy* **2020**, *45*, 2486–2503. [CrossRef]
14. Abánades, A.; Rubbia, C.; Salmieri, D. Technological challenges for industrial development of hydrogen production based on methane cracking. *Energy* **2012**, *46*, 359–363. [CrossRef]
15. Tyrer, D. Production of Hydrogen. USA Patent 1803221, 28 April 1931.
16. Steinberg, M. Fossil fuel decarbonization technology for mitigating global warming. *Int. J. Hydrog. Energy* **1999**, *24*, 771–777. [CrossRef]
17. Serban, M.; Lewis, M.A.; Marshall, C.L.; Doctor, R.D. Hydrogen Production by Direct Contact Pyrolysis of Natural Gas. *Energy Fuels* **2003**, *17*, 705–713. [CrossRef]
18. Geißler, T.; Plevan, M.; Abanades, A.; Heinzel, A.; Mehravaran, K.; Rathnam, R.; Rubbia, C.; Salmieri, D.; Stoppel, L.; Stuckrad, S.; et al. Experimental investigation and thermo-chemical modeling of methane pyrolysis in a liquid metal bubble column reactor with a packed bed. *Int. J. Hydrog. Energy* **2015**, *40*, 14134–14146. [CrossRef]
19. Catalan, L.J.; Rezaei, E. Modelling the hydrodynamics and kinetics of methane decomposition in catalytic liquid metal bubble reactors for hydrogen production. *Int. J. Hydrog. Energy* **2022**, *47*, 7547–7568. [CrossRef]
20. Borgogna; Iaquaniello, G.; Biagioni, V.; Murmura, M.A.; Annesini, M.C.; Cerbelli, S. Estimate of the Height of Molten Metal Reactors for Methane Cracking. *Chem. Eng. Trans.* **2022**, *96*, 427–432.
21. Upham, D.C.; Agarwal, V.; Khechfe, A.; Snodgrass, Z.R.; Gordon, M.J.; Metiu, H.; McFarland, E.W. Catalytic molten metals for the direct conversion of methane to hydrogen and separable carbon. *Science* **2017**, *358*, 917–921. [CrossRef]
22. Paxman, D.; Trottier, S.; Flynn, M.R.; Kostiuk, L.; Secanell, M. Experimental and numerical analysis of a methane thermal decomposition reactor. *Int. J. Hydrog. Energy* **2017**, *42*, 25166–25184. [CrossRef]
23. Trommer, D.; Hirsch, D.; Steinfeld, A. Kinetic investigation of the thermal decomposition of CH4 by direct irradiation of a vortex-flow laden with carbon particles. *Int. J. Hydrog. Energy* **2004**, *29*, 627–633. [CrossRef]
24. Ojha, A.; Al Dahhan, M. Investigation of local gas holdup and bubble dynamics using four-point optical probe technique in a split-cylinder airlift reactor. *Int. J. Multiph. Flow* **2018**, *102*, 1–15. [CrossRef]
25. Kulkarni, A.A.; Joshi, J.B. Bubble Formation and Bubble Rise Velocity in Gas−Liquid Systems: A Review. *Ind. Eng. Chem. Res.* **2005**, *44*, 5873–5931. [CrossRef]
26. Rodat, S.; Abanades, S.; Coulié, J.; Flamant, G. Kinetic modelling of methane decomposition in a tubular solar reactor. *Chem. Eng. J.* **2009**, *146*, 120–127. [CrossRef]

Article

Study on the Hydrodynamic Performance of a Countercurrent Total Spray Tray under Sloshing Conditions

Jinliang Tao, Guangwei Zhang, Jiakang Yao, Leiming Wang and Feng Wei *

National-Local Joint Engineering Laboratory for Energy Conservation in Chemical Process Integration and Resources Utilization, School of Chemical Engineering and Technology, Hebei University of Technology, Tianjin 300130, China

* Correspondence: weifeng@hebut.edu.cn; Tel.: +86-135-120-346-32

Abstract: In this paper, a new type of total spray tray (TST) with gas–liquid countercurrent contact is proposed to solve the problem of slight operation flexibility and poor sloshing resistance in towers under offshore conditions. Its hydrodynamic performance indicators, such as pressure drop, weeping, entrainment, and liquid level unevenness, were experimentally studied under rolling motion. A tower with an inner diameter of 400 mm and tray spacing of 350 mm was installed on a sloshing platform to simulate offshore conditions. The experimental results show that the rolling motion affected the hydrodynamic performance of the tray under experimental conditions. When the rolling amplitude did not exceed 4°, the degree of fluctuation of the hydrodynamic performance was small, and the tray could still work stably. With increasing rolling amplitude, the TST wet plate pressure drop, weeping, and liquid level unevenness fluctuations also increased. When the rolling amplitude reached 7°, the maximum fluctuation of the wet plate pressure drop was 8.9% compared to that in the static state, and the plate hole kinetic energy factor, as the TST reached the lower limit of weeping, increased rapidly from 6.2 at rest to 7.8 under the experimental conditions. It can be seen that the TST still exhibits good hydrodynamic performance under rolling motion.

Keywords: pressure drop; countercurrent total spray tray; sloshing platform; hydrodynamic performance; offshore conditions

Citation: Tao, J.; Zhang, G.; Yao, J.; Wang, L.; Wei, F. Study on the Hydrodynamic Performance of a Countercurrent Total Spray Tray under Sloshing Conditions. *Processes* **2023**, *11*, 355. https://doi.org/10.3390/pr11020355

Academic Editors: Elio Santacesaria, Riccardo Tesser and Vincenzo Russo

Received: 22 November 2022
Revised: 9 January 2023
Accepted: 19 January 2023
Published: 22 January 2023

1. Introduction

As the most widely used critical common technology in the chemical industry, distillation is widely used in petroleum, natural gas, chemical, pharmaceutical, and environmental protection, and other industries. Furthermore, it occupies a considerable proportion of industrial production [1]. In the 21st century, with the increasing depletion of onshore oil and gas resources and the continuous breakthrough of offshore oil and gas exploration and exploitation technologies, the natural gas industry has begun to extend to the sea [2,3]. The distillation tower is the core equipment of the natural gas pretreatment process, and its operation has a significant impact on the gas purification effect, gas quality, and economic benefits [4]. Therefore, the development of offshore distillation and the realization of stable and efficient distillation towers on offshore platforms have become inevitable trends in the development of the distillation industry [5–9]. However, due to the influence of ocean wind and waves, floating devices will produce three angular motions of rolling, pitching, and yawing, and three displacement motions of swaying, surging, and heaving. The movement of offshore platforms is shown in Figure 1. The sloshing that has the greatest influence on the hydrodynamic performance of a traditional distillation tower is rolling (pitching) [10–12]. Therefore, it is of great significance to understand the performance of the tower under rolling motion.

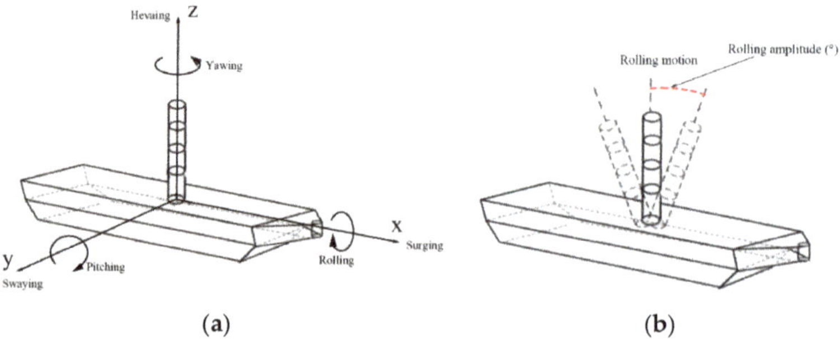

Figure 1. Movement of offshore platform. (**a**) Movement of offshore platform; (**b**) rolling motion.

Research on the influence of sloshing on equipment has mainly concentrated on the design calculation and load analysis of ships and tanks [13–16], while research on its effect on towers is relatively rare. Ma [4] carried out an experimental study on bubble cap trays and valve trays on a sloshing platform. The results showed that the performance of the trays decreased significantly with increasing rolling amplitude, and they could not work when the rolling amplitude exceeded 3°. Cheng [17] studied the hydrodynamic performance of LBJ (low backmixing jet) and DLJ (double-layer jet) trays under sloshing conditions and found that their ability to resist sloshing was improved compared to that of traditional trays. Zhang et al. [18] studied the influence of offshore sloshing on a packed column and found that the liquid accumulated obviously near the wall on the tilt side. The flow field parameters in the column changed significantly after the inclination exceeded 3°. Weedman et al. [19] studied the performance of several different packed columns under tilting conditions. Due to the high length–diameter ratio of the distillation column, the separation efficiency decreased rapidly under slight tilting conditions. Di et al. [20] studied the effects of ship motion on the mass transfer area in structured packed columns for offshore gas production. The results confirmed that the mass transfer area will decrease under any typical ship motion. Yang et al. [21] proposed a small air separation plant with a dual-column distillation process and carried out experiments under offshore conditions, providing engineering guidance for the design of cryogenic distillation columns for offshore applications. China University of Petroleum [22–28] conducted an experimental study on a packed column and plate tower on a sloshing platform and analyzed the influence of sloshing on the distribution and flow of gas and liquid in the tower. It was found that rolling motion was the most influential form of sloshing with regard to the hydrodynamic performance of the tower, and the arrangement of partitions could reduce the influence of sloshing on the liquid in the tower.

It can be seen that offshore conditions seriously affect the hydrodynamic performance of packed column and plate towers. The reason for this is that the tilt scale of the tower has a significant amplification effect on the uneven liquid distribution in the tower. For a plate tower with gas–liquid cross-flow contact, the liquid on the tray flows horizontally and cannot be blocked in the flow direction. Thus, the unevenness of liquid on the plate becomes severe with an increasing diameter of the tower under the tilt state. For a packed column with gas–liquid countercurrent contact, the liquid in the column flows vertically downward; by increasing the height of the column, the liquid will deviate to one side of the column during falling into the tilt state. Therefore, the traditional onshore distillation tower cannot maintain high performance under harsh conditions such as sea waves and typhoons [28].

Now, people mainly use different types of packed columns to solve the effect of offshore sloshing on distillation columns. Compared to a packed column, a plate tower has the advantages of a large operating range, suitability for large tower diameters, and convenient maintenance. However, it is impossible to block the liquid in the direction of liquid flow, which limits the application of traditional plate towers in the sea. If the plate tower can be made to resist sloshing, the plate tower will have greater advantages under some working conditions. Therefore, it is of great significance to study the applicability of plate towers under offshore conditions.

To solve the bottlenecks in existing plate towers—their low operational flexibility and poor resistance to sloshing under sloshing conditions—we propose a new type of total spray tray (TST) [29–31] with gas–liquid countercurrent contact and with the liquid flow in the tower guided by a three-dimensional space barrier. Under rolling conditions, the hydrodynamic performance of this TST was studied experimentally. The variation law of the hydrodynamic performance of the tower plate under offshore conditions is analyzed. Its operational flexibility range is clarified. Its ability to adapt to offshore sloshing conditions is explored, providing technical support for the design optimization of tower equipment on offshore platforms.

2. Materials and Methods

2.1. TST Space Barrier Drainage Principle

The TST consists of three parts: a tray with a plate hole, a spray tube above the plate hole, and a liquid sealing cap. The structure is shown in Figure 2.

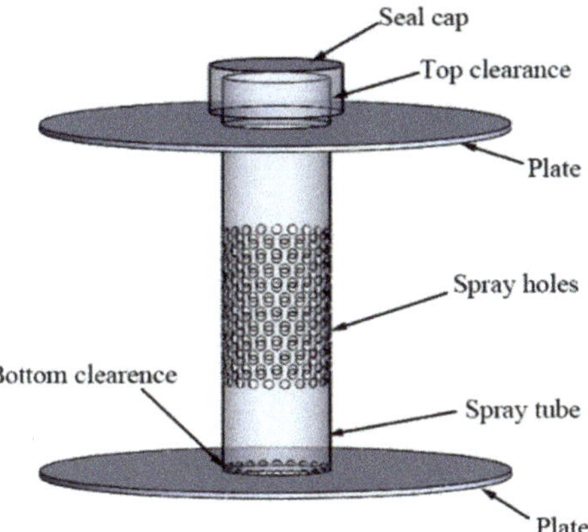

Figure 2. The structure of the TST.

By setting a layer of tray every 350 mm, the liquid will not accumulate obviously near the wall with increasing tower height in the sloshing state. At the same time, because of the gas–liquid countercurrent contact, the liquid does not need to cross the whole tray. The problem with the traditional plate tower where the partitions cannot be increased in the flow direction to reduce the sloshing effect can be solved here by increasing the partitions. The effect of the partitions is shown in Figure 3.

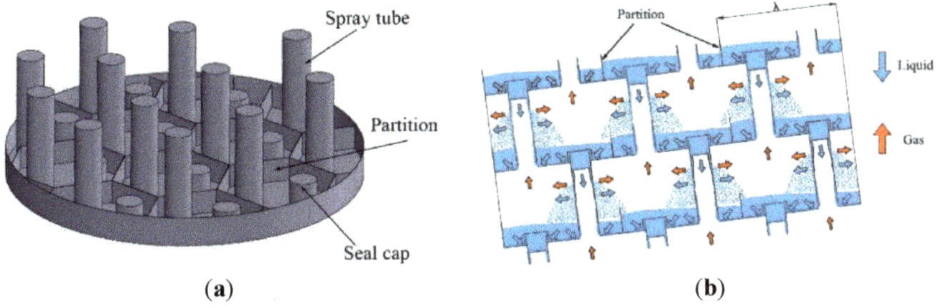

Figure 3. The effect of the partitions. (**a**) Form of the partitions; (**b**) working principles of the TST.

2.2. Experimental Setup and Process

The spray tube of the TST has the function of dropping liquid. Thus, the maximum characteristic of the TST is that there is no independent liquid-receiving plate or liquid drop area on the tray, giving the TST the features of structural symmetry in all horizontal directions. Therefore, this paper only studied the influence of the most influential sloshing (rolling) on the hydrodynamic performance of the TST. To simplify the experimental setup, only one barrier unit was analyzed in this experiment, and the diameter of the tower used was 400mm, that is, λ = 400 mm. Mobil presented a model wave experiment of the FPSO device in 1998, which showed that under severe sea conditions, the hull roll did not exceed 6°. Han et al. [32] pointed out that the maximum amplitude of rolling or pitching is 5.15° under the once-in-a-century combination of wind and waves along the coast of China. Through the relevant research, it is found that the sloshing period is mostly between 6 s–20 s with regard to the influence of offshore sloshing on the tower. Therefore, combined with experimental conditions, the rolling experiment was carried out at 0–7° and for rolling periods of 8s, 12s, 16s, and 20s. The gas flow velocity and liquid flow were selected from the normal operating conditions in the static state. The experimental conditions are shown in Table 1.

Table 1. Experimental conditions.

Condition	Value
Liquid flow (m^3/h)	2.2
F_0 ((m/s)(kg/m^3)$^{0.5}$)	6–15
F_T ((m/s)(kg/m^3)$^{0.5}$)	1.16, 2.01, 2.42
Rolling amplitude (°)	0, 1, 2, 3, 4, 5, 6, 7
Rolling period (s)	8, 12, 16, 20

The experimental setup is shown in Figure 4. It consists of a tower, a blower, a circulating pump, a measuring device, and a sloshing platform. The experimental tower used in the experiment was composed of organic glass and PPR (pentatricopeptide repeats) material, with a diameter of 400 mm, a tray spacing of 350 mm, and two fixed TSTs with a diameter of 90 mm. On the spray tube side wall of the TST, 231 holes with a diameter of 8 mm were opened, and 12 half-holes with a diameter of 8 mm were opened on the bottom. The experimental tower was placed on a circulating water tank with a diameter of 600 mm. The highest plate was used to collect entrained liquid. The lowest plate was used to collect weeping liquid. The experimental tower and the circulating water tank were connected to the sloshing platform to slosh together with the sloshing platform. The sloshing platform is shown in Figure 5.

The TST hydrodynamic experiment was carried out under ambient temperature and pressure using an air–water system. During the experiment, gas from an air blower, driven by a frequency conversion motor with 5 kW rated power, was introduced into the bottom of the tower. The velocity of the gas was measured using a pitot tube flowmeter. After the air contacted the liquid phase from bottom to top, it was removed from the vent after passing through the entrainment collector. The water was pumped out by a circulating pump from a circulating water tank and entered from the top of the tower after the rotameter measured the flow. The sloshing platform was controlled by electric machinery to achieve different amplitudes and periods of rolling. The tray pressure drop was measured using a ZCYB-1000 electronic differential pressure gauge, which has an accuracy margin of 1 Pa. After the operation reached stability, the rolling origin was selected as the starting point to start timing, and the pressure drop value was recorded every quarter-period. At the same time, the weeping and entrainment rates were calculated by collecting the weeping liquid and entrained liquid droplets within a certain time in a graduated cylinder. The height of clear liquid could be read from the ruler at the detection point of the tower wall, with an accuracy margin of 1 mm.

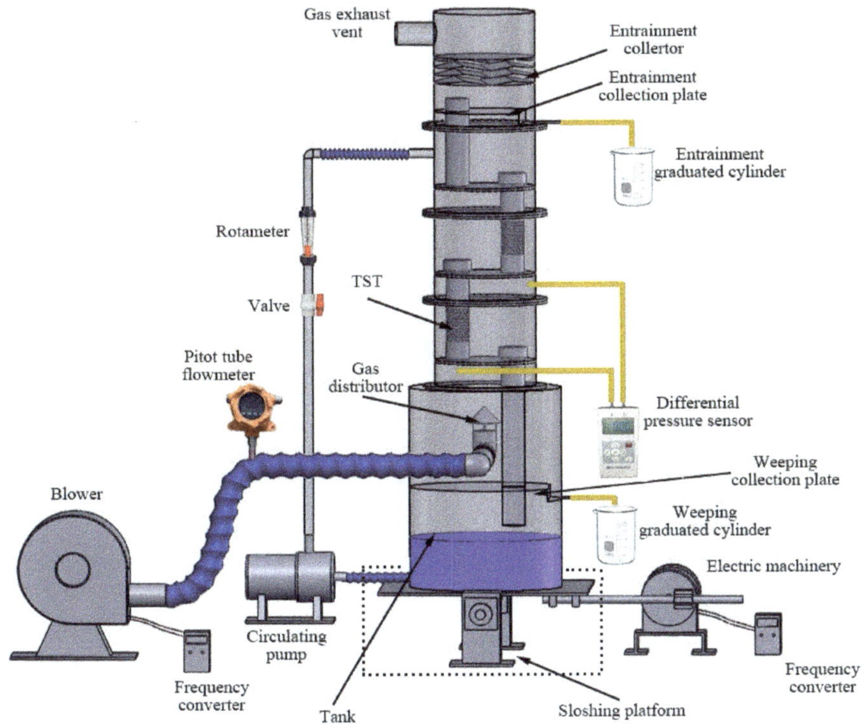

Figure 4. Experimental setup.

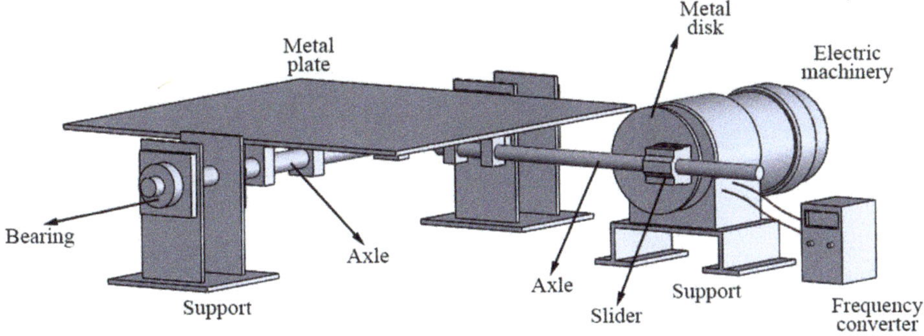

Figure 5. Sloshing platform.

3. Results and Discussion

The offshore rolling motion mainly affects the hydrodynamic performance of the tray. The applicability of the plate tower under offshore conditions is indicated by the hydrodynamic performance of the tray. The following hydrodynamic performance indices are analyzed in this paper.

3.1. Pressure Drop

Plate pressure drop includes dry plate pressure drop and wet plate pressure drop, which are directly related to energy consumption in the operation process. A low pressure drop of the tray means that the fluid flowing through the tray loses less energy. Pressure drop is an important indicator for evaluating the performance of the tray [33,34]. The pressure drop (ΔP) is calculated via Formula (1):

$$\Delta P = P_b - P_t \tag{1}$$

where P_b is the pressure at the bottom of the tray and P_t is the pressure at the top of the tray.

3.1.1. Dry Plate Pressure Drop

The dry plate pressure drop refers to the energy loss caused by the gas passing through all of the components on the tray when there is no liquid flow, which reflects the influence of the tray structure on the performance [28]. Energy loss in dry pressure drop during TST operation is mainly caused by the gas passing through the plate holes and spray holes.

The change in dry plate pressure drop under different rolling amplitudes of the TST was experimentally analyzed under a rolling period of 8 s and $F_0 = 15$. The experimental results are shown in Figure 6a. At times 0T, 0.5 T, and 1T, the tray was horizontal, and the dry plate pressure drop of the tray was similar to that in the static state. In addition, in a rolling period, the pressure drop at other times was larger than that in the static state. The pressure drop reached the maximum value in the first half-period and the second half-period at 0.25 T and 0.75 T. It can be seen from the figure that when the rolling amplitude was 0–4°, the pressure drop changed little compared to that in the static state, and the maximum pressure drop increased by 20 Pa compared to that in the static state. When the rolling amplitude exceeded 4°, the fluctuation degree of the pressure drop increased significantly. When the rolling amplitude was 7°, the maximum pressure drop increased by 50 Pa. The reason for this is that, due to the influence of rolling, the gas coming out of the plate hole was no longer parallel to the spray tube. This gas directly impacted the inclined spray tube at an angle, making it more likely to form vortices and lose more energy, resulting in an increased pressure drop. With increasing rolling amplitude, the influence is more obvious. It can be seen that rolling had the adverse effect of increasing the TST dry plate pressure drop, but the effect was small when the rolling amplitude was within 4°.

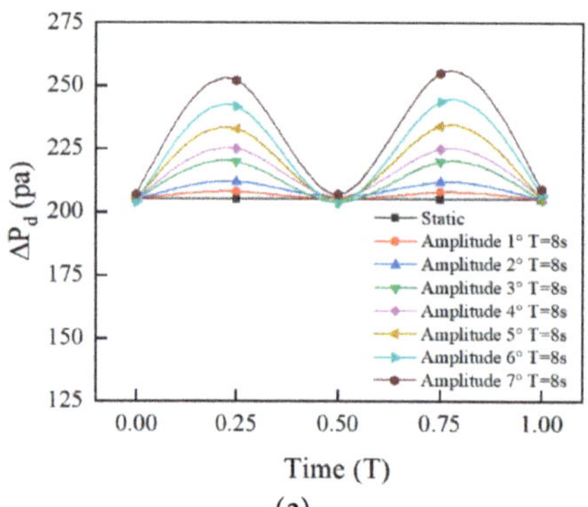

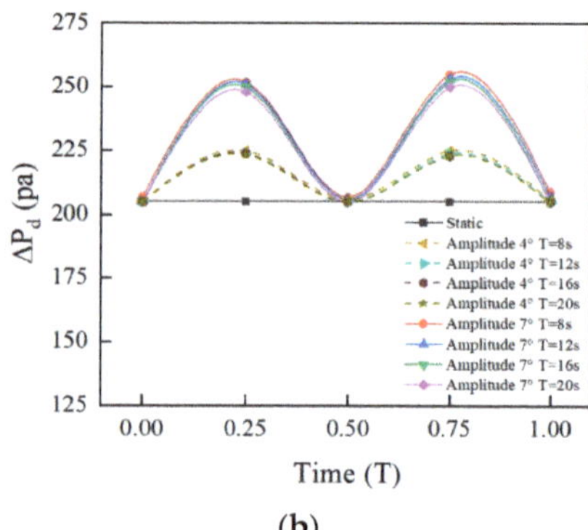

Figure 6. The dry plate pressure drop under rolling motion. (**a**) Different rolling amplitudes; (**b**) different rolling periods.

For rolling amplitudes of 4° and 7°, the dry plate pressure drop under different rolling periods was analyzed. The experimental results are shown in Figure 6b. It can be seen that the curve trends of the dry plate pressure drop under different rolling periods were consistent, and the change was small across different rolling periods. When the periods were 8 s and 20 s, the maximum pressure drop difference was only 2%; the pressure drop can thus be considered to be unaffected by the rolling period.

3.1.2. Wet Plate Pressure Drop

The wet plate pressure drop of the TST is different from that of other bubbling trays due to its special structure and gas–liquid flow mode. The wet plate pressure drop of the TST includes two parts: one is the energy loss of gas through the tray structure; the other is the energy lost when the gas contacts the liquid through the spray tube. The wet plate pressure drop of the tray is an important index to evaluate the hydraulic performance of a tower, and it represents the energy lost by the gas phase passing through the tray. According to these data, the tower structure can be improved, which is of great significance to the optimization of the tray structure [35].

Figure 7 shows the change in wet plate pressure drop under rolling motion when $F_0 = 8.74$, $V_L = 2.2 \ m^3/h$, and $T = 8 \ s$. It can be seen from Figure 7a that when rolling occurred, the pressure drop fluctuated with the rolling and reached maximum fluctuation values in the first half-cycle and the second half-cycle at about 0.25 T and 0.75 T. When the rolling amplitude was not more than 4°, the wet plate pressure drop fluctuated little compared to that in the static state, and when the rolling amplitude reached 4°, the pressure drop at 0.25 T and 0.75 T changed by 2.3% and 2.7%, respectively, compared to that in the static state. After 4°, the pressure drop fluctuated obviously with increased rolling amplitude. When the rolling amplitude reached 7°, the pressure drops at 0.25 T and 0.75 T changed by 8.1% and 8.9%, respectively, compared to that in the static state. The reason for this is that the rolling motion causes a fluctuation in the clear liquid layer on the tray, and weeping may occur during this process, which leads to fluctuations in the pressure drop. At 0.25 T, the spray tube sloshes to the lowest position, and the pressure drop increases due to the increase in the liquid level around the spray tube. At 0.75 T, the spray tube sloshes to the highest position and the liquid level at the spray tube is the lowest, resulting in the lowest pressure drop. The larger the rolling amplitude, the greater the pressure drop fluctuation, and the more unstable the working state of the tray.

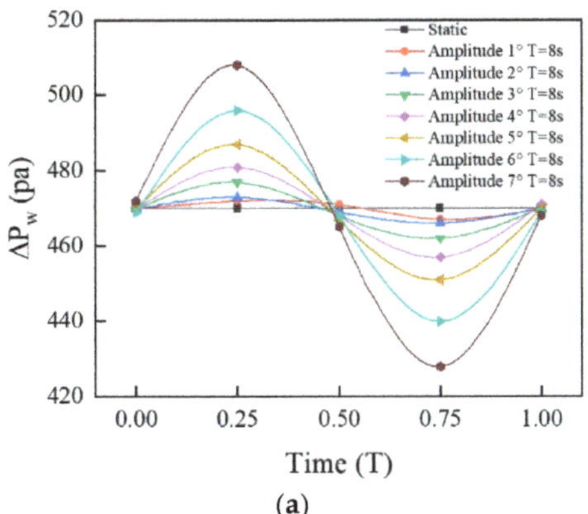

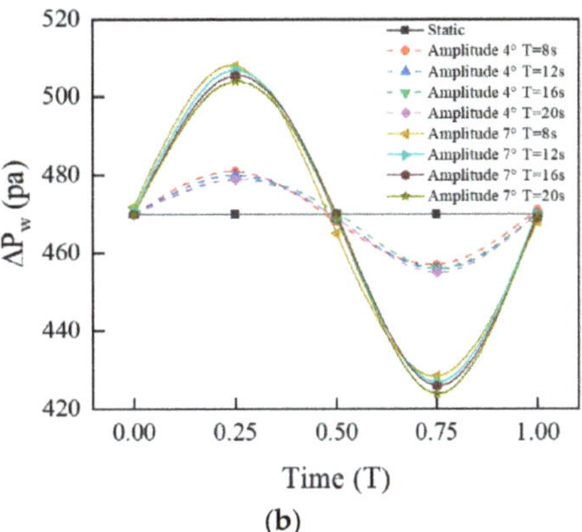

Figure 7. The wet plate pressure drop under rolling motion. (**a**) Different rolling amplitudes; (**b**) different rolling periods.

Figure 7b shows that the degree of pressure drop fluctuation under different periods differed little, and the difference between the maximum pressure drop and the minimum pressure drop at 0.25 T and 0.75 T was only about 1%.

We can see that the wet plate pressure drop under rolling conditions fluctuated with rolling, and the smaller the fluctuation, the stronger the ability to resist sloshing. A rolling amplitude within 4° had little effect on the wet plate pressure drop. When it reached 7°, the fluctuation was controlled at 10%, indicating that the TST can still maintain a good pressure drop distribution under rolling conditions. Further, the change was small under different rolling periods, showing little effect due to the rolling period.

3.2. Weeping

When the rising gas velocity is low, the rising gas' power in the riser is not enough to support the liquid. The liquid directly drops from the riser, which is called weeping. Weeping will affect the plate's gas–liquid contact and reduce the tower plate's efficiency. At the same time, serious weeping will make the plate unable to accumulate fluid, resulting in abnormal operation [29,31]. It is generally considered that the weep rate should not exceed 10%. Therefore, the gas velocity at a weep rate of 10% is called the weep point gas velocity in the industry. The gas velocity at the weeping point is the lower limit of the normal operating range of the tray. The weep rate (e_L) is calculated via Formula (2).

$$e_L = \frac{V_W}{V_L} \tag{2}$$

In the formula, V_W and V_L are the volume flow rates of the weeping liquid and the feed liquid, respectively.

Figure 8 shows the weeping under rolling motion with different F_0 at $V_L = 2.2 \ \mathrm{m^3/h}$. It can be seen from Figure 8a that there was no weeping under the three conditions in the static state, and the weep rate increased with increased rolling amplitude. The growth rate was flat when $F_0 = 8.74$ and $F_0 = 10.05$, and the weep rate was still less than 1% when the rolling amplitude reached 7°. The tray operated well. When $F_0 = 7.86$, the weep rate increased significantly with increased rolling amplitude, and the weep rate reached 7% when the rolling amplitude reached 7°, which is a significant change from the static state. It can be seen that the rolling motion had the adverse effect of increasing the weep rate. At low gas velocity, the rolling motion changes the height of the clear liquid layer and the distribution of the airflow, resulting in the local airflow kinetic energy being insufficient to support the liquid gravity, causing weeping. Moreover, when the rolling amplitude exceeded 4°, the liquid level unevenness increased and the pressure drop fluctuated significantly, so the weep rate increased significantly compared to that from before.

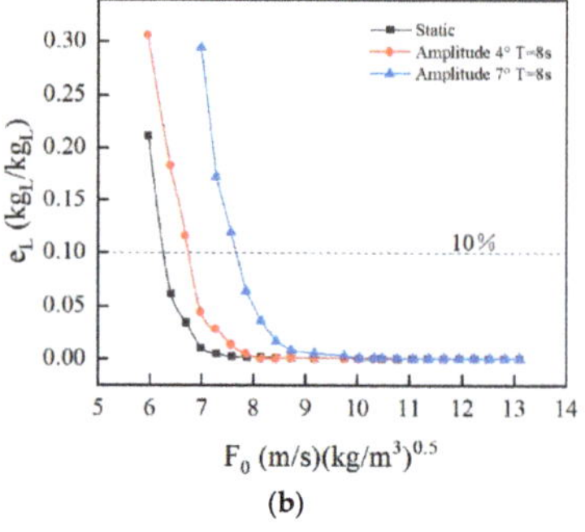

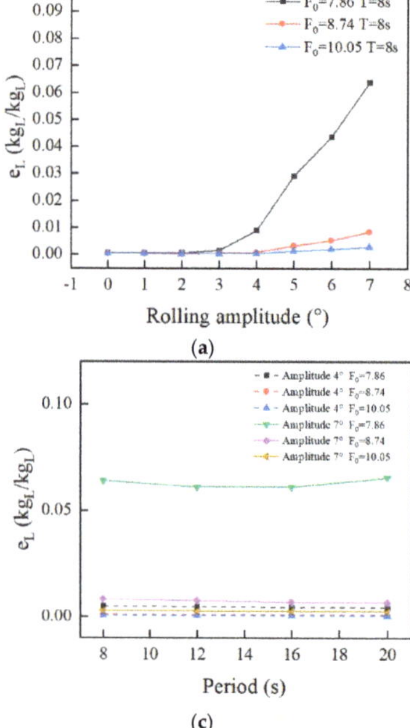

Figure 8. Weeping under rolling motion. (**a**) Different rolling amplitudes; (**b**) different gas velocities; and (**c**) different rolling periods.

It can be seen from Figure 8b that with increased rolling amplitude, the weep rate increased, and the lower operating limit of the tower increased. In the static state, the weep rate reached 10% at F_0 = 6.3. When the rolling amplitude was 4°, the weep rate reached 10% at F_0 = 6.75. Under a rolling amplitude of 7°, the weep rate reached 10% when F_0 = 7.7. We can see that the rolling motion reduced the normal operating range of the tower plate. When the rolling amplitude was 4°, the lower operating limit of the tower increased by about 7.5% compared to that in the static state, having little effect on the tower. When the rolling amplitude was 7°, the lower operating limit of the tower increased by about 22% compared to that in the static state. However, the weep rate can still be well controlled under the appropriate plate hole kinetic energy factor. Figure 8c shows that as the rolling period changed in the range of 8 s ~ 20 s, the differences between the weep rates of each period were small and could be ignored.

3.3. Entrainment

When the gas flow velocity is low, weeping occurs, making the tray unable to operate normally. Conversely, when the gas velocity is too large, some small droplets will be carried by the gas to the upper tray, which is called entrainment. Excessive entrainment will affect the efficiency of the tower [36]. In industrial production, remedial measures must be taken when entrainment reaches 5% [37]. Entrainment is calculated via Formula (3):

$$e_v = \frac{M_e}{M_G} \qquad (3)$$

where M_e is the mass rate of the liquid lifted to the foam capture tray by the gas and M_G is the mass rate of the gas.

In order to better study the entrainment of the tray, experimental analysis was carried out with different F_T at V_L = 2.2 m³/h. It can be seen from Figure 9a that with increasing rolling amplitude, the entrainment tended to decrease, and the larger the gas velocity, the more obvious the change trend. When F_T = 1.16, the tray had no entrainment under the rolling condition as it was under the static state, and when F_T = 2.42, the entrainment decreased slightly with increased rolling amplitude.

It can be seen from Figure 9b that the entrainment was positively correlated with F_T and decreased with increased rolling amplitude. When F_T = 2.42, the entrainment was reduced by about 5% compared to that in the static state when the rolling amplitude was 4°, and it was reduced by 9% when the rolling amplitude was 7°. The reason for this is that, due to the influence of rolling, in the process of gas–liquid injection, part of the gas is sprayed downward and part of the gas is sprayed upward. The gas and liquid sprayed obliquely downward will directly fall on the tray, and due to the effect of gravity, the liquid-carrying rate of this part of the gas is relatively higher. Rolling increases the collision between the droplets and the tower components so that some of the small droplets converge into large droplets after collision and fall directly. The higher the gas velocity, the greater the collision's severity. Therefore, rolling causes the entrainment to slightly decrease.

Figure 9c shows that when F_T was low, the rolling period did not affect the entrainment. When F_T was high, entrainment increased with increased period length. When the rolling amplitudes were 4° and 7° under the condition of F_T = 2.42, the entrainment ratios in the 20 s period increased by 4.1% and 5.3%, respectively, compared to that in the 8 s period. The reason for this is that, with an increase in the rolling period length, the collision intensity between droplets decreases, which makes the entrainment rise.

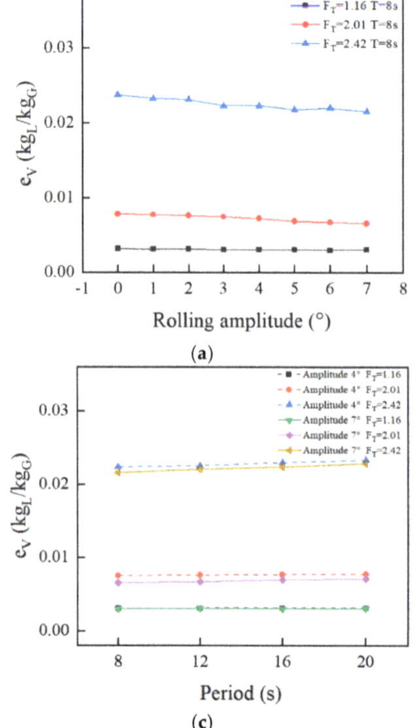

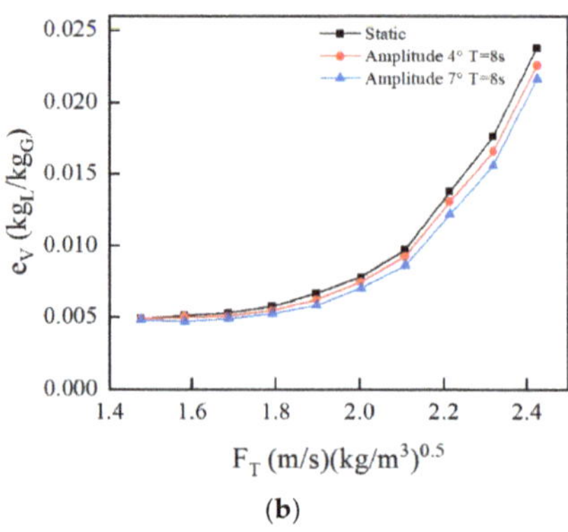

Figure 9. Entrainment under rolling motion. (**a**) Different rolling amplitudes; (**b**) different gas velocities; and (**c**) different rolling periods.

3.4. Liquid Level Unevenness

When the tower is tilted due to rolling motion, the free liquid level on the plate will be different, and the liquid level at each position will change at any time with the rolling. In this paper, the degree of liquid level unevenness at a certain time is called the liquid level unevenness. The greater the liquid level unevenness is, the greater the pressure drop fluctuation of the tray is, and the easier it is to cause weeping and other adverse effects. Eight test points were taken on the tower wall in the experiment, as shown in Figure 10.

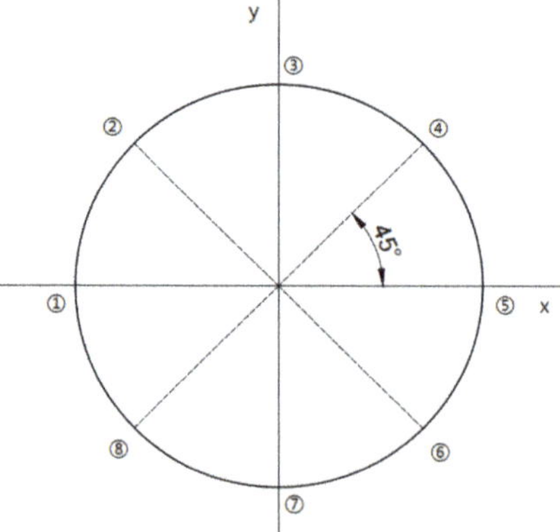

Figure 10. The locations of monitoring points.

In this experiment, the diameter of the tower and the frequency of rolling were small, so it can be considered that the free liquid level on the tower plate was still in the horizontal state and did not fluctuate during the rolling. The liquid layer on the tray had a clear free liquid level, which could be read directly. We reduced the observation error by measuring the liquid level under multiple periods and calculating the average value. After measuring the liquid level at the eight points, the liquid level unevenness was calculated according to Formula (4).

$$M_f = \left[\frac{1}{n} \sum_{i=1}^{n} \left(\frac{h_i - \bar{h}}{\bar{h}} \right)^2 \right]^{0.5} \tag{4}$$

In the formula, n represents the number of measuring points on the tower wall; hi represents the liquid level at point i on the tower wall, mm; and $\bar{h}$ represents the average liquid level height on the tray, mm.

Figure 11 shows the influence of different rolling amplitudes and rolling periods on the liquid level unevenness when $F_0 = 8.74$ and $V_L = 2.2 m^3/h$. Figure 11a shows that the liquid level unevenness under different rolling amplitudes fluctuated periodically with time, reaching upper and lower half-period maxima at around 0.25 T and 0.75 T. The fluctuation in amplitude of liquid level unevenness increased with increased rolling amplitude. Figure 11b shows that the trend of level unevenness on the tower plate was consistent under different rolling periods, and the difference between each period was small enough to be ignored. It can be seen that rolling motion had adverse effects on free surface fluctuation, but there was no sharp change in the free surface in the experiment, and the tray could still work normally.

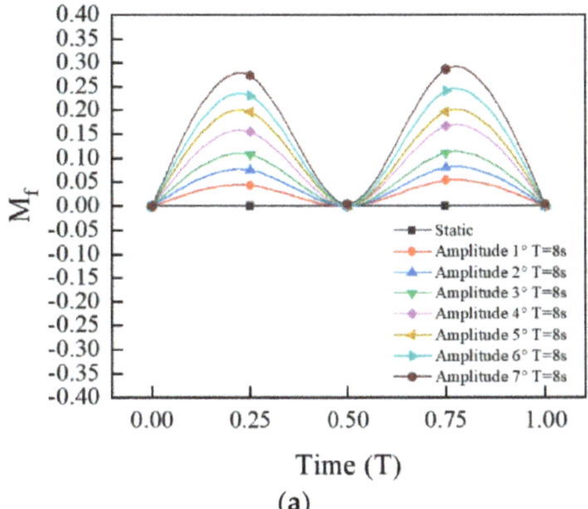

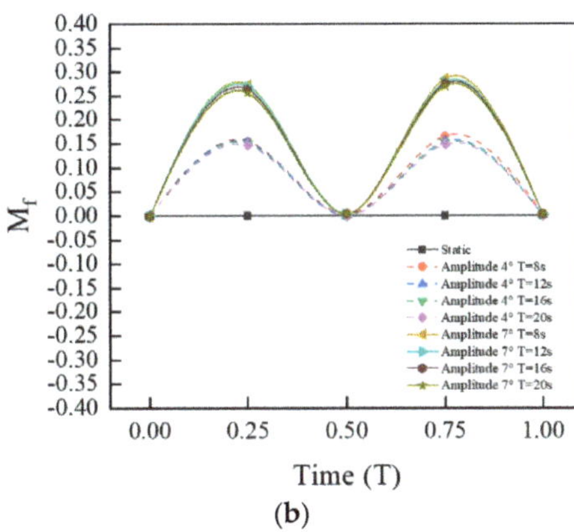

Figure 11. Liquid level unevenness under rolling motion. (**a**) Different rolling amplitudes; (**b**) different rolling periods.

4. Conclusions

In this paper, a new type of total spray tray (TST) with gas–liquid countercurrent contact was proposed to solve the problem of poor resist sloshing ability in existing towers under offshore conditions. Its hydrodynamic performance was experimentally studied under rolling motion to evaluate the influence of offshore conditions on the TST. The following conclusions were obtained under the experimental conditions:

Rolling caused adverse effects such as hydrodynamic performance fluctuation of the tray. When the rolling amplitude did not exceed 4°, the fluctuation was small. As the rolling amplitude exceeded 4°, the influence of rolling on the TST gradually increased.

The dry plate pressure drop of the TST fluctuated with the rolling motion. When the rolling amplitude was 4°, the dry plate pressure drop fluctuated by a maximum of 9% compared to that in the static state, and the fluctuation was 22% when the rolling amplitude was 7°. The fluctuation amplitude of wet plate pressure drop increased with increased rolling amplitude. When the rolling amplitude was 4°, the maximum fluctuation of wet plate pressure drop was 2.7% compared to that in the static state, and when the rolling amplitude was 7°, the fluctuation was 8.9%.

Rolling induced weeping, reducing the normal range of the tray. When the rolling amplitude was 4°, the lower limit of operation of the tray was 7.5% higher than that in the static state, which had little effect on the tower. At 7°, the lower limit of operation of the tray was 22% higher than that in the static state. However, under the condition of an appropriate kinetic energy factor, the weep rate could still be well controlled within 10%.

Entrainment decreased slightly with an increase in the rolling amplitude, which shows that the rolling motion had little effect on the entrainment. The fluctuation in liquid level unevenness increased with increased rolling amplitude. However, there was no serious liquid level fluctuation at large amplitudes, and the tower could still operate stably.

The difference in the hydrodynamic performance of the TST in different periods was very small, so different rolling periods can be considered to have little effect on the performance of the tray.

At the same time, it can be seen that increasing the gas velocity within the appropriate range can reduce the adverse effects such as weeping caused by sloshing in practical applications.

Author Contributions: Conceptualization, methodology, J.T.; data curation, writing—original draft, software, G.Z.; validation, J.Y.; software, L.W.; writing—review and editing, F.W. All authors have read and agreed to the published version of the manuscript.

Funding: This research received no external funding.

Institutional Review Board Statement: Not applicable.

Informed Consent Statement: Not applicable.

Data Availability Statement: Not applicable.

Conflicts of Interest: The authors declare no conflict of interest.

Nomenclature

e_L	$e_L = \frac{V_w}{V_L}$ relative weeping
e_v	$e_v = \frac{L_2 \rho_L}{M_v}$ gas entrainment
F_0	$F_0 = u_0 \sqrt{\rho_G}$ plate hole kinetic energy factor $((m/s)(kg/m^3)^{0.5})$
F_T	$F_T = u_T \sqrt{\rho_G}$ empty tower kinetic energy factor $((m/s)(kg/m^3)^{0.5})$
h_i	liquid level at point i on the tower wall (mm)
$\bar{h}$	average liquid level on the tray (mm)
L_w	volume of liquid per unit time (m^3/h)
M_e	mass rate of the liquid (kg/h)
M_f	unevenness of liquid level
M_G	mass rate of the gas (kg/h)
n	number of measuring points on the tower wall
ΔP	pressure drop (Pa)
P_d	pressure at the bottom of the tray (Pa)
P_t	pressure at the top of the tray (Pa)
ΔP_d	dry pressure drop across the tray (Pa)
ΔP_w	wet pressure drop across the tray (Pa)
u_0	velocity of the gas in the plate holes (m/s)
u_T	velocity of the gas in the empty column (m/s)
T	period (s)
V_L	volume flow rates of the weeping liquid (m^3/h)
V_w	volume flow rates of the feed liquid (m^3/h)

Greek symbols

θ	deviation angle of the column axis from the vertical axis ($^\circ$)
ρ_G	density of the gas (kg/m^3)
ρ_L	density of the liquid (kg/m^3)
λ	partition spacing (mm)

References

1. Hailun, R.; Dengchao, A.; Taoyue, Z.; Hailong, L.; Xingang, L. Distillation technology research progress and industrial application. *Chem. Ind. And. Eng. Prog.* **2016**, *35*, 1606–1626.
2. Chengzao, J.; Yongfeng, Z.; Xia, Z. Prospects of and challenges to natural gas industry development in China. *Nat. Gas Ind.* **2014**, *34*, 8–18. [CrossRef]
3. Wenhua, Z. Numerical and Experimental Study on Hydrodynamics of an FLNG System. Ph.D. Thesis, Shanghai Jiao Tong University, Shanghai, China, 2014.
4. Ma, P. Studies on Adaptation of Plate Column in FLNG Unit. Master's Thesis, China University of Petroleum (EastChina), Dongying, China, 2017.
5. Jiwei, S. Key Technology Research in FLNG General Design. *Shipbuild. China* **2015**, *56*, 81–86.
6. Bin, X.; Xichong, Y.; Xuliang, H.; Yan, L. Research status of FLNG and its application prospect for deep water gas field development in South China Sea. *China Offshore Oil Gas* **2017**, *29*, 127–134.
7. Bin, X.; Shisheng, W.; Xichong, Y.; Xia, H. FLNG/FLPG engineering models and their economy evaluation. *Nat. Gas Ind.* **2012**, *32*, 99–102+119–120.
8. Scott, E.B.; Lane, M.K. SS: Floating Offshore LNG: Offshore LNG Value Chain Optimization. In Proceedings of the Offshore Technology Conference, Houston, TX, USA, 4–7 May 2009.
9. Wijngaarden, W.V.; Meek, H.J.; Schier, M. The Generic LNG FPSO—A Quick & Cost-Effective Way to Monetize Stranded Gas Fields. In Proceedings of the SPE Asia Pacific Oil and Gas Conference and Exhibition, Perth, Australia, 20–22 October 2008.
10. Gu, Y.; Ju, Y. LNG-FPSO: Offshore LNG solution. *Front. Energy Power Eng. China* **2008**, *2*, 249–255. [CrossRef]
11. Chun, Z. Studies on Floating LNG Pretreatment Technology of Liwan Gas Field in the South China Sea. Master's Thesis, China University of Petroleum, Dongying, China, 2011.
12. Jianfeng, T.; Qiang, C.; Haojie, Z.; Wengang, Y.; Qingyan, X.; Zelin, S. Effect of shaking on pressure drop of structured packing absorption column. *J. China Univ. Pet. Ed. Nat. Sci.* **2018**, *42*, 142–148.
13. Hafez, K.; El-Kot, A.-R. Comparative analysis of the separation variation influence on the hydrodynamic performance of a high speed trimaran. *J. Mar. Sci. Appl.* **2011**, *10*, 377–393. [CrossRef]
14. Piller, M.; Nobile, E.; Hanratty, T.J. DNS study of turbulent transport at low Prandtl numbers in a channel flow. *J. Fliud Mech.* **2002**, *458*, 419–441. [CrossRef]
15. Chengsheng, W.; Decai, Z.; Bo, L.; Lei, G. CFD Computation of Ship Motions and Added Resistance for a High Speed Trimaran in Regular Heading Waves. *Shipbuild. China* **2010**, *51*, 1–10.
16. Shuxi, T. Sloshing Load Analysis for FLNG Tank Design. Master's Thesis, Dalian University of Technology, Dalian, China, 2012.
17. Cheng, Q. Study on Performance Evaluation of Two New Types of Tray on FLNG. Master's Thesis, China University of Petroleum (EastChina), Dongying, China, 2018.
18. Zhang, M.; Li, Y.; Li, Y.; Han, H.; Teng, L. Numerical simulations on the effect of sloshing on liquid flow maldistribution of randomly packed column. *Appl. Therm. Eng.* **2017**, *112*, 585–594. [CrossRef]
19. Weedman, J.A.; Dodge, B.F. Rectification of Liquid Air in a Packed Column. *Ind. Eng. Chem.* **2002**, *39*, 732–744. [CrossRef]
20. Di, X.; Ma, J.; Huang, Y. Mass-transfer area in a pilot-scale structured-packing column under different types of ship motion. *Chem. Eng. Sci.* **2019**, *203*, 302–311. [CrossRef]
21. Meng, Y.; Wang, S.; Zhang, Y.; Chen, S.; Hou, Y.; Chen, L. Experimental evaluation of the performance of a cryogenic distillation system under offshore conditions. *Chem. Eng. Sci.* **2022**, *263*, 118084. [CrossRef]
22. Bin, H. Studies on Decarbonisation Performance of Packed Absorption Tower under Sloshing Conditions. Master's Thesis, China University of Petroleum (EastChina), Dongying, China, 2016.
23. Fan, Y. Studies on Distributing Performance of ladder liquid distributor Under Sloshing Conditions. Master's Thesis, China University of Petroleum (EastChina), Dongying, China, 2016.
24. Zelin, S. Studies on Flow Performance of Structured Packing in Absorption Tower under the shaking Condition. Master's Thesis, China University of Petroleum (EastChina), Dongying, China, 2016.
25. Kai, Z. Studies on Decarbonization Tower Design Optimization Under Sloshing Conditions. Master's Thesis, China University of Petroleum (EastChina), Dongying, China, 2015.
26. Jianfeng, T.; Jian, C.; Yunfei, X.; Wengang, Y.; Xinpeng, J.; Weiming, Z. Research on orifices diameter of calandria liquid distributor used in offshore deacidification tower with different spray densities. *Chem. Ind. Eng. Prog.* **2017**, *36*, 1192–1201.
27. Jianfeng, T.; Xinming, J.; Junyi, Z.; Wengang, Y.; Jian, C.; Haojie, Z. Compatibility test of gas distributor in the FLNG packed tower. *Oil Gas Storage Transp.* **2018**, *37*, 822–830.
28. Jianfeng, T.; Bin, H.; Xinming, J.; Yihuai, H.; Zelin, S.; Qiang, C. Fluid distribution performance of packed tower under coupling sloshingworking conditions. *J. China Univ. Pet. Ed. Nat. Sci.* **2017**, *41*, 130–137.

29. Yanli, L.; Jinliang, T.; Bo, L.; Zhicheng, S.; Feng, W. Study on hydrodynamics performance and mass transfer efficiency of total spray tray. *Mod. Chem. Ind.* **2018**, *38*, 200–204.
30. Tao, J.L.; Shi, Z.C.; Ling, Y.L.; Wei, F. Hydrodynamic Characteristics in the Counter-Flow Total Spray Tray. *Chem. Eng. Technol.* **2019**, *42*, 1199–1204. [CrossRef]
31. Ran, W.; Jinliang, T.; Yanli, L.; Feng, W.; Jidong, L. Total spray tray (TST) for distillation columns: A new generation tray with lower pressure drop. *Chem. Ind. Chem. Eng. Q.* **2017**, *23*, 523–527. [CrossRef]
32. Xuliang, H.; Bin, X.; Xiaosong, Z.; Jingrui, Z. Numerical Simulation and Experimental Study on Hydrodynamic Performance of FLNG with Liquid Tanks. *Shipbuild. China* **2016**, *57*, 87–97.
33. Wang, H.; Niu, X.; Li, C.; Li, B.; Yu, W. Combined trapezoid spray tray (CTST)—A novel tray with high separation efficiency and operation flexibility. *Chem. Eng. Process.* **2017**, *112*, 38–46. [CrossRef]
34. Tang, M.; Zhang, S.; Wang, D.; Liu, Y.; Zhang, Y.; Wang, H.; Yang, K. Hydrodynamics of the tridimensional rotational flow sieve tray in a countercurrent gas-liquid column. *Chem. Eng. Process.* **2019**, *142*, 107568. [CrossRef]
35. Zhang, M.; Zhang, B.Y.; Zhao, H.K.; Zhao, Y.; Sun, J.; Ren, Z.Q.; Li, Q.S. Hydrodynamics and mass transfer performance of flow-guided jet packing tray. *Chem. Eng. Process.* **2017**, *120*, 330–336. [CrossRef]
36. Jaćimović, B.M. Entrainment effect on tray efficiency. *Chem. Eng. Sci.* **2000**, *55*, 3941–3949. [CrossRef]
37. Kister, H.Z.; Haas, J.R. Entrainment from sieve trays in the froth regime. *Ind. Eng. Chem. Res.* **2002**, *27*, 2331–2341. [CrossRef]

Article

Effects of S and Mineral Elements (Ca, Al, Si and Fe) on Thermochemical Behaviors of Zn during Co-Pyrolysis of Coal and Waste Tire: A Combined Experimental and Thermodynamic Simulation Study

Yaxin Lan [1], Shuangling Jin [1,*], Jitong Wang [2], Xiaorui Wang [1], Rui Zhang [1], Licheng Ling [2] and Minglin Jin [1,*]

1   School of Materials Science and Engineering, Shanghai Institute of Technology, Shanghai 201418, China
2   State Key Laboratory of Chemical Engineering, East China University of Science and Technology, Shanghai 200237, China
*   Correspondence: jinshuangling@sit.edu.cn (S.J.); jml@sit.edu.cn (M.J.)

Abstract: The transformation behaviors of Zn during co-pyrolysis of waste tires and coal were studied in a fixed-bed reaction system. The effects of pyrolysis temperature and the Zn content of coal mixture on the Zn distributions in the pyrolytic products (coke, tar and gas) were investigated in detail. It is found that the relative percentages of Zn in the pyrolytic products are closely related to the contents of S and mineral elements (Ca, Al, Si and Fe) in the coal. The thermodynamic equilibrium simulations conducted using FactSage 8.0 show that S, Al and Si can interact with Zn to inhibit the volatilization of Zn from coke. The reaction sequence with Zn is S > Al > Si, and the thermal stability of products is in the order of $ZnS > ZnAl_2O_4 > Zn_2SiO_4$. These results provide insights into the migration characteristics of Zn during co-pyrolysis of coal and waste tires, which is vital to the prevention and control of Zn emissions to reduce the environmental burden.

Keywords: coal; waste tire; co-pyrolysis; Zn; thermochemical behaviors

Citation: Lan, Y.; Jin, S.; Wang, J.; Wang, X.; Zhang, R.; Ling, L.; Jin, M. Effects of S and Mineral Elements (Ca, Al, Si and Fe) on Thermochemical Behaviors of Zn during Co-Pyrolysis of Coal and Waste Tire: A Combined Experimental and Thermodynamic Simulation Study. *Processes* **2022**, *10*, 1635. https://doi.org/10.3390/pr10081635

Academic Editors: Elio Santacesaria, Riccardo Tesser and Vincenzo Russo

Received: 26 July 2022
Accepted: 15 August 2022
Published: 18 August 2022

Publisher's Note: MDPI stays neutral with regard to jurisdictional claims in published maps and institutional affiliations.

1. Introduction

With the rapid development of the automobile industry, the disposal of increasing non-biodegradable waste tires has brought a great burden on the natural environment. After the removal of steel, carcass and textiles, the tire material generally consists of synthetic and natural rubber, carbon black, and inorganic components such as ZnO and SiO_2 [1–4]. The carbon content of waste tires exceeds 80% and the ash content is comparable to that of coal [4,5]. The moisture content of waste tires is relatively low in comparison with other alternative energy sources such as municipal solid waste (MSW) or biomass [4,5]. In addition, the waste tires possess a high calorific value of 30–40 MJ/kg, which is larger than those of coal and other solid fuels [1–6]. Considering these characteristics, combustion, gasification and pyrolysis are proposed as the potential approaches to utilizing waste tires [7–27]. The pyrolysis of waste wires in an inert atmosphere can produce 10–30% of gas, 38–55% of pyrolytic oil and 33–38% of char, all of which are valuable products [17–21]. The pyrolysis gas is composed of methane, ethane, butadiene, hydrogen and other hydrocarbons with a high calorific value (37–42 MJ/kg), which can be used as a source of energy for the pyrolysis process itself and other required processes [19–23]. The tire pyrolysis oil (TPO) is a complex of C5-C20 hydrocarbons containing paraffins, olefins and aromatic compounds with a high calorific value of 41–44 MJ/kg, which can be employed as a substitute for diesel fuel to reduce fossil fuel consumption and as a high value-added chemical source for producing benzene, toluene, xylenes, isoprene and limonene [20–23]. The produced char has a calorific value of 30–40 MJ/kg and can be used as a fuel [23]. Additionally, the char can be reused as a low-quality carbon black in the tire industry because it contains high contents of ash (12–16%), S (1.8–4%) and Zn (3–5%) and as high-quality activated carbon via activation using steam or carbon dioxide as activation agents for adsorption, catalysis and electrochemical applications [24–27].

Apart from classic pyrolysis, the co-pyrolysis of waste tires with coal for metallurgical coke production is considered an economical route for recycling waste products in the cokemaking process, decreasing coal consumption and reducing the cost of waste disposal without causing an apparent deterioration in coke quality under suitable conditions [28–32]. W.R. Leeder studied the effect of particle size of waste tires on the quality of coke and found that 5 wt.% of finely pulverized tires can be incorporated into coal blends without weakening the coke quality, but the coke yield is reduced because of the higher volatile matter of tires [28]. C. Barriocanal et al. compared the yields and characteristics of pyrolysis products from blends of coal and tire wastes in a fixed bed (FB) and a rotary oven (RO) [30]. It was found that the char yields obtained in the two configurations were similar and the RO promoted the production of gas while the FB produced a higher amount of oil. The pore characteristics of chars obtained from the two ovens were similar and the chars generated from the waste tires are mainly mesoporous whereas that from the coal contained a larger amount of macro- and micropores [31–33]. The oil produced in the RO was more aromatic and contained a smaller number of oxygen functional groups due to their higher residence time in the hot zone of the reactor. A.M. Fernández et al. found that the decrease in the fluidity of industrial coal blends after the addition of tire wastes and the ash composition of tires contributed to the deterioration in coke quality [34,35]. They found that the presence of Zn-bearing phases in the tire wastes increased the coke reactivity and proposed that a small number of waste tires (2 wt.%) should be incorporated into the industrial coal blends to guarantee coke quality and good blast furnace performance.

ZnO, as an activator for sulfur vulcanization, is widely used in tire rubber manufacturing with a weight content of 1–2% [36,37]. Zn is also known as a deleterious element for the production and life of the blast furnace [35–40], which is present in the blast furnace as a component of sinter charge or the coke in the form of an oxide (ZnO) and a sulphide (ZnS) and can be reduced by CO or carbon to metallic Zn [38,39]. The melting and boiling points of metallic Zn are 420 °C and 910 °C, respectively. Therefore, the metallic Zn is easily vaporized in the high-temperature regions and then condenses in the low-temperature zones or is re-oxidized to ZnO by CO_2 and water vapor [40]. Consequently, reduction, vaporization, condensation, oxidation and circulation of Zn could occur in the blast furnace [38,41]. The harmful effects of Zn on blast furnaces have been investigated by several studies [42,43]. The Zn vapor can flow into the air holes of the refractory lining, then the deposition and/or the oxidation of Zn can give rise to internal stress, volume expansion and material damage [42]. The coke acts as a fuel, a reductant, structural support and a carburizer in the blast furnace [43]. The penetration and deposition of Zn in the coke pores can weaken the coke strength and accelerate the pulverization of coke. The Zn vapor can also react with the primary minerals near the coke pores to form new Zn-bearing compounds, resulting in the accumulation of Zn in the coke [43]. Meanwhile, Zn can catalyze the gasification reaction of coke, thus the high content of Zn will increase the coke reactivity index (CRI) and decrease the coke strength after reaction (CSR) [43].

As mentioned above, the presence of Zn in the coke is detrimental to the quality of coke and the production of a blast furnace. Actually, the volatilization and condensation of Zn vapor may also be harmful to the refractory materials of the coke oven during the co-pyrolysis of coal with waste tires. Therefore, it is essential to study the transformation behaviors of Zn during co-pyrolysis of waste tire and coal. The main mineral components in coal are SiO_2, Al_2O_3, Fe_2O_3 and CaO [44]. Although the reactions between ZnO, carbon and S have been studied during the pyrolysis of waste tires alone [45–50], the impacts of inorganic components (Ca, Al, Si, Fe) originating from the coal on the Zn conversion and Zn distribution in the pyrolytic products during co-pyrolysis remain poorly understood. Considering these problems, the co-pyrolysis of coal and waste tires was carried out in a fixed bed reactor, and the Zn contents in the pyrolytic products (coke, tar and gas) with the different addition ratios of waste tires (or ZnO) at typical temperatures were studied.

The mineral compositions of cokes were analyzed by XRD measurement. Meanwhile, the thermodynamic equilibrium calculations were conducted using FactSage 8.0 to simulate the thermochemical conversion behaviors and phase distributions of Zn and other mineral elements (Ca, Al, Si and Fe) under different co-pyrolysis conditions. Additionally, the consistency between the experimental results and the thermodynamic equilibrium analysis was verified [51–53].

2. Materials and Methods

2.1. Materials

An industrial coal blend provided by an iron-making plant in China was used as the base coal, and three kinds of waste rubber powders (WT-1, WT-2 and WT-3) gained from the tire recycling industry were employed as additives, which were obtained by the grinding of tread rubbers from truck (WT-1) and car (WT-2) tires and sidewall rubber (WT-3), respectively. Table 1 provides the proximate and elemental analyses of base coal and tire wastes. The moisture content of sample was obtained by drying at 105 °C to constant mass. The ash content of the sample was obtained by calcining in air at 800 °C to constant mass. The sample was heated at 900 °C for 7 min in air to measure the volatile matter content. The contents of C, H, O, N and S elements were determined on an elemental analyzer (Elementar-vario EL cube). The contents of Zn, Al, Ca, Fe, Mg, Si and Ti elements in coal and WT-2 were measured using inductively coupled plasma atomic emission spectrometry (ICP-AES) on Agilent 720 spectrometer, as listed in Table 2. It is shown that the chemical compositions of three tire wastes are similar, and their H, O, S and Zn contents are higher than the base coal. WT-2 was selected as the additive to study the Zn distributions in the pyrolytic products by fixed-bed pyrolysis experiments and the transformation behaviors of Zn via thermodynamic simulations using FactSage 8.0.

Table 1. Characteristics of coal and tire wastes.

Amples	Proximate Analysis			Elemental Analysis					Zn (wt.% [a])
	Moisture (wt.%)	Ash (wt.% [a])	VM [b] (wt.% [a])	C (wt.% [c])	N (wt.% [c])	H (wt.% [c])	S (wt.% [c])	O (wt.% [c])	
Coal	1.85	7.36	26.24	89.14	1.60	1.78	0.92	6.56	0.01
WT-1	1.35	7.00	62.02	87.93	0.62	3.81	1.67	5.97	2.17
WT-2	1.40	9.24	60.36	85.82	0.60	4.31	1.88	7.39	2.13
WT-3	1.50	9.79	67.68	83.94	0.60	4.28	2.07	9.11	2.24

[a] Dry basis. [b] Volatile matter. [c] Dry ash-free basis.

Table 2. Mineral compositions of coal and WT-2.

Samples	Al (wt.%)	Ca (wt.%)	Fe (wt.%)	Mg (wt.%)	Si (wt.%)	Ti (wt.%)
Coal	1.23	0.27	0.33	0.025	1.51	0.053
WT-2	0.038	0.29	0.13	0.042	1.01	0.006

2.2. Experimental

The co-pyrolysis experiments were carried out on a temperature-controlled fixed-bed reactor system, which includes a heating, cooling and gas adsorption section, as illustrated in Figure 1. The coal was mixed with 1, 3 and 5 wt.% of WT-2 powders by ball milling, corresponding to 0.02, 0.06 and 0.1 wt.% of Zn in the hybrid, respectively. To further expand the scope of the study, the 1 and 5 wt.% of ZnO powders were also blended with the coal, related to 0.8 and 4 wt.% of Zn in the mixture, respectively. Then the blends were put into the quartz tube of fixed-bed reactor and heated to 700, 900 and 1050 °C and kept for 2 h with a heating rate of 5 °C/min in nitrogen, respectively. For comparison, the separate pyrolysis of coal or WT-2 was also conducted under the same condition. The produced tar was found to be condensed on the wall of the quartz tube and in the tar collector. To analyze the Zn content in tar, the quartz tube and tar collector were calcined in muffle furnace at 800 °C for 2 h and the obtained ash was dissolved in HCl aqueous solution (37.5 wt.%), which was subjected to inductively coupled plasma atomic emission spectrometry (ICP-AES) measurement. Additionally, the HCl aqueous solution in the gas absorber was also analyzed by ICP-AES to determine the amount of Zn vaporized into gas. Zn content in coke was determined by subtraction method.

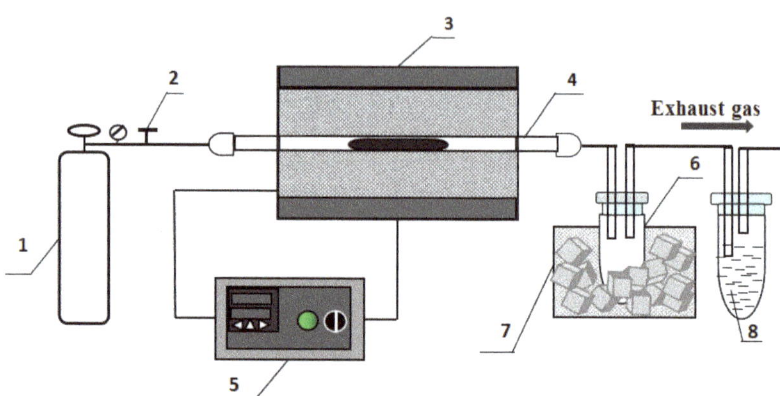

Figure 1. Schematic diagram of fixed-bed experimental device. 1—N_2 cylinder, 2—flowmeter, 3—electric heating furnace, 4—quartz tube, 5—program temperature controller, 6—coal tar collector, 7—ice water, 8—HCl solution.

2.3. Materials Characterization

The phase compositions of samples were analyzed by X-ray diffraction (XRD) on a Rigaku Smartlab 9 kw diffractometer with a Cu Kα radiation (λ = 0.15406 nm) in the 2θ range of 5–90° with a scanning rate of 10°/min. The thermal decomposition behaviors of samples were analyzed with thermogravimetric and differential scanning calorimetry (TG-DSC) on a Chi 449F3 thermogravimeter from room temperature to 1100 °C at a heating rate of 10 °C/min under a nitrogen flow.

2.4. Thermodynamic Equilibrium Simulation

In order to better understand the conversion behaviors of Zn during co-pyrolysis process, the thermodynamic equilibrium simulations were carried out using FactSage 8.0 based on the principle of Gibbs free energy minimization. The contents of C, H, O, N, S and mineral elements (Ca, Al, Si and Fe) of coal were used as the inputs. The amounts of S, Ca, Al, Si and Fe are 0.24, 0.07, 0.46, 0.53 and 0.06 mol per kg of coal, respectively. Additionally, the amount of Zn is 0.0019, 0.0052, 0.0117, 0.0182, 0.12 and 0.62 mol in per kg of coal when the Zn content is 0.01, 0.02, 0.06, 0.1, 0.8 and 4 wt.%, respectively. Equilibrium calculations were performed at the temperature range of 100–1500 °C with an interval of 100 °C in the nitrogen atmosphere under a pressure of 1 atm.

3. Results and Discussion

3.1. Phase Compositions of Coal and Tire Wastes

The phase compositions of coal and three tire wastes were analyzed by XRD, as shown in Figure 2a. The main compounds in coal are kaolinite $Al_2Si_2O_5(OH)_4$ (JCPDS 29-1488), quartz SiO_2 (JCPDS 99-0088), goethite $FeO(OH)$ (JCPDS 99-0055) and gypsum $CaSO_4(H_2O)_2$ (JCPDS 06-0047). Calcite containing magnesium $(Ca, Mg)CO_3$ (JCPDS 43-0697) is found in WT-2 and limestone $CaCO_3$ (JCPDS 86-0174) exists in WT-3. $CaSO_4$ (JCPDS 74-2421) and ZnO (JCPDS 99-0111) are detected in all three tire wastes. $CaSO_4$ is commonly used in rubber modification, which can effectively improve the mechanical performance and thermal stability of rubber [54]. The XRD patterns of ashes obtained by the calcination of coal and tire wastes at 800 °C for 1 h are displayed in Figure 2b. SiO_2, $CaSO_4$ and Fe_2O_3 (JCPDS 85-0599) are detected in the ash of coal. The ashes of three tire wastes are composed of SiO_2, $CaSO_4$, ZnO and Zn_2SiO_4 (JCPDS 37-1485). Zn_2SiO_4 is formed by the reaction of ZnO and SiO_2 at high temperatures [55].

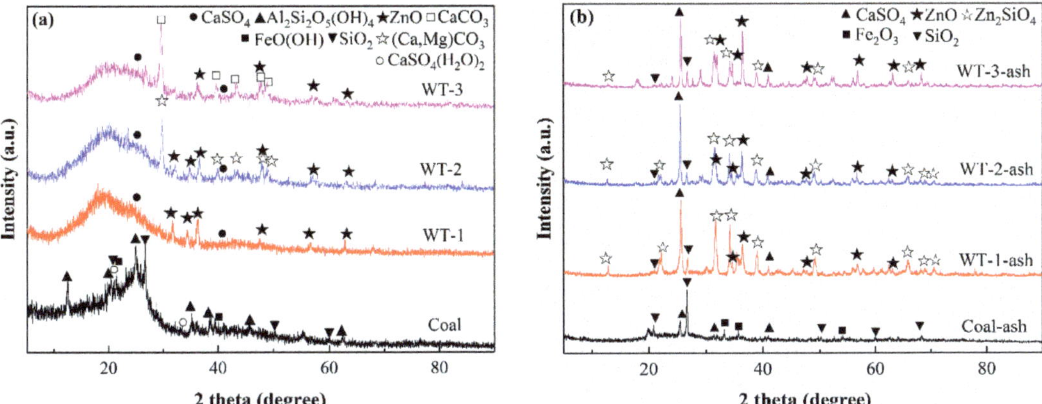

Figure 2. XRD patterns of base coal and three tire wastes (**a**) and their ashes (**b**).

3.2. TG-DSC Analysis

The influences of waste tires and ZnO additives on the thermal stability of coal were examined by TG-DSC measurements. The TG-DTG and DSC curves of coal, WT-2 and coal containing 5 wt.% of WT-2 or 5 wt.% of ZnO were displayed in Figure 3. The TG curve of coal can be divided into three stages, including dehydration, primary pyrolysis and secondary pyrolysis processes, giving rise to a total weight loss of 29.2% from room temperature to 1100 °C. The first stage from room temperature to 200 °C is attributed to the removal of physically adsorbed water, corresponding to a peak at 74 °C in the DTG curve and a small endothermic peak at around 82 °C in the DSC curve. The second weight loss takes place at 400–600 °C; meanwhile, a peak at 479 °C in the DTG curve and a broad endothermic peak at 482 °C in the DSC curve are observed, which is related to the primary pyrolysis of coal with the breaking and recombining of organic functional groups accompanied by the evolution of gas products such as CO_2, CO, light aliphatics, CH_4, H_2O and so on [56–58]. The third stage occurs above 700 °C, which is assigned to the secondary pyrolysis of condensed carbon matrix with the release of CO and H_2, corresponding to a broad endothermic peak at 600–1000 °C. The decomposition of waste tire WT-2 starts at 200 °C and a sharp DTG peak at 382 °C is observed to exhibit a weight loss of 68.9% up to 1100 °C. The low thermal stability of WT-2 should be due to the high volatile content and easily breakable molecular bonds of rubber [35,59]. The co-pyrolysis of coal with 5 wt.% of WT-2 displays an additional peak at 384 °C in the DTG curve and the weight loss increases by 1.7% up to 1100 °C compared with that of coal, indicating that more volatile matter is released after the addition of waste tire [60]. In comparison with coal, the DTG curve for a mixture of coal and 5 wt.% of ZnO shows an extra broad peak around 800 °C, and an exothermic peak at 830 °C in the DSC curve is detected, which is attributed to the carbothermal reduction of ZnO.

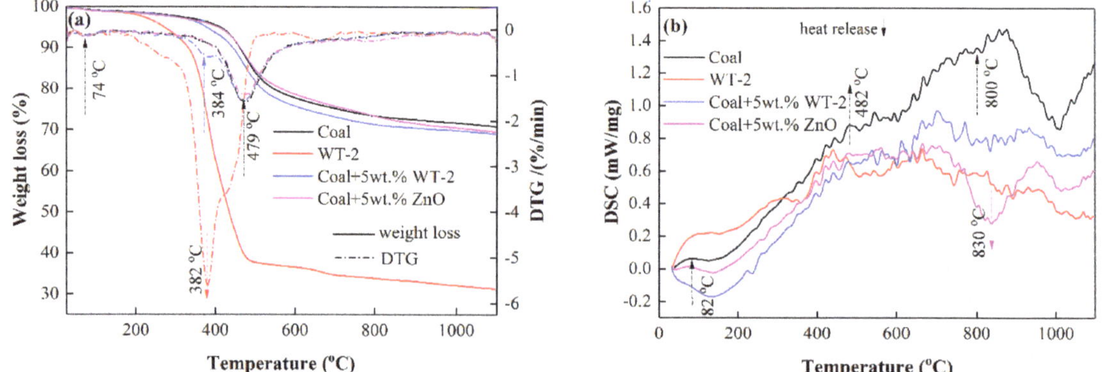

Figure 3. TG-DTG (**a**) and DSC (**b**) curves of coal, WT-2, coal containing 5 wt.% of WT-2 or 5 wt.% of ZnO.

3.3. Phase Compositions of Cokes

The XRD patterns of cokes generated by the separate pyrolysis of coal and WT-2 are displayed in Figure 4. As shown in Figure 4a, SiO_2 is detected in the coke obtained by the pyrolysis of coal at 700, 900 and 1050 °C. Additionally, CaS (JCPDS 08-0464) is found in the coke generated at 900 and 1050 °C. When the WT-2 pyrolyzes alone at 700 °C, the sphalerite ZnS (JCPDS 77-2100), ZnO, CaS and SiO_2 are detected in the resultant coke. With the temperature rising, the diffraction peaks of ZnO disappear in the coke obtained at 900 °C and those of ZnS cannot be detected in the coke formed at 1050 °C, indicating that the ZnS is more stable than ZnO in the coke, but ZnS can also be reduced by carbon to metallic Zn vapor at high temperature.

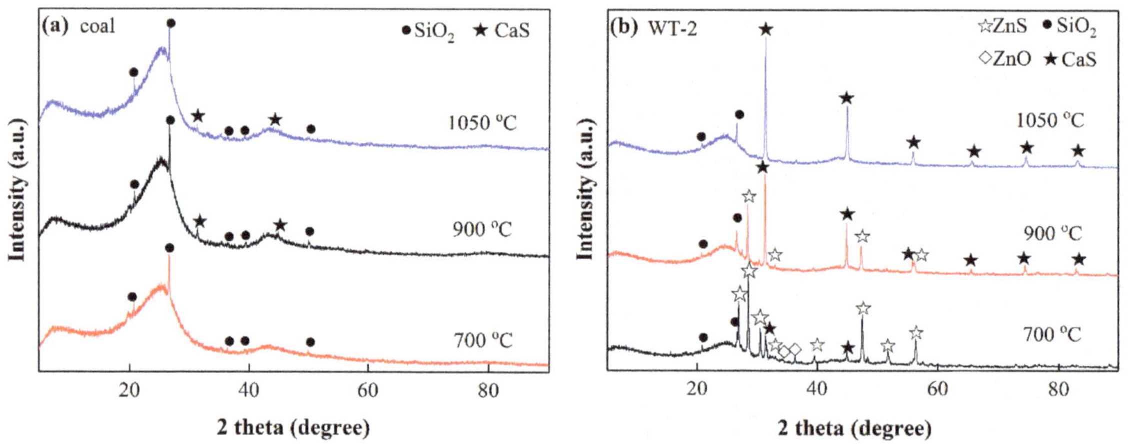

Figure 4. XRD patterns of cokes obtained by the separate pyrolysis of coal (**a**) and WT-2 (**b**) at different temperatures.

Figure 5 shows the XRD patterns of coke produced by pyrolysis of coal containing 1, 3 and 5 wt.% of WT-2, 1 and 5 wt.% of ZnO, corresponding to 0.02, 0.06, 0.1, 0.8 and 4 wt.% of Zn content in the blends, respectively. When the Zn content is low (0.06 and 0.1 wt.%), ZnS is only detected in the cokes obtained at 700 °C, and as the Zn content increases to 0.8 and 4 wt.%, ZnS exists in the cokes formed at 700–1050 °C. ZnS cannot be detected in the coke obtained by the pyrolysis of coal containing 0.02 wt.% of Zn, which should be due to the ultra-low content of Zn and the resultant ultra-fine crystallite size of ZnS. As the

Zn content increases to 4 wt.%, ZnO and ZnS coexist in the coke produced at 700 °C. The diffraction peaks of ZnO vanish and those of ZnAl$_2$O$_4$ (JCPDS 73-1961) appear at 900 °C, whereas the ZnAl$_2$O$_4$ phase disappears and the diffraction peaks of ZnS become weak at 1050 °C. The above results show that Zn preferentially combines with S to generate ZnS, and when the Zn content exceeds the molar of S (4 wt.% of Zn), excess ZnO can react with Al$_2$O$_3$ to form ZnAl$_2$O$_4$, which is also not stable at 1050 °C and can be further reduced to metallic Zn vapor by carbon. Moreover, the CaS phase can be detected in the cokes formed at 700, 900 and 1050 °C when the Zn content is low (0.02, 0.06 and 0.1 wt.%), which can only be observed in the cokes obtained at 900 and 1050 °C as the Zn content increases to 0.8 and 4 wt.%. This suggests that the S preferentially binds with Zn rather than Ca because CaS can only be formed once some ZnS decomposes.

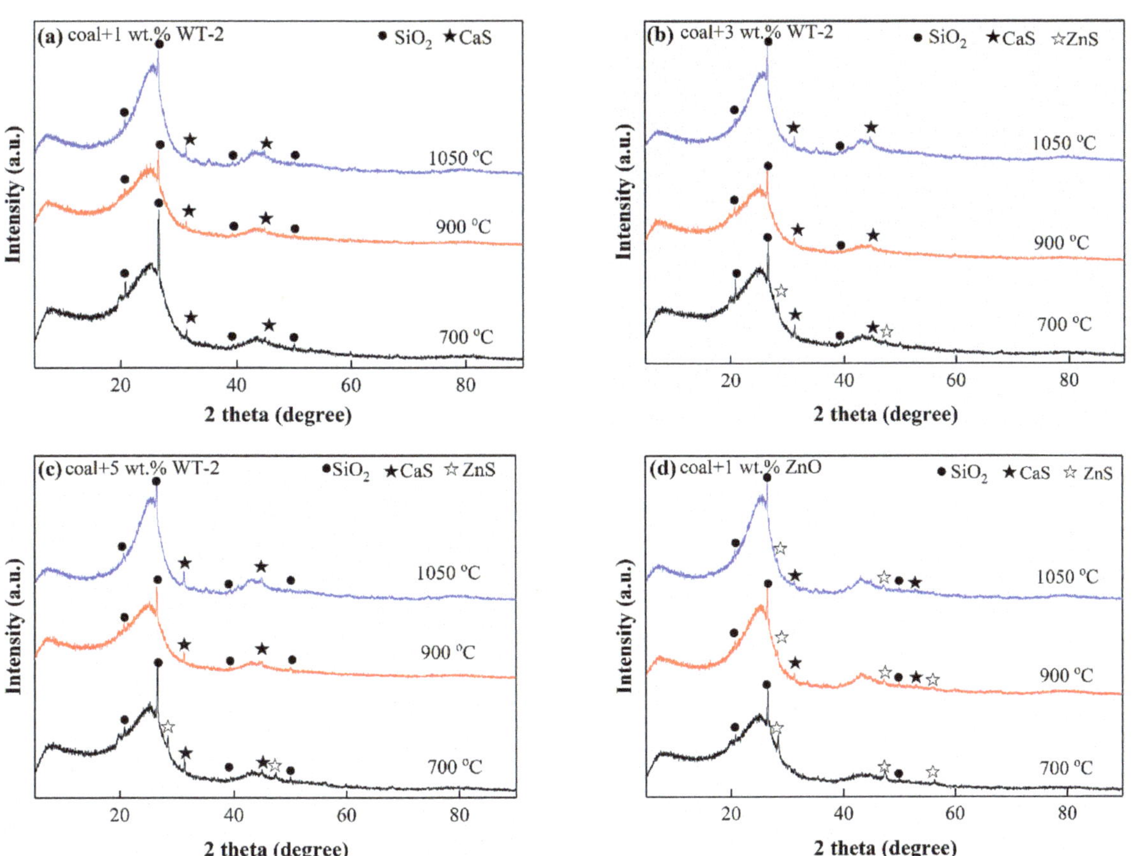

Figure 5. *Cont.*

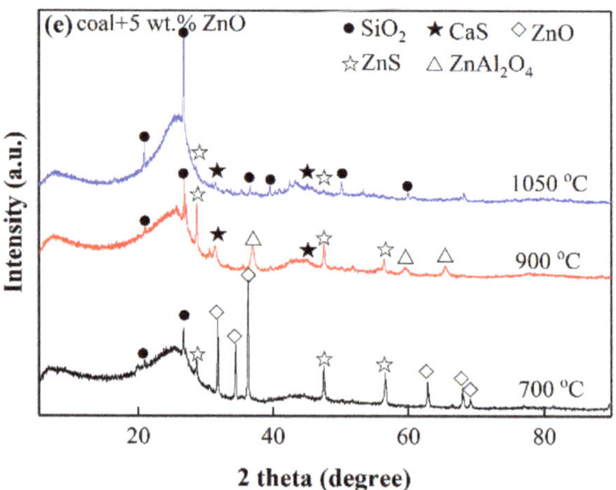

Figure 5. XRD patterns of coke generated by pyrolysis of coal containing 1 wt.% (**a**), 3 wt.% (**b**) and 5 wt.% (**c**) of WT-2, 1 wt.% (**d**) and 5 wt.% (**e**) of ZnO.

3.4. Migration of Zn during Pyrolysis

The mass distributions of Zn in the pyrolytic products (coke, tar and gas) under different pyrolysis conditions are listed in Table S1, and the corresponding relative percentages (%) and contents (g/g coke) of Zn are displayed in Figure 6. It should be emphasized that the results for separate pyrolysis of coal and co-pyrolysis of coal with 1 wt.% of WT-2 are not included due to the extremely low content of Zn in the pyrolytic products and the resultant large detection error. As shown in Figure 6a, c and e, the relative percentages of Zn in the coke decrease with the temperature increasing. The Zn residual rate in the coke produced at 700 °C is 97.4%, 97.3% and 96.8% as the Zn content is 0.06, 0.1 and 0.8 wt.% in the coal mixture, which declines to 90.3% when the Zn content is 4 wt.%. When the co-pyrolysis temperature rises to 900 and 1050 °C, most of the Zn escapes from the coke and enters into the tar. Moreover, the residual rate of Zn in the coke obtained at 900 °C when the Zn content is 0.8 and 4 wt.% is higher than that of coke produced from the coal containing 0.06 and 0.1 wt.% of Zn. Additionally, the residual rate of Zn in the coke obtained at 1050 °C from the coal owning 4 wt.% of Zn is slightly lower than that of coke produced from the coal containing 0.06, 0.1 and 0.8 wt.% of Zn. The Zn contents in the pyrolytic products were also calculated based on the per gram coke, as shown in Figure 6b,d,f. It is shown that the Zn content of obtained coke increases with the increase of Zn content in the coal at the same pyrolysis temperature and decreases with the increase of pyrolysis temperature under the same Zn content of coal. The Zn content of coke formed at 700 °C is highest as the coal containing 4 wt.% of Zn, reaching 0.047 g/g coke and the largest amount of Zn (0.051 g/g coke) migrates into the tar produced at 1050 °C from the coal owning 4 wt.% of Zn. When the WT-2 pyrolyzes alone, the Zn residual rate in the coke obtained at 700 °C is 88.3% and then decreases to 40.2% and 0.05% at 900 and 1050 °C, respectively (Figure 6a). Compared with the co-pyrolysis of coal and WT-2, the coke yield by separate pyrolysis of WT-2 is almost reduced by half (Table S1), which results in the enhancement of Zn content calculated based on the per gram coke in the pyrolytic products (Figure 6b,d,f).

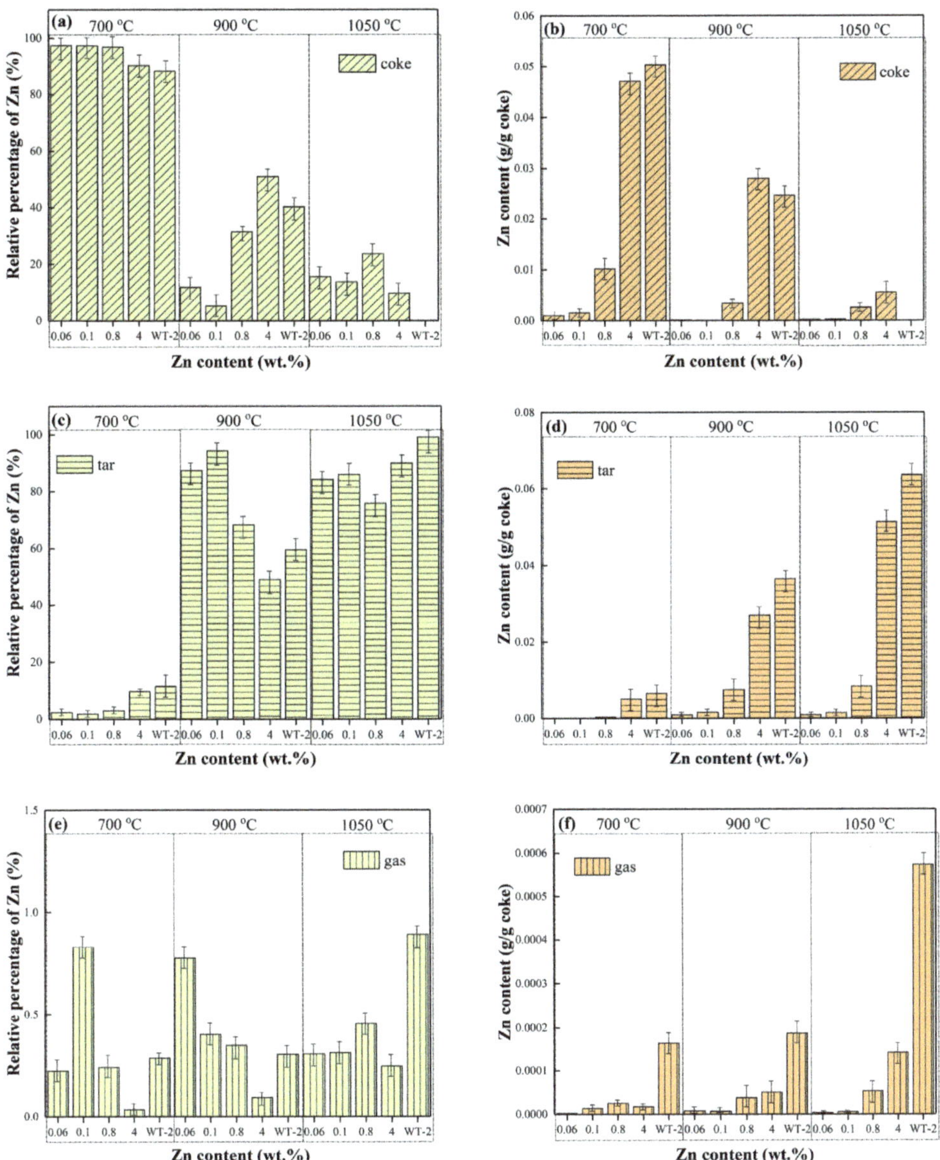

Figure 6. The relative percentages (%) (**a,c,e**) and contents (g/g coke) (**b,d,f**) of Zn in the coke, tar and gas under different pyrolysis conditions.

To sum up, when the coal containing 0.06, 0.1 and 0.8 wt.% of Zn (the molar ratio of Zn to S < 1) pyrolyzes at 700 °C, Zn is fixed in coke in the form of ZnS, resulting in a 97% Zn residual rate in coke. When the molar of Zn exceeds that of S (4 wt.% of Zn), ZnO and ZnS coexist in the coke produced at 700 °C and partial ZnO is reduced to gaseous Zn by carbon, causing a slight decrease in Zn residual rate in the coke (90.3%). It is noted that the Zn residual rate in the coke obtained by separate pyrolysis of WT-2 at 700 °C is 88.3%, which is lower than that of co-pyrolysis of coal with WT-2, although the molar ratio of Zn to S in the WT-2 is 0.64. This should be attributed to the large crystallite size of ZnO in the WT-2, which cannot be completely transformed into ZnS at 700 °C, as shown in Figure 4b, resulting in the coexistence of ZnO and ZnS in the coke obtained at 700 °C. As the pyrolysis temperature increases, ZnS is reduced to metallic Zn vapor by carbon, so the residual rate of Zn in coke decreases and the percentage of Zn in tar increases. In addition, the residual rate of Zn in coke is related to the Zn content of coal. As the Zn content of coal is 0.8 wt.%, a large amount of ZnO is reduced slowly by carbon with the temperature rising and leads to a higher Zn residual rate in the cokes obtained at 900 and 1050 °C. When the Zn content of coal is 4 wt.%, excess ZnO reacts with Al_2O_3 to generate $ZnAl_2O_4$ at 900 °C to enhance the Zn residual rate in coke to 50%. Additionally, $ZnAl_2O_4$ can also be reduced by carbon to generate gaseous Zn at 1050 °C, giving rise to a reduction of Zn residual rate in coke of 10%. Based on the above experimental results, the relative percentage distribution of Zn in the pyrolytic products is closely related to the content of S and mineral elements in the coal. FactSage 8.0 is further used to calculate the thermochemical conversion behaviors and phase distributions of Zn at different temperatures in detail.

3.5. Thermodynamic Equilibrium Analysis

3.5.1. The Interactions of Zn with S

The effect of S on the thermodynamic equilibrium distributions of Zn in the coke was investigated, as shown in Figure 7. When only ZnO exists in the coal, ZnO is reduced by carbon to generate metallic Zn vapor at the temperature range of 500–700 °C with 0.01 wt.% of Zn content (0.0019 mol/kg coal), as displayed in Figure 7a. As the Zn content is increased to 4 wt.% (0.62 mol/kg coal), the stability of ZnO is improved, which decomposes at the temperature range of 600–900 °C (Figure 7b). It is found that the generated amount of CO after the addition of ZnO is 0.62 mol more than that of the base coal (Figure 7c), so the reaction of ZnO and carbon produces CO gas, as described in Reaction (1). When the Zn content is 0.01 wt.% (0.0019 mol/kg coal), the presence of S (0.24 mol/kg coal) suppresses the volatilization of Zn due to the formation of ZnS, which is further reduced to gaseous Zn by carbon at 700–1000 °C (Figure 7d). ZnO and ZnS coexist in the coke as the Zn content is 4 wt.% (0.62 mol/kg coal). The gaseous Zn is produced at 600 °C by the reduction of ZnO and ZnS, which starts to decompose at 900 °C. Some ZnS undergo crystal phase transformation at 1000–1200 °C, from sphalerite ZnS(s) to wurtzite ZnS(s2) and are completely reduced to Zn vapor at 1200 °C (Figure 7e). The thermodynamic equilibrium distributions of Zn with the Zn contents of 0.02 wt.% (0.0052 mol/kg coal), 0.1 wt.% (0.0182 mol/kg coal) and 0.8 wt.% (0.12 mol/kg coal) are also calculated and displayed in Figure S1, further demonstrating that the thermal stabilities of ZnO and ZnS are enhanced with the increase of Zn content.

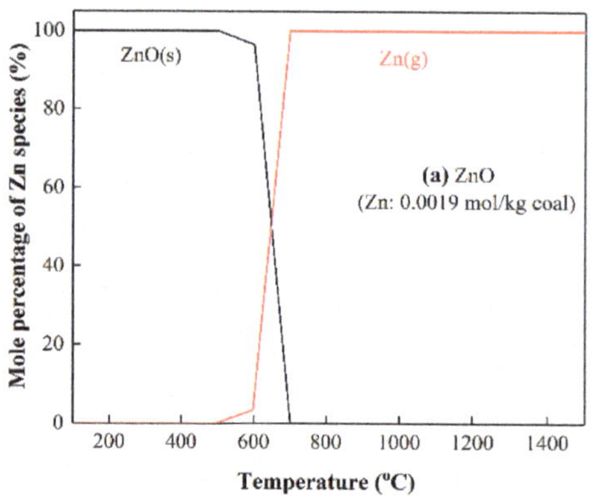

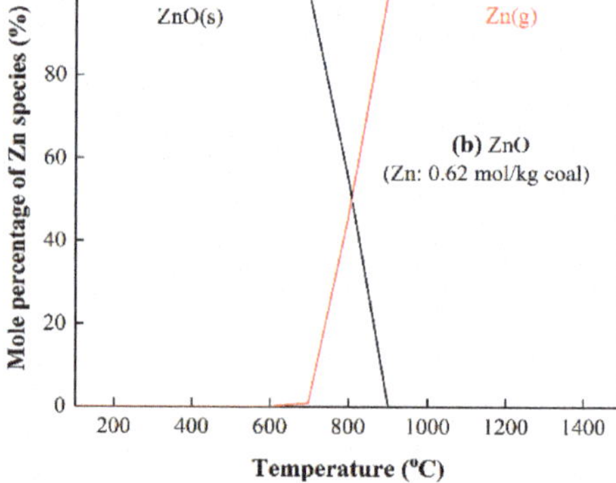

Figure 7. *Cont.*

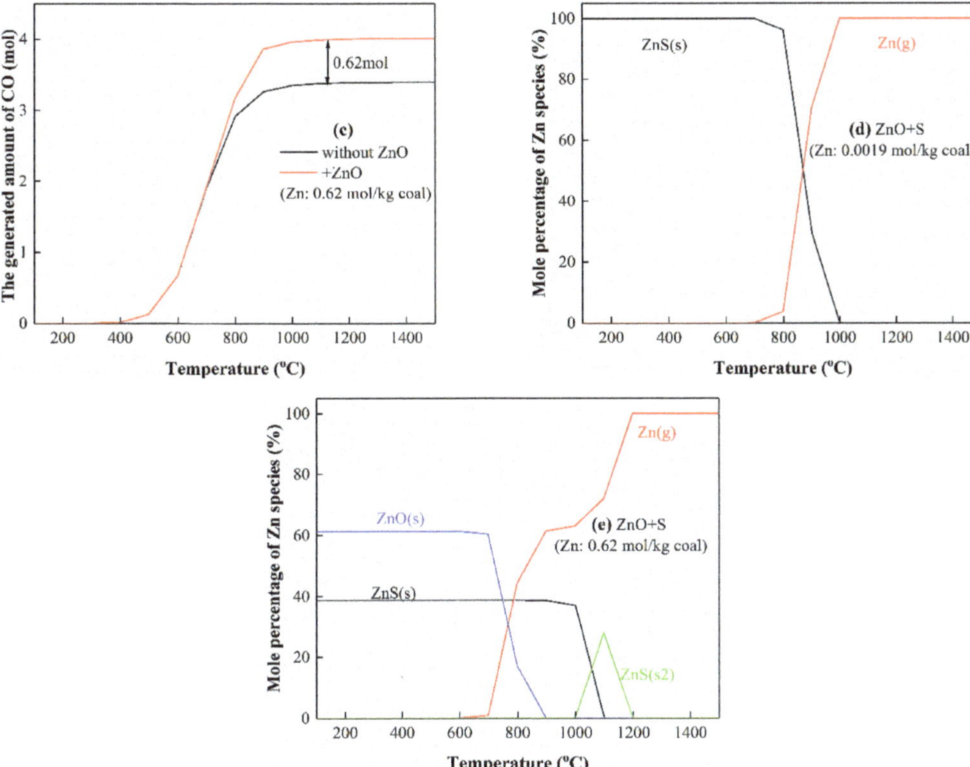

Figure 7. The thermodynamic equilibrium distributions of Zn (**a–e**) and generated amount of CO (**c**) when S and Zn coexist in the coal.

$$ZnO + C \rightarrow Zn(g) + CO(g) \tag{1}$$

3.5.2. The Interactions of Zn with S and Ca

The effect of the co-existence of S and Ca on the thermodynamic equilibrium distribution of Zn was analyzed, as shown in Figure 8. When the Zn content is low (Zn/S/Ca = 0.0019/0.24/0.07), the Zn incorporates with S to form ZnS, which decomposes into gaseous Zn from 700 to 900 °C (Figure 8a). Additionally, Ca also combines with S to generate CaS at 400–500 °C, which is stable before 1500 °C (Figure 8b). When the amount of Zn is high (Zn/S/Ca = 0.62/0.24/0.07), CaO can react with ZnS at 800–900 °C to form CaS and gaseous Zn (Figure 8c,d). Meanwhile, 0.07 mol of CO gas is generated in this process (Figure 8e), as described in Reaction (2). The above results indicate that the presence of Ca reduces the stability of ZnS and promotes the volatilization of Zn from coke. Consistent with the XRD results, CaS can be detected in the coke generated at 700 °C when the Zn content is low (0.0019 mol/kg coal), while CaS is only formed in the coke obtained at 900 °C when the Zn content is high (0.62 mol/kg coal).

$$CaO + ZnS + C \rightarrow CaS + Zn(g) + CO(g) \tag{2}$$

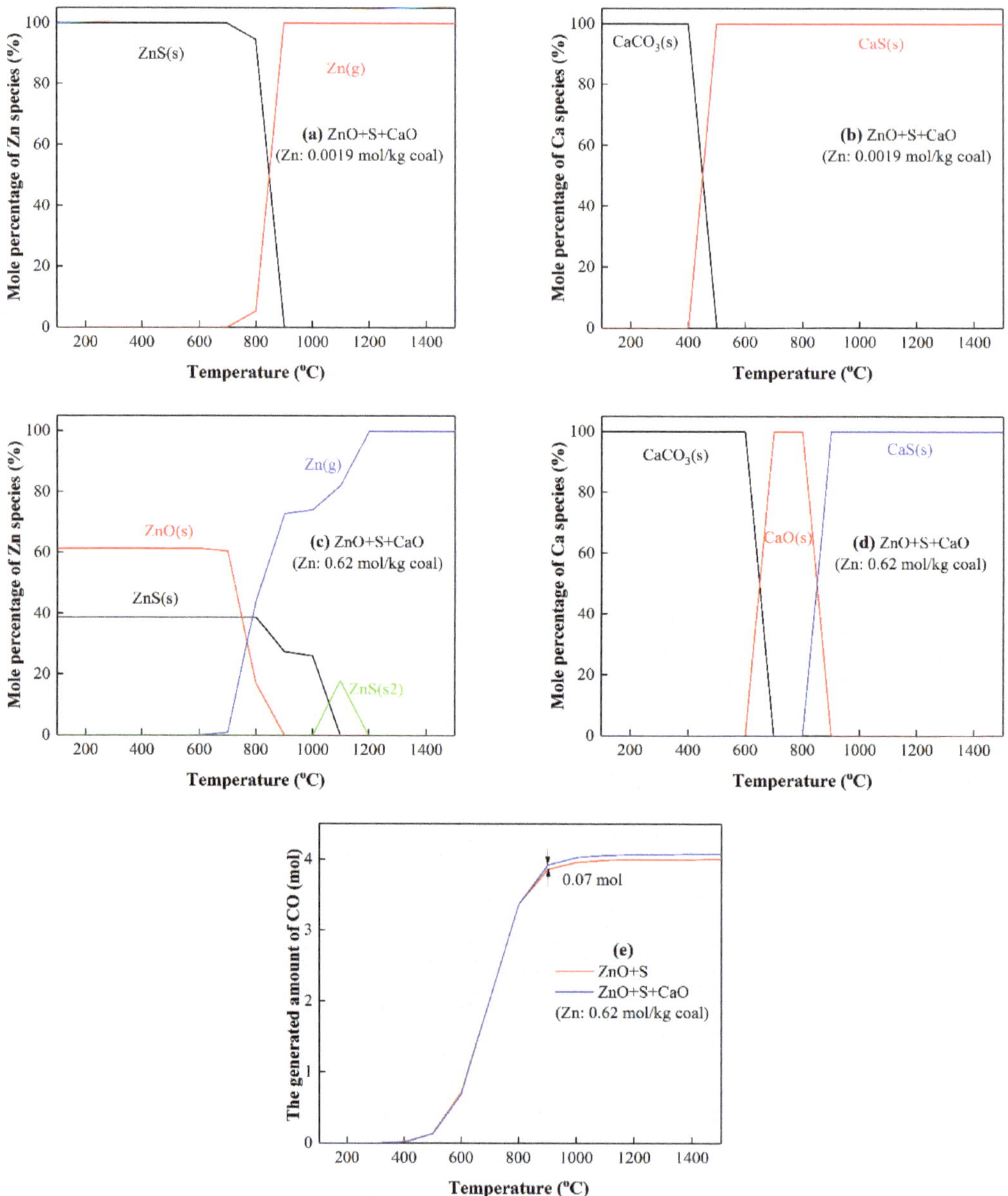

Figure 8. The thermodynamic equilibrium distributions of Zn (**a,c**), Ca (**b,d**) and generated amount of CO (**e**) when S, Zn and Ca coexist in the coal.

3.5.3. The Interactions of Zn with S and Fe

The influence of the co-existence of S and Fe on the thermodynamic equilibrium distributions of Zn is displayed in Figure 9. When the Zn content is low (Zn/S/Fe = 0.0019/0.24/0.06), ZnS is formed and reduced by carbon to gaseous Zn at the temperature range of 700–1000 °C (Figure 9a). Fe combines with S to form FeS_2 (pyrite), FeS_2 is transformed to FeS(s) (monoclinic pyrrhotite) at 200–300 °C, FeS(s) is converted to FeS(s2) (hexagonal pyrrhotite) at 300–400 °C, and FeS(s2) reacts with carbon to form Fe_3C (cementite) at 1300–1500 °C (Figure 9b). When the Zn content is high ($Zn/S/Fe = 0.62/0.24/0.06$), Fe combines with partial Zn to form $ZnFe_2O_4$ (Figure 9c), as indicated in Reaction (3). $ZnFe_2O_4$ decomposes into ZnO and Fe_3O_4 at 400–500 °C. Fe_3O_4 is continuously reduced to FeO and elemental Fe by carbon at 500–700 °C, then Fe reacts with carbon at 800–900 °C to form Fe_3C. Fe_3C is further transformed to FeS at 1000–1100 °C, meanwhile, the ZnS is decomposed to metallic Zn gas, finally, FeS is converted to Fe_3C at 1300–1500 °C (Figure 9d). The above results demonstrate that the presence of Fe has no effect on the interaction of Zn and S. In addition, when Fe and Ca coexist in coal, compared with the case where only Fe or Ca exists, the phase distributions of Zn and Ca do not change within 100–1500 °C, but the stability of FeS becomes worse. The FeS is transformed to Fe_3C from 1200 °C, as shown in Figure S2.

$$ZnO + Fe_2O_3 \rightarrow ZnFe_2O_4 \tag{3}$$

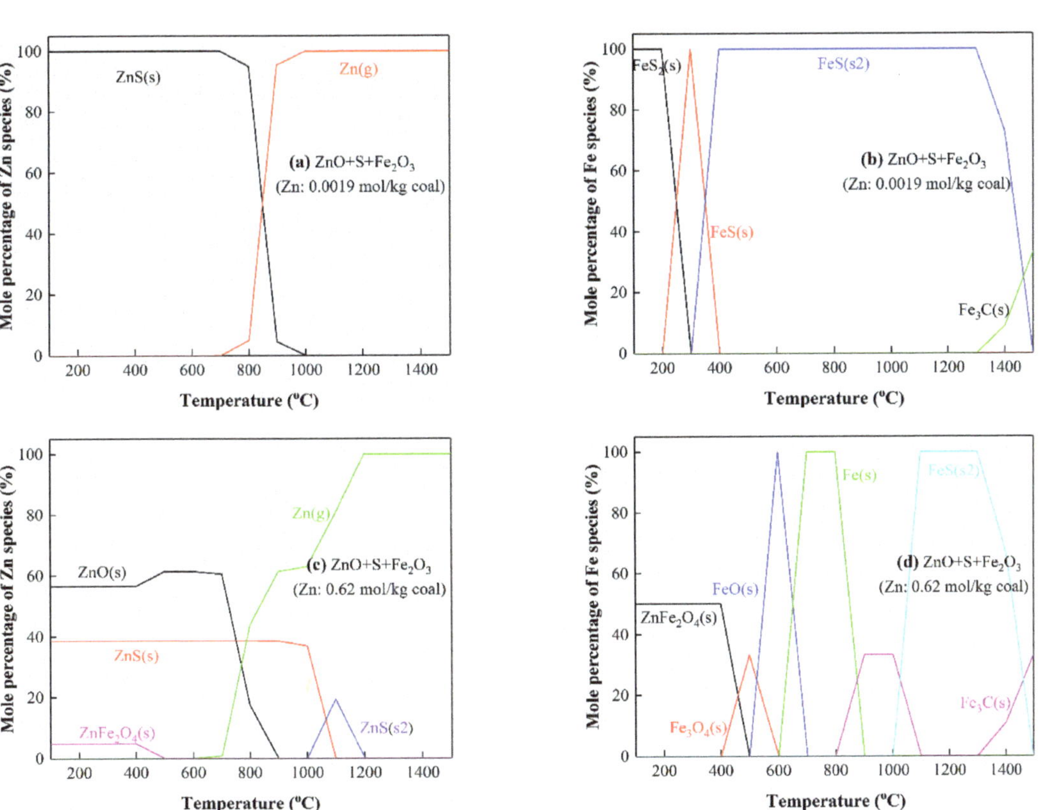

Figure 9. The thermodynamic equilibrium distributions of Zn (**a**,**c**) and Fe (**b**,**d**) when S, Zn and Fe coexist in the coal.

3.5.4. The Interactions of Zn with Si

The thermodynamic equilibrium distributions of Zn and Si as Zn and Si coexist in the coal are displayed in Figure 10. When the Zn content is low ($Zn/Si = 0.0019/0.53$), partial Si combines with Zn to form Zn_2SiO_4 (Reaction (4)), which is further reduced to gaseous Zn by carbon at 600–800 °C (Reaction (5)), as shown in Figure 10a. $SiO_2(s)$ (α-quartz) is converted to $SiO_2(s2)$ (β-quartz) at 500–600 °C, $SiO_2(s2)$ is transformed to $SiO_2(s3)$ (tridymite) at 800–900 °C and partial $SiO_2(s3)$ reacts with carbon to form $SiO(g)$ and $CO(g)$ at 1300–1500 °C (Reaction (6)), and $SiC(s)$ and $CO(g)$ at 1400–1500 °C (Reaction (7)), respectively (Figure 10b). When the Zn content is high ($Zn/Si = 0.62/0.53$), the stability of

Zn$_2$SiO$_4$ is improved. It reacts with carbon to generate SiO$_2$(s2) and Zn(g) at 700–800 °C and SiO$_2$(s3) and Zn(g) at 800–900 °C (Reaction (5)), respectively (Figure 10c,d). The phase transitions of Si at the temperature range of 900–1500 °C are the same as the condition of low Zn content (Figure 10d). When S and Si coexist in the system, as shown in Figure S3, Zn preferentially reacts with S to form ZnS. When the Zn content is high (molar ratio of Zn/S > 1), the excessive Zn combines with Si to generate Zn$_2$SiO$_4$, which is converted to SiO$_2$(s2) and Zn(g) at 700–800 °C and SiO$_2$(s3) and Zn(g) at 800–900 °C. In addition, SiO$_2$(s3) can react with S and carbon to generate SiS(g) and CO(g) at 1200 °C (Reaction (8)). The generated amount of CO during the above processes is shown in Figure 10e.

$$2ZnO + SiO_2 \rightarrow Zn_2SiO_4 \tag{4}$$

$$Zn_2SiO_4 + 2C \rightarrow SiO_2 + 2Zn(g) + 2CO(g) \tag{5}$$

$$SiO_2 + C \rightarrow SiO(g) + CO(g) \tag{6}$$

$$SiO_2 + 3C \rightarrow SiC + 2CO(g) \tag{7}$$

$$SiO_2 + S + 2C \rightarrow SiS(g) + 2CO(g) \tag{8}$$

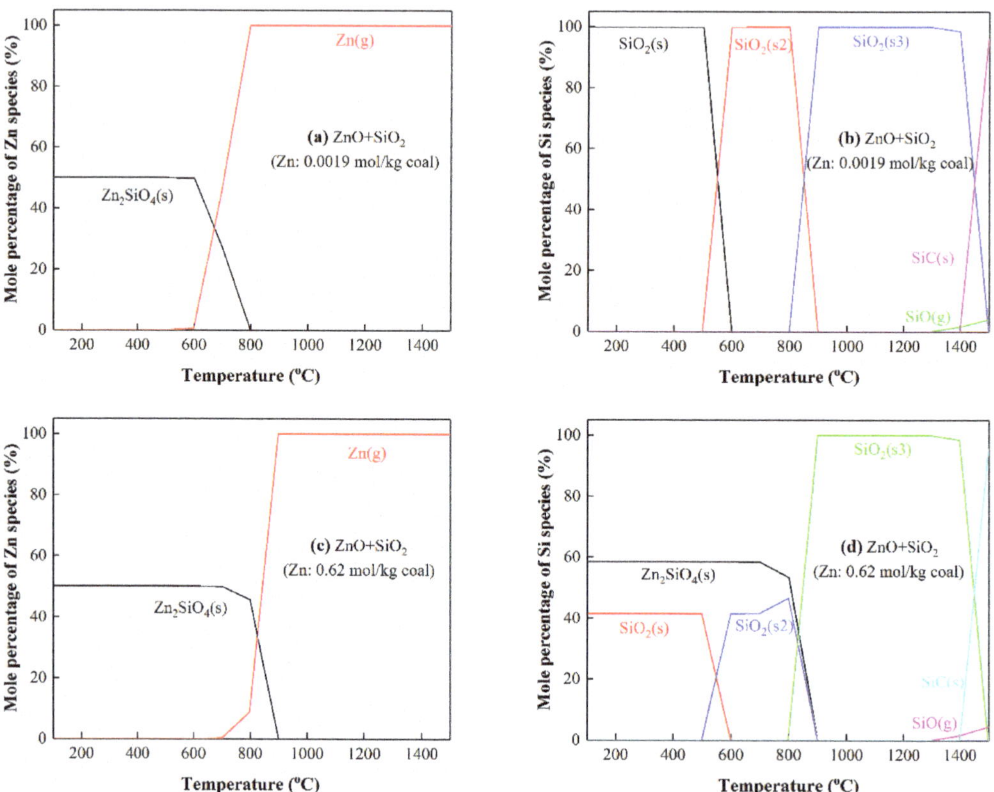

Figure 10. *Cont.*

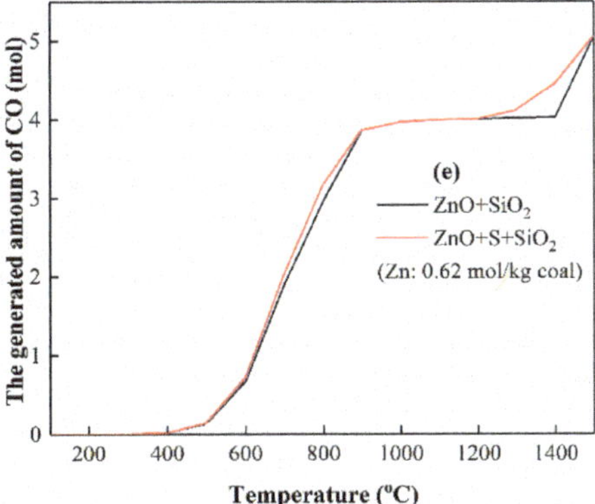

Figure 10. The thermodynamic equilibrium distributions of Zn (**a**,**c**), Si (**b**,**d**) and generated amount of CO (**e**) when Zn and Si coexist in the coal.

3.5.5. The Interactions of Zn with Al

The interactions of Zn and Al were also investigated, as shown in Figure 11. When the Zn content is low (Zn/Al = 0.0019/0.46), partial Al combines with Zn to form $ZnAl_2O_4$(s) (Reaction (9)), and $ZnAl_2O_4$ is reduced to gaseous Zn and Al_2O_3(s) by carbon at 600–800 °C (Reaction (10)), as shown in Figure 11a,b. When the Zn content is high (Zn/Al = 0.62/0.46), Zn exists as ZnO(s) and $ZnAl_2O_4$(s), which are reduced to Zn(g) by carbon at 600–900 and 900–1000 °C (Figure 11b,c), respectively, suggesting that the stability of $ZnAl_2O_4$ is higher than that of ZnO and that the existence of Al inhibits the volatilization of Zn from coke. The reaction of $ZnAl_2O_4$ and carbon can also produce CO gas, and the generated amount of CO is shown in Figure 11e, which is the same as the condition that only ZnO exists. When S and Al coexist in the system, as shown in Figure S4, Zn preferentially reacts with S to form ZnS. When the Zn content is high, Zn exists in the form of ZnS, $ZnAl_2O_4$ and ZnO, which are decomposed at the temperature range of 900–1200, 800–1000 and 600–800 °C, respectively. Consistent with the XRD results, $ZnAl_2O_4$ can be detected in the coke obtained at 900 °C when the Zn content is high (molar ratio of Zn/S > 1).

$$ZnO + Al_2O_3 \rightarrow ZnAl_2O_4 \tag{9}$$

$$ZnAl_2O_4 + C \rightarrow Al_2O_3 + Zn(g) + CO(g) \tag{10}$$

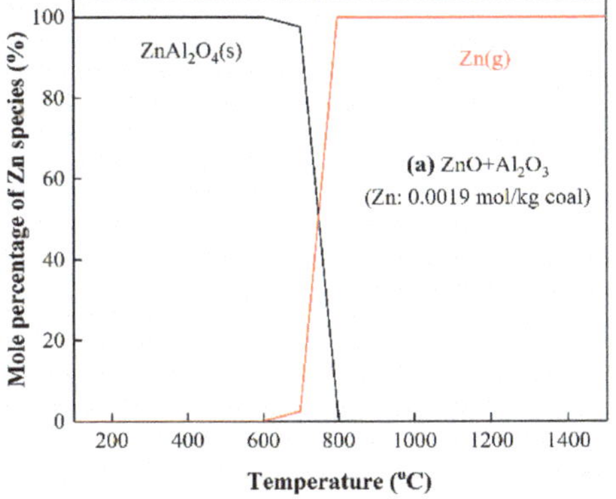

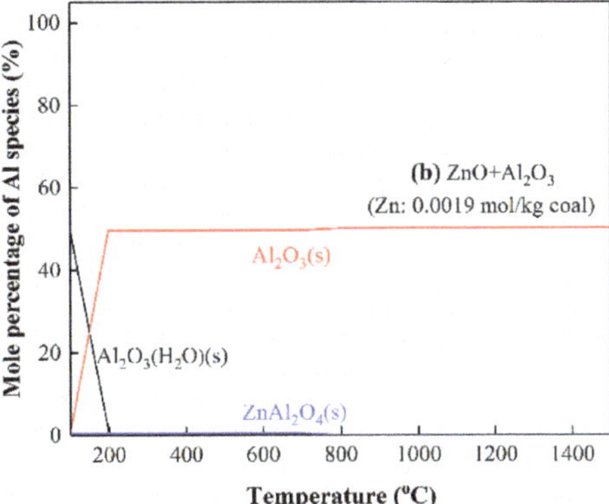

Figure 11. *Cont.*

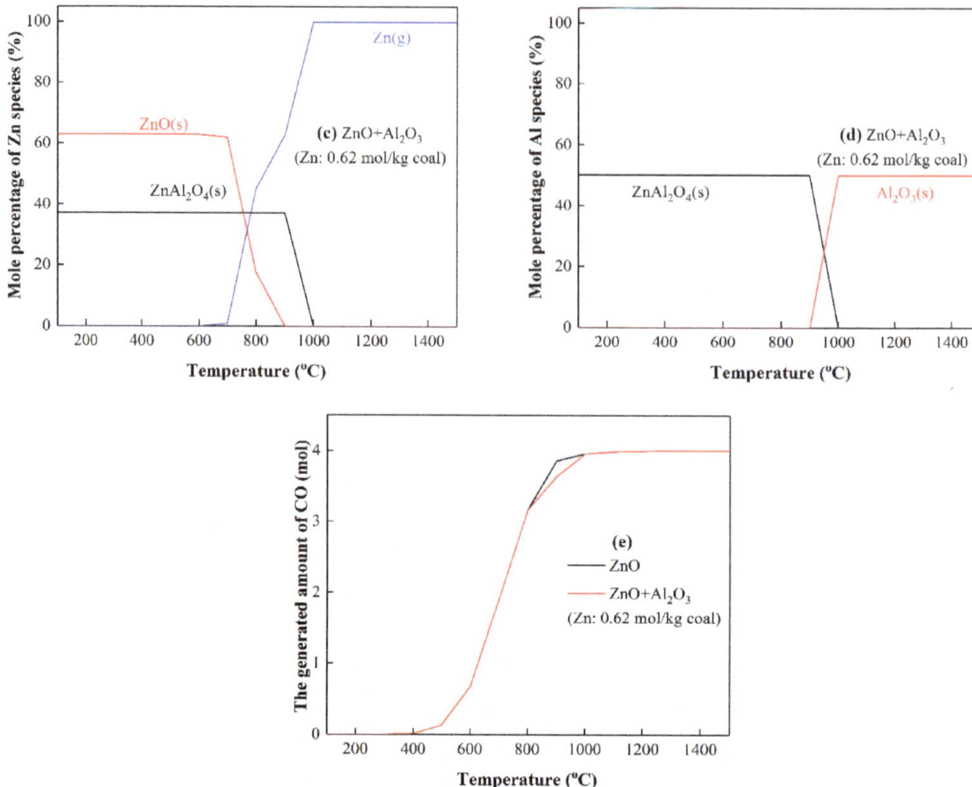

Figure 11. The thermodynamic equilibrium distributions of Zn (**a,c**), Al (**b,d**) and generated amount of CO (**e**) when Zn and Al coexist in the coal.

3.5.6. The Interactions of Zn with Si and Al

The effect of the co-existence of Si and Al on the thermodynamic equilibrium distributions of Zn was also investigated, as shown in Figure 12. When the Zn content is low (Zn/Si/Al = 0.0019/0.53/0.46mol), Zn combines with Al to form $ZnAl_2O_4(s)$ (Figure 12a), indicating that Al is easier to react with Zn than Si. The Si and residual Al exists as $(Al_2O_3)(SiO_2)_2(H_2O)_2(s)$ (kaolinite) and $SiO_2(s)$, and $(Al_2O_3)(SiO_2)_2(H_2O)_2(s)$ decomposes into $Al_2SiO_5(s)$ (kyanite) and $SiO_2(s)$ at 100–200 °C. Then $Al_2SiO_5(s)$ (kyanite) transforms to $Al_2SiO_5(s2)$ (andalusite) at 200–300 °C, and $SiO_2(s)$ (α-quartz) converts into $SiO_2(s2)$ (β-quartz) at 500–600 °C (Figure 12c). The $Al_2SiO_5(s2)$ (andalusite) decomposes into $Al_6Si_2O_{13}(s)$ and $SiO_2(s2)$ at 700–800 °C (Figure 12c,e), as described in reaction (11). Additionally, $Al_6Si_2O_{13}(s)$ reacts with carbon to form $Al_2O_3(s)$, $SiC(s)$ and $CO(g)$ at 1400–1500 °C (reaction (12)). When Zn content is high (Zn/Si/Al = 0.62/0.53/0.46), Zn combines with Al to form $ZnAl_2O_4(s)$ and partial Si to generate $Zn_2SiO_4(s)$ (Figure 12b). Zn_2SiO_4 is reduced by carbon to generate $SiO_2(s2)$ and gaseous Zn at 700–900 °C (Figure 12b,d). Additionally, $ZnAl_2O_4$ reacts with $SiO_2(s3)$ and carbon to generate $Al_6Si_2O_{13}(s)$, $Zn(g)$ and $CO(g)$ at 900–1000 °C (Figure 12b,d,f), as described in reaction (13). The generated amount of CO in the reactions (12) and (13) is shown in Figure 12g. When S is present in the system, Zn preferentially reacts with S to form ZnS and the excess Zn combines with Al and Si to generate $ZnAl_2O_4$ and Zn_2SiO_4, respectively, as shown in Figure S5. There is no change in the phase distributions of Al and Si except for the formation of $SiS(g)$ at 1200 °C. The standard-state Gibbs free energies $(\Delta_r G_m^\theta)$ as a function of temperature for reactions (1)–(13) are calculated by FactSage 8.0 and listed in Table S2 and Figure S6. When the Gibbs free energy is negative, the reaction is spontaneous. It is shown that the favored temperature ranges for different reactions are consistent with the generation sequence of corresponding products discussed above.

$$3Al_2SiO_5 \rightarrow Al_6Si_2O_{13} + SiO_2 \tag{11}$$

$$Al_6Si_2O_{13} + 6C \rightarrow 3Al_2O_3 + 2SiC + 4CO(g) \tag{12}$$

$$3ZnAl_2O_4 + 2SiO_2 + 3C \rightarrow Al_6Si_2O_{13} + 3Zn(g) + 3CO(g) \tag{13}$$

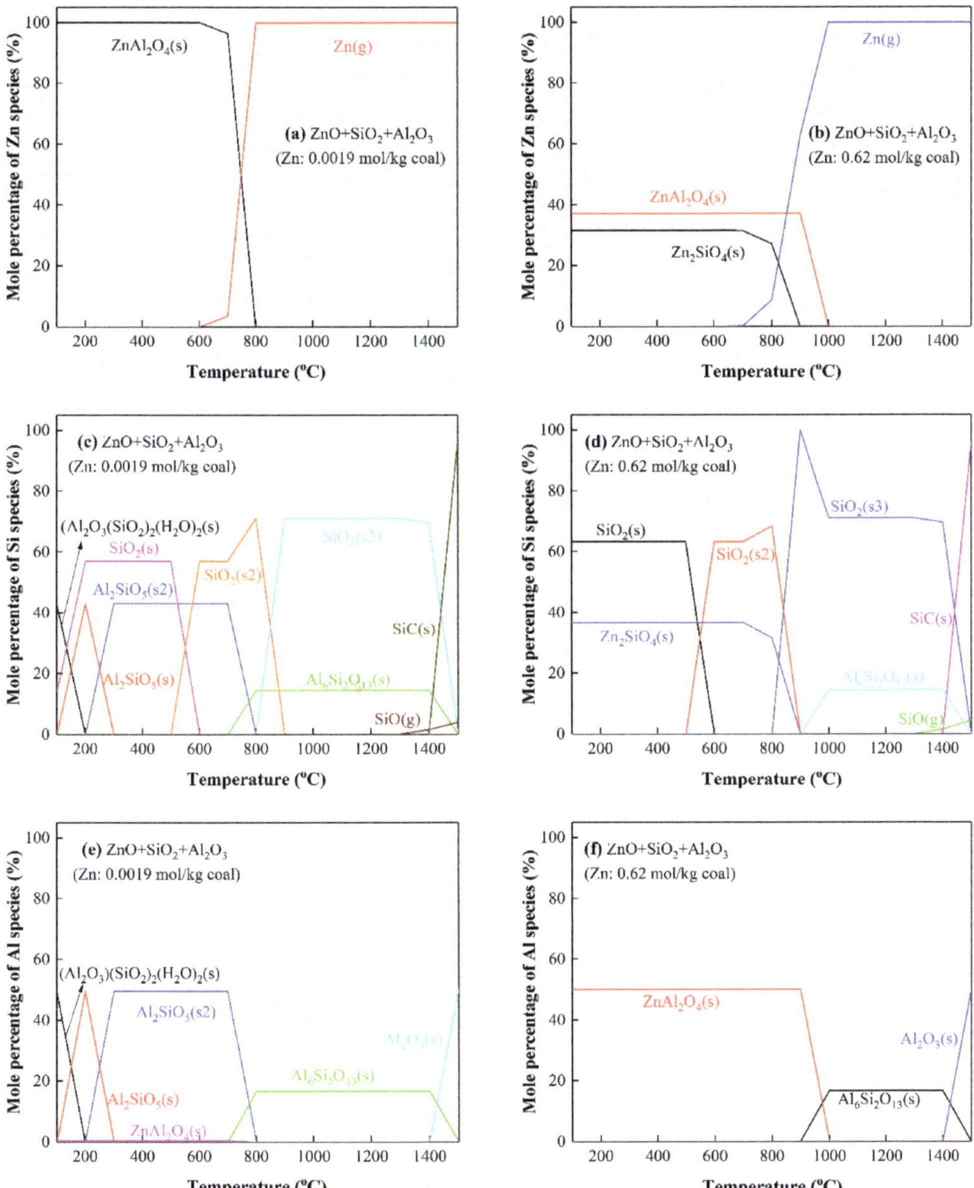

Figure 12. *Cont.*

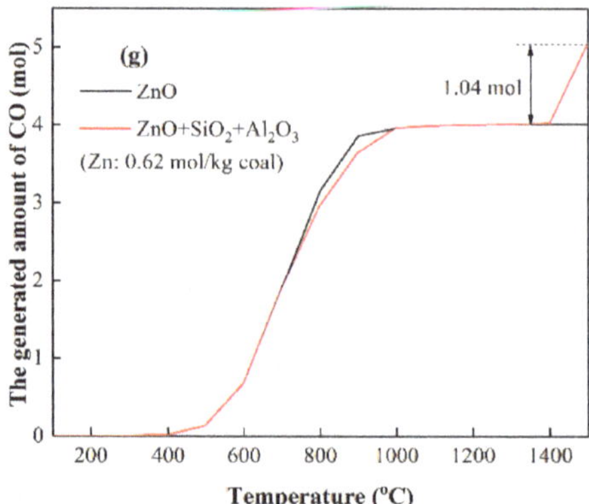

Figure 12. The thermodynamic equilibrium distributions of Zn (**a,b**), Si (**c,d**), Al (**e,f**) and generated amount of CO (**g**) when Zn, Si and Al coexist in the coal.

3.5.7. The Interactions of Zn with Ca and Al

The effect of the co-existence of Ca and Al on the phase distributions of Zn was investigated, as shown in Figure 13. When the Zn content is low (Zn/Ca/Al = 0.0019/0.07/0.46), Zn combines with Al to form $ZnAl_2O_4$, which decomposes to gaseous Zn at 600–800 °C (Figure 13a). $CaAl_4O_7(s)$ and $CaAl_{12}O_{19}(s)$ are formed by the interaction between Ca and Al at 400 °C, which exists stably at high temperatures (Figure 13c,e). When the Zn content is high (Zn/Ca/Al = 0.62/0.07/0.46), Zn exists as ZnO and $ZnAl_2O_4$, and partial $ZnAl_2O_4$ begins to decompose at 600–700 °C (Figure 13b), and the generated Al species combines with Ca to form $Ca_3Al_2O_6$ (Figure 13d,f). $Ca_3Al_2O_6$ transforms into $CaAl_4O_7$ at 800–900 °C, and part of $CaAl_4O_7$ converts into $CaAl_{12}O_{19}$ at 900–1000 °C. The above results show that there is a competitive relationship between Ca and Zn in the interaction of Al. Below 600 °C, Zn preferentially reacts with Al to generate $ZnAl_2O_4$. With the increase in temperature, Ca begins to plunder Al to form Ca aluminate. When S is also present in the above system (Figure S7), ZnS is generated as the Zn content is low (Zn/S/Ca/Al = 0.0019/0.24/0.07/0.46). $CaAl_{12}O_{19}$ and CaS are formed at 400–500 °C, and $CaAl_{12}O_{19}$ begins to decompose and reacts with S to form CaS and Al_2O_3 at 700–800 °C. When the Zn content is high (Zn/S/Ca/Al = 0.62/0.24/0.07/0.46), Zn exists as ZnS, $ZnAl_2O_4$ and ZnO. $ZnAl_2O_4$ decomposes into ZnO at 600–700 °C and gaseous Zn at 700–1000 °C, respectively. Meanwhile, $Ca_3Al_2O_6$ is generated at 600–700 °C, and partial $Ca_3Al_2O_6$ transforms into $CaAl_2O_4$ at 700–800 °C, then the residual $Ca_3Al_2O_6$ and $CaAl_2O_4$ converts into $CaAl_4O_7$ at 800–900 °C. Subsequently, part of $CaAl_4O_7$ turns into $CaAl_{12}O_{19}$ at 900–1000 °C, and the remaining $CaAl_4O_7$ and $CaAl_{12}O_{19}$ combine with S to form CaS and Al_2O_3 at 1000–1100 °C.

3.5.8. The Interactions of Zn with S, Ca, Al, Si and Fe

Finally, when Zn, Ca, Al, Si and Fe are simultaneously present in the coal, their phase distributions are displayed in Figure 14 and Figure S8. As the Zn content is low (Zn/S/Ca/Al/Si/Fe = 0.0019/0.24/0.07/0.46/0.53/0.06), Zn still interacts with S to generate ZnS, which turns into gaseous Zn at 700–1000 °C (Figure 14a). The reactions among Ca, Si and Al are complex. $CaAl_2Si_2O_7(OH)_2(H_2O)$, $(CaO)_2(Al_2O_3)_2(SiO_2)_8(H_2O)_7$ and $(Al_2O_3)(SiO_2)_2(H_2O)_2$ all transform into $CaAl_4Si_2O_{10}(OH)_2$ and $Al_2SiO_5(s)$ (kyanite) at 100–200 °C, and $CaAl_4Si_2O_{10}(OH)_2$ decomposes into $CaAl_2Si_2O_8$ and $Al_2SiO_5(s2)$ (andalusite) at 200–300 °C (Figure 14b,c). Meanwhile, the $Al_2SiO_5(s)$ (kyanite) also transforms into $Al_2SiO_5(s2)$ (andalusite) at 200 °C, and $Al_2SiO_5(s2)$ (andalusite) converts into $Al_6Si_2O_{13}(s)$ at 700–800 °C (Figure 14c). A small amount of $CaAl_2Si_2O_8$ decomposes and CaS is generated from 1400 °C (Figure 14b–d). At 1400–1500 °C, $Al_6Si_2O_{13}$ transforms into Al_2O_3 and SiC under the action of carbon (Figure 14c,d). The phase distribution of Fe at high temperatures is affected by the presence of Si. Fe_3C turns into FeSi at 1400–1500 °C (Figure 14e). When the Zn content is high (Zn/S/Ca/Al/Si/Fe = 0.62/0.24/0.07/0.46/0.53/0.06), Zn exists as ZnS, $ZnAl_2O_4$ and Zn_2SiO_4, as shown in Figure S8. Two-step decompositions at 200–500 and 800–1000 °C are observed for $ZnAl_2O_4$, in which the $ZnAl_2O_4$ converts into Zn_2SiO_4 and gaseous Zn, respectively (Figure S8a). $Ca_3Fe_2Si_3O_{12}$ and $CaFeSi_2O_6$ are formed at 100–200 and 100–300 °C, respectively. $Ca_3Fe_2Si_3O_{12}$ transforms into $CaFeSi_2O_6$ and $Ca_3Al_2Si_3O_{12}$ at 200–300 °C with a little decrease in $ZnAl_2O_4$ (Figure S8b,c). Then $Ca_3Al_2Si_3O_{12}$ converts into $CaAl_2Si_2O_8$ at 300–400 °C along with consumption of $ZnAl_2O_4$ and $SiO_2(s)$ (Figure S8c,d). Subsequently, $CaFeSi_2O_6$ turns into $CaAl_2Si_2O_8$ and Fe_2SiO_4 at 400–500 °C, accompanied by the decomposition of $ZnAl_2O_4$. Fe_2SiO_4 decomposes into elemental Fe and $SiO_2(s2)$ at 700–800 °C (Figure S8d,e). The elemental Fe combines with carbon to produce Fe_3C at 800–900 °C, which then transforms to FeS at 1000–1100 °C. The FeS decomposes and Fe_3C is formed again at 1300–1400 °C, which finally converts into FeSi at 1400–1500 °C.

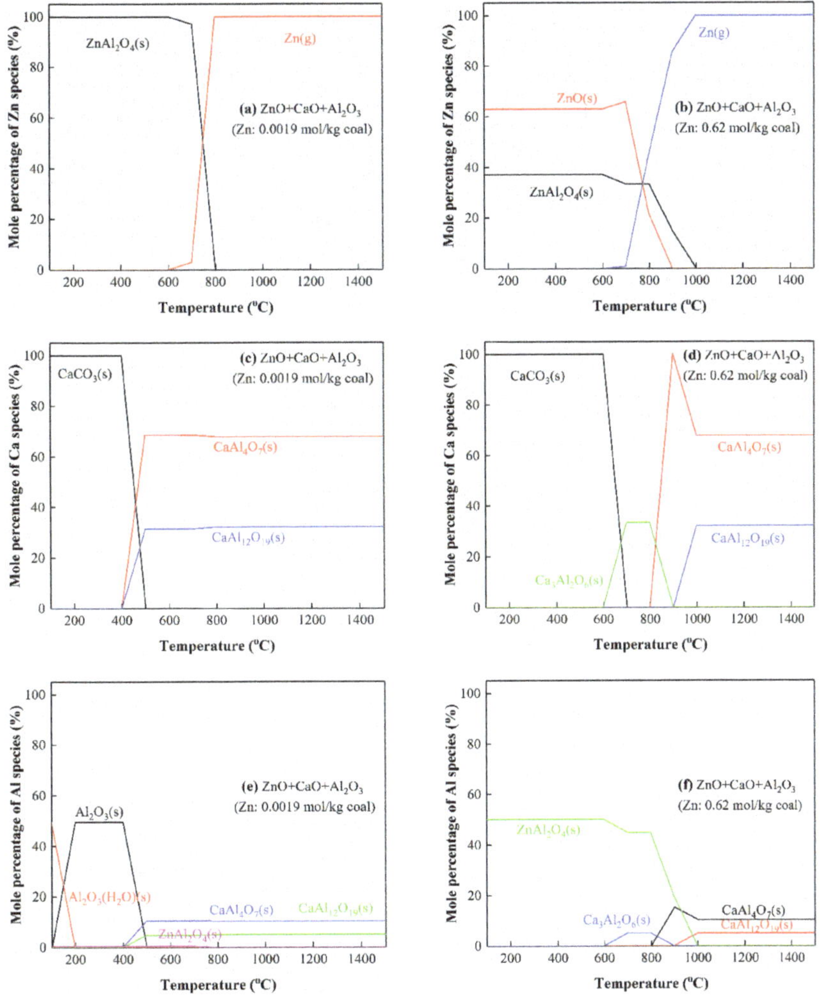

Figure 13. Thermodynamic equilibrium distributions of Zn (**a,b**), Ca (**c,d**) and Al (**e,f**) when Zn, Ca and Al coexist in the coal.

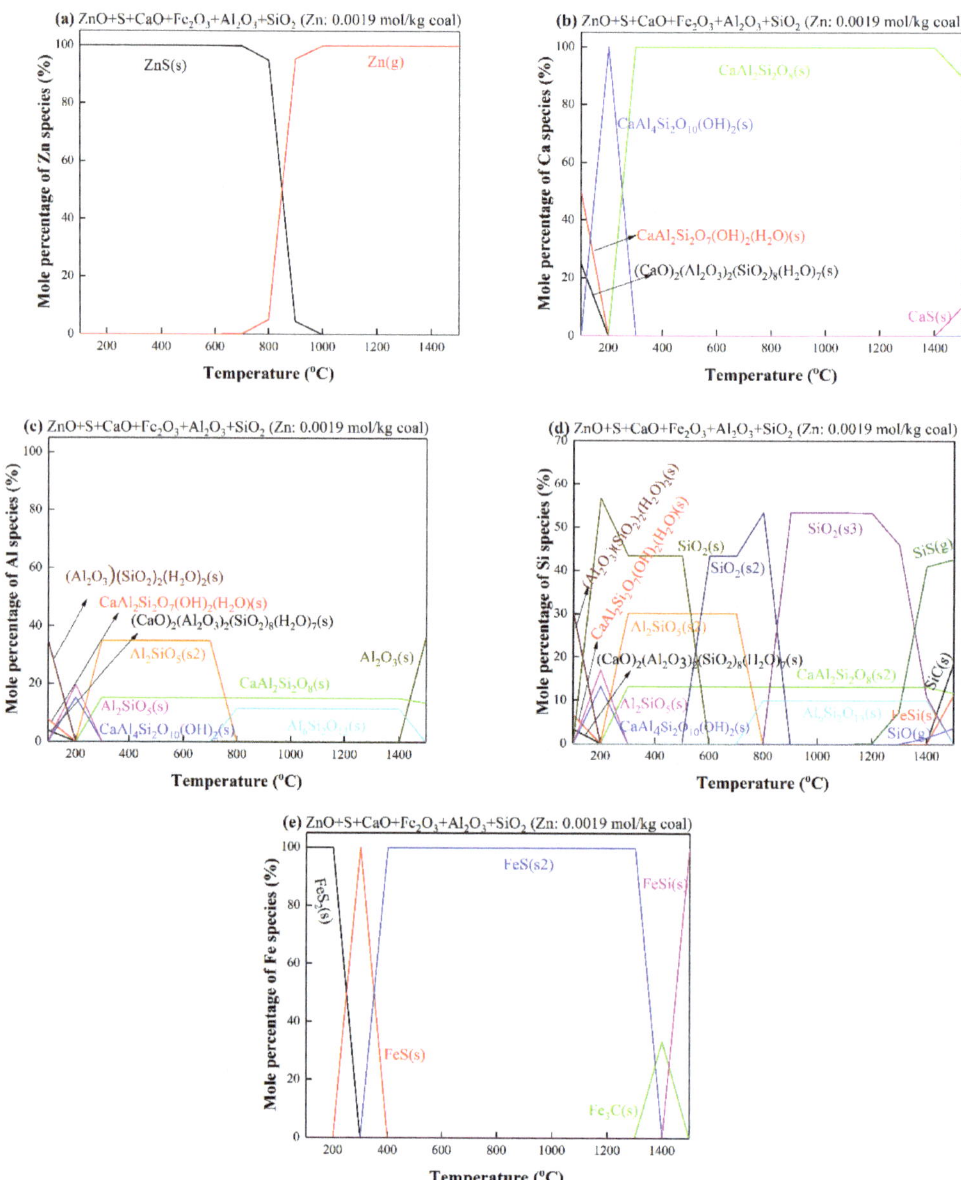

Figure 14. Thermodynamic equilibrium distributions of Zn (**a**), Ca (**b**), Al (**c**), Si (**d**) and Fe (**e**) in the coal when the Zn content is low (Zn/S/Ca/Al/Si/Fe = 0.0019/0.24/0.07/0.46/0.53/0.06).

Consistent with the XRD results, when the Zn content is low (molar ratio of Zn/S < 1), ZnS and SiO_2 are the main minerals in the obtained coke, and ZnS, $ZnAl_2O_4$ and SiO_2 can be detected when the Zn content is high (molar ratio of Zn/S > 1). In addition, CaS is present in the coke whatever the Zn content, which is not the situation as predicted by the thermodynamic equilibrium simulations. It is deduced that there is limited contact of Ca with Al and Si species in the coal, so the Ca preferentially combines with S dispersed in the coal matrix to generate CaS. Moreover, ZnO is found in the coke obtained at 700 °C, while the theoretically predicted Zn_2SiO_4 is not detected, which should be due to the physical isolation of Zn from Si in the coal or the crystallite size of generated Zn_2SiO_4 being too small to be detected by the XRD measurement. Based on the above analysis, the Zn species in the blends of coal and waste tires can migrate to the gas products in the form of gaseous Zn via a carbothermal reduction reaction during pyrolysis. The formation temperature of gaseous Zn is dependent on the contents of Zn, S and mineral elements in coal. The S, Al and Si can interact with Zn to inhibit the volatilization of Zn from coke. The reaction sequence with Zn is S > Al > Si, and the thermal stability of products is in the order of $ZnS > ZnAl_2O_4 > Zn_2SiO_4$. The decomposition of ZnS, $ZnAl_2O_4$ and Zn_2SiO_4 can occur in the temperature range of 700–1200, 600–1000 and 600–900 °C, respectively. Moreover, Fe and Ca can also bind with S to form metal sulfides, but their interactions are weaker than that of Zn with S and ZnS can be preferentially formed. In actual industrial production, the central temperature of coke cake produced in the coke oven is controlled at 950–1050 °C to ensure sufficient strength, so most of the Zn species can escape from the coke, which may be detrimental to the refractory bricks of the coke oven. Additionally, the deposition of Zn also increases the risk of clogging of ascension pipe that is used for the discharge of coke oven crude gas.

Although the transformation mechanisms described above are concentrated on the co-pyrolysis process of waste tires and coal, the corresponding rules are also suitable to understand the thermochemical behaviors of Zn in systems containing C, S, Ca, Al, Si and Fe under an inert atmosphere. Particularly, except for Zn, the contents of S and other inorganic components such as Ca, Al, Si and Fe are appreciable in the waste tires [34,61–63], which is also verified by our analysis results (see Table 2). Therefore, the conclusions from this study can also provide insights into the migration characteristics of Zn during the pyrolysis of waste tires alone, which is vital to the prevention and control of Zn emission to reduce the environmental burden.

4. Conclusions

In this paper, the transformation behaviors of Zn during co-pyrolysis of waste tires and coal were studied in a fixed-bed reactor system. It is shown that the relative percentage of Zn in the pyrolytic products (coke, tar and gas) obtained at different temperatures is closely related to the content of S and mineral elements (Ca, Al, Si and Fe) in the coal. When the molar ratio of Zn to S is less than 1, ZnS is formed in the coke obtained at 700 °C, resulting in ca. 97% of Zn residual rate in the coke. As the pyrolysis temperature increases to 900 and 1050 °C, ZnS is reduced to metallic Zn vapor by carbon, so the residual rate of Zn in coke decreases and the percentage of Zn in the tar increases. When the molar of Zn exceeds that of S, ZnO and ZnS coexist in the coke produced at 700 °C and partial ZnO is reduced to gaseous Zn by carbon, causing a small decrease in Zn residual rate in coke. Excess ZnO can react with Al_2O_3 to generate $ZnAl_2O_4$ at 900 °C, resulting in the enhancement of the Zn residual rate in coke (50%). Additionally, $ZnAl_2O_4$ can also be reduced by carbon to generate gaseous Zn at 1050 °C, giving rise to a reduction of the Zn residual rate in coke to 10%. The thermodynamic equilibrium simulations show that the formation temperature of gaseous Zn is dependent on the contents of Zn, S and mineral elements in coal. The S, Al and Si can interact with Zn to inhibit the volatilization of Zn from coke in the sequence of S > Al > Si, and the thermal stability of products is in the order of $ZnS > ZnAl_2O_4 > Zn_2SiO_4$. The decompositions of ZnS, $ZnAl_2O_4$ and Zn_2SiO_4 into Zn vapor occur in the temperature range of 700–1200, 600–1000 and 600–900 °C, respectively. Based on the above analysis, most Zn species can escape from the coke during the industrial cokemaking process, which is harmful to the refractory materials of the coke oven and brings the risk of blockage of ascension pipe due to the deposition of Zn. The Zn transformation mechanisms studied in this work are applicable not only for the co-pyrolysis of coal and waste tires but also for the pyrolysis of waste tires alone.

Supplementary Materials: The following supporting information can be downloaded at: https://www.mdpi.com/article/10.3390/pr10081635/s1, Figure S1: Thermodynamic equilibrium distributions of Zn when S and Zn coexist in the coal; Table S1: Mass distributions of Zn in the pyrolytic products; Figure S2: The thermodynamic equilibrium distributions of Zn (a, b), Ca (c, d) and Fe (e, f) when S, Zn, Ca and Fe coexist in the system; Figure S3: The thermodynamic equilibrium distributions of Zn (a, c) and silicon (b, d) when S, Zn and Si coexist in the system; Figure S4: The thermodynamic equilibrium distributions of Zn (a, c) and Al (b, d) when S, Zn and Al coexist in the system; Figure S5: Thermodynamic equilibrium distributions of Zn (a, b), Si (c, d) and Al (e, f) when S, Zn, Si and Al coexist in the system; Table S2: Standard Gibbs free energies of typical reactions; Figure S6: The standard-state Gibbs free energies of reactions 1−13 as a function of temperature; Figure S7: Thermodynamic equilibrium distributions of Zn (a, b), Ca (c, d) and Al (e, f) when S, Zn, Ca and Al coexist in the system; Figure S8: Thermodynamic equilibrium distributions of Zn (a), Ca (b), Al (c), Si (d) and Fe (e) in the coal when the Zn content is high (Zn/S/Ca/Al/Si/Fe =0.62/0.24/0.07/0.46/0.53/0.06).

Author Contributions: Conceptualization, S.J. and M.J.; methodology, S.J. and Y.L.; software, S.J. and Y.L.; validation, S.J., Y.L. and X.W.; formal analysis, Y.L., J.W., X.W. and R.Z.; investigation, Y.L., J.W., X.W. and R.Z.; resources, J.W. and L.L.; data curation, S.J. and Y.L.; writing—original draft preparation, Y.L.; writing—review and editing, S.J.; visualization, S.J. and Y.L.; supervision, L.L. and M.J.; project administration, M.J.; funding acquisition, L.L. All authors have read and agreed to the published version of the manuscript.

Funding: This research was funded by the National Natural Science Foundation of China [No. U1710252].

Institutional Review Board Statement: Not applicable.

Informed Consent Statement: Not applicable.

Data Availability Statement: The data used to support the findings of this study are available from the corresponding author upon request.

Acknowledgments: This work was financially supported by the National Natural Science Foundation of China (No. U1710252).

Conflicts of Interest: The authors declare no conflict of interest.

References

1. Elbaba, I.F.; Wu, C.; Williams, P.T. Catalytic pyrolysis-gasification of waste tire and tire elastomers for hydrogen production. *Energy Fuels* **2010**, *24*, 3928–3935. [CrossRef]
2. Lehmann, C.M.B.; Rostam-Abadi, M.; Rood, M.J.; Sun, J. Reprocessing and reuse of waste tire rubber to solve air-quality related problems. *Energy Fuels* **1998**, *12*, 1095–1099. [CrossRef]
3. Pan, D.L.; Jiang, W.T.; Guo, R.T.; Huang, Y.; Pan, W.G. Thermogravimetric and kinetic analysis of Co-combustion of waste tires and coal blends. *ACS Omega* **2021**, *6*, 5479–5484. [CrossRef] [PubMed]
4. RMurillo; Navarro, M.V.; López, J.M.; Aylón, E.; Callén, M.S.; García, T.; Mastral, A.M. Kinetic model comparison for waste tire char reaction with CO_2. *Ind. Eng. Chem. Res.* **2004**, *43*, 7768–7773. [CrossRef]
5. Levendis, Y.A.; Atal, A.; Carlson, J.B. On the correlation of CO and PAH emissions from the combustion of pulverized coal and waste tires. *Environ. Sci. Technol.* **1998**, *32*, 3767–3777. [CrossRef]
6. Zhang, Y.S.; Wu, C.; Nahil, M.A.; Williams, P. Pyrolysis-catalytic steam reforming/gasification of waste tires for production of carbon nanotubes and hydrogen. *Energy Fuels* **2015**, *29*, 3328–3334. [CrossRef]
7. Levendis, Y.; Atal, A.; Carlson, J.; Dunayevskiy, Y.; Vouros, P. Comparative study on the combustion and emissions of waste tire crumb and pulverized coal. *Environ. Sci. Technol.* **1996**, *30*, 2742–2754. [CrossRef]
8. Zaharia, M.; Sahajwalla, V.; Kim, B.C.; Khanna, R.; Saha-Chaudhury, N.; O'Kane, P.; Dicker, J.; Skidmore, C.; Knights, D. Recycling of rubber tires in electric arc furnace steelmaking: Simultaneous combustion of metallurgical coke and rubber tyres blends. *Energy Fuels* **2009**, *23*, 2467–2474. [CrossRef]
9. Díaz-Bautista, M.A.; Alvarez-Rodríguez, R.; Clemente-Jul, C.; Mastral, A.M. AFBC of coal with tyre rubber. Influence of the co-combustion variables on the mineral matter of solid by-products and on Zn lixiviation. *Fuel* **2013**, *106*, 10–20. [CrossRef]
10. Kříž, V.; Brožová, Z.; Přibyl, O.; Sýkorová, I. Possibility of obtaining hydrogen from coal/waste-tyre mixture. *Fuel Processing Technol.* **2008**, *89*, 1069–1075. [CrossRef]
11. Straka, P.; Bučko, Z. Co-gasification of lignite/waste-tyre mixture in a moving bed. *Fuel Processing Technol.* **2009**, *90*, 1202–1206. [CrossRef]
12. Oni, B.A.; Sanni, S.E.; Olabode, O.S. Production of fuel-blends from waste tyre and plastic by catalytic and integrated pyrolysis for use in compression ignition (CI) engines. *Fuel* **2021**, *297*, 120801. [CrossRef]

13. Murugan, S.; Ramaswamy, M.C.; Nagarajan, G. A comparative study on the performance, emission and combustion studies of a DI diesel engine using distilled tyre pyrolysis oil-diesel blends. *Fuel* **2008**, *87*, 2111–2121. [CrossRef]
14. Frigo, S.; Seggiani, M.; Puccini, M.; Vitolo, S. Liquid fuel production from waste tyre pyrolysis and its utilisation in a diesel engine. *Fuel* **2014**, *116*, 399–408. [CrossRef]
15. Bičáková, O.; Straka, P. Co-pyrolysis of waste tire/coal mixtures for smokeless fuel, maltenes and hydrogen-rich gas production. *Energy Convers. Manag.* **2016**, *116*, 203–213. [CrossRef]
16. Mastral, A.M.; Murillo, R.; Callén, M.S.; García, T. Evidence of coal and tire interactions in coal-tire coprocessing for short residence times. *Fuel Processing Technol.* **2001**, *69*, 127–140. [CrossRef]
17. Aydın, H.; İlkılıç, C. Optimization of fuel production from waste vehicle tires by pyrolysis and resembling to diesel fuel by various desulfurization methods. *Fuel* **2012**, *102*, 605–612. [CrossRef]
18. Song, Z.L.; Yan, Y.C.; Xie, M.M.; Lv, X.; Yang, Y.Q.; Liu, L.; Zhao, X.Q. Effect of steel wires on the microwave pyrolysis of tire powders. *ACS Sustain. Chem. Eng.* **2018**, *6*, 13443–13453. [CrossRef]
19. Leung, D.Y.C.; Yin, X.L.; Zhao, Z.L.; Xu, B.Y.; Chen, Y. Pyrolysis of tire powder: Influence of operation variables on the composition and yields of gaseous product. *Fuel Processing Technol.* **2002**, *79*, 141–155. [CrossRef]
20. Lopez, G.; Olazar, M.; Aguado, R.; Elordi, G.; Amutio, M.; Artetxe, M.; Bilbao, J. Vacuum pyrolysis of waste tires by continuously feeding into a conical spouted bed reactor. *Ind. Eng. Chem. Res.* **2010**, *49*, 8990–8997. [CrossRef]
21. Osorio-Vargas, P.; Shanmugaraj, K.; Herrera, C.; Campos, C.H.; Torres, C.C.; Castillo-Puchi, F.; Arteaga-Pérez, L.E. Valorization of waste tires via catalytic fast pyrolysis using palladium supported on natural halloysite. *Ind. Eng. Chem. Res.* **2021**, *60*, 18806–18816. [CrossRef]
22. Campuzano, F.; Cardona-Uribe, N.; Agudelo, A.F.; Sarathy, S.M.; Martínez, J.D. Pyrolysis of waste tires in a twin-auger reactor using CaO: Assessing the physicochemical properties of the derived products. *Energy Fuels* **2021**, *35*, 8819–8833. [CrossRef]
23. Onay, O.; Koca, H. Determination of synergetic effect in co-pyrolysis of lignite and waste tyre. *Fuel* **2015**, *150*, 169–174. [CrossRef]
24. Ren, Q.Q.; Wu, Z.Y.; Hu, S.; He, L.M.; Su, S.; Wang, Y.; Jiang, L.; Xiang, J. Sulfur self-doped char with high specific capacitance derived from waste tire: Effects of pyrolysis temperature. *Sci. Total Environ.* **2020**, *741*, 140193. [CrossRef] [PubMed]
25. Xu, J.Q.; Yu, J.X.; He, W.Z.; Huang, J.W.; Xu, J.S.; Li, G.M. Replacing commercial carbon black by pyrolytic residue from waste tire for tire processing: Technically feasible and economically reasonable. *Sci. Total Environ.* **2021**, *793*, 148597. [CrossRef] [PubMed]
26. Xu, J.Q.; Yu, J.X.; He, W.Z.; Huang, J.W.; Xu, J.S.; Li, G.M. Recovery of carbon black from waste tire in continuous commercial rotary kiln pyrolysis reactor. *Sci. Total Environ.* **2021**, *772*, 145507. [CrossRef]
27. Xu, J.Q.; Yu, J.X.; Xu, J.L.; Sun, C.L.; He, W.Z.; Huang, J.W.; Li, G.M. High-value utilization of waste tires: A review with focus on modified carbon black from pyrolysis. *Sci. Total Environ.* **2020**, *742*, 140235. [CrossRef]
28. Leeder, W.R. Additions of discarded automobile tyres to coke-oven blends. *Fuel* **1974**, *53*, 283–284. [CrossRef]
29. Chaala, A.; Roy, C. Production of coke from scrap tire vacuum pyrolysis oil. *Fuel Processing Technol.* **1996**, *46*, 227–239. [CrossRef]
30. Acevedo, B.; Barriocanal, C.; Alvarez, R. Pyrolysis of blends of coal and tyre wastes in a fixed bed reactor and a rotary oven. *Fuel* **2013**, *113*, 817–825. [CrossRef]
31. Han, M.C.; He, H.W.; Kong, W.K.; Dong, K.; Wang, B.Y.; Yan, X.; Wang, L.M.; Ning, X. High-performance electret and antibacterial polypropylene meltblown nonwoven materials doped with boehmite and ZnO nanoparticles for air filtration. *Fibers Polym.* **2022**, *23*, 1947–1955. [CrossRef]
32. Cheng, Z.L.; Guo, Z.G.; Fu, P.; Yang, J.; Wang, Q.W. New insights into the effects of methane and oxygen on heat/mass transfer in reactive porous media. *Int. Commun. Heat Mass Transf.* **2021**, *129*, 105652. [CrossRef]
33. Liu, Y.; Zhang, Z.L.; Liu, X.; Wang, L.; Xia, X.H. Ore image classification based on small deep learning model: Evaluation and optimization of model depth, model structure and data size. *Miner. Eng.* **2021**, *172*, 107020. [CrossRef]
34. Fernandez, A.M.; Barriocanal, C.; Gupta, S.; French, D. Effect of blending carbon-bearing waste with coal on mineralogy and reactivity of cokes. *Energy Fuels* **2014**, *28*, 291–298. [CrossRef]
35. Fernández, A.M.; Barriocanal, C.; Díaz-Faes, E. Recycling tyre wastes as additives in industrial coal blends for cokemaking. *Fuel Processing Technol.* **2015**, *132*, 173–179. [CrossRef]
36. Qin, X.; Xu, H.S.; Zhang, G.G.; Wang, J.D.; Wang, Z.; Zhao, Y.Q.; Wang, Z.Y.; Tan, T.W.; Bockstaller, M.R.; Zhang, L.Q.; et al. Enhancing the performance of rubber with nano ZnO as activators. *ACS Appl. Mater. Interfaces* **2020**, *12*, 48007–48015. [CrossRef]
37. Rhodes, E.P.; Ren, Z.Y.; Mays, D.C. Zinc leaching from tire crumb rubber. *Environ. Sci. Technol.* **2012**, *46*, 12856–12863. [CrossRef]
38. Esezobor, D.E.; Balogun, S.A. Zinc accumulation during recycling of iron oxide wastes in the blast furnace. *Ironmak. Steelmak.* **2006**, *33*, 419–425. [CrossRef]
39. Zhang, F.; An, S.L.; Luo, G.P.; Zhang, S.Z. Behaviours of zinc and lead in blast furnace of Baotou iron and steel group Co. *Adv. Mater. Res.* **2011**, *194*, 306–309. [CrossRef]
40. Stepin, G.M.; Mkrtchan, L.S.; Dovlyadov, I.V.; Borshchevskii, I.K. Problems related to the presence of Zinc in Russian blast-furnace smelting and ways of solving them. *Metallurgist* **2001**, *45*, 382–390. [CrossRef]
41. Trinkel, V.; Mallow, O.; Thaler, C.; Schenk, J.; Rechberger, H.; Fellner, J. Behavior of chromium, nickel, lead, zinc, cadmium, and mercury in the blast furnace-a critical review of literature data and plant investigations. *Ind. Eng. Chem. Res.* **2015**, *54*, 11759–11771. [CrossRef]
42. Yang, X.F.; Chu, M.S.; Shen, F.M.; Zhang, Z.M. Mechanism of zinc damaging to blast furnace tuyere refractory. *Acta Metall. Sin.-Engl.* **2009**, *22*, 454–460. [CrossRef]

43. Li, K.J.; Zhang, J.L.; Liu, Z.J.; Wang, T.Q.; Ning, X.J.; Zhong, J.B.; Xu, R.S.; Wang, G.W.; Ren, S.; Yang, T.J. Zinc accumulation and behavior in tuyere coke. *Metall. Mater. Trans. B* **2014**, *45*, 1581–1588. [CrossRef]

44. Zeng, X.W.; Zheng, S.; Zhou, H.C. The phenomena of secondary weight loss in high-temperature coal pyrolysis. *Energy Fuels* **2017**, *31*, 10178–10185. [CrossRef]

45. Jones, I.; Preciado-Hernandez, J.; Zhu, M.M.; Zhang, J.; Zhang, Z.Z.; Zhang, D.K. Utilisation of spent tyre pyrolysis char as activated carbon feedstock: The role, transformation and fate of Zn from the perspective of production. *Waste Manag.* **2021**, *126*, 549–558. [CrossRef]

46. Darmstadt, H.; Roy, C.; Kaliaguine, S. Characterization of pyrolytic carbon blacks from commercial tire pyrolysis plants. *Carbon* **1995**, *33*, 1449–1455. [CrossRef]

47. Mis-Fernandez, R.; Azamar-Barrios, J.A.; Rios-Soberanis, C.R. Characterization of the powder obtained from wasted tires reduced by pyrolysis and thermal shock process. *J. Appl. Res. Technol.* **2008**, *6*, 95–104. [CrossRef]

48. Ilnicka, A.; Okonski, J.; Cyganiuk, A.W.; Patyk, J.; Lukaszewicz, J.P. Zinc regarding the utilization of waste tires by pyrolysis. *Pol. J. Environ. Stud.* **2016**, *25*, 2683–2687. [CrossRef]

49. Cataldo, F. On the characterisation of carbon black from tire pyrolysis. *Fuller. Nanotub. Carbon Nanostructures* **2020**, *28*, 368–376. [CrossRef]

50. Zhou, Q.Q.; Yang, S.S.; Wang, H.T.; Liu, Z.Y.; Zhang, L. Selective deoxygenation of biomass volatiles into light oxygenates catalysed by S-doped, nanosized zinc-rich scrap tyre char with in-situ formed multiple acidic sites. *Appl. Catal. B Environ.* **2021**, *282*, 119603. [CrossRef]

51. Wang, B.; Li, W.; Ma, C.; Yang, W.; Pudasainee, D.; Gupta, R.; Sun, L. Synergistic effect on the co-gasification of petroleum coke and carbon-based feedstocks: A state-of-the-art review. *J. Energy Inst.* **2022**, *102*, 1–13. [CrossRef]

52. Liu, Y.; Zhang, Z.L.; Liu, X.; Wang, L.; Xia, X.H. Efficient image segmentation based on deep learning for mineral image classification. *Adv. Powder Technol.* **2021**, *32*, 3885–3903. [CrossRef]

53. Zhang, X.Y.; Ma, F.N.; Dai, Z.X.; Wang, J.; Chen, L.; Ling, H.; Soltanian, M.R. Radionuclide transport in multi-scale fractured rocks: A review. *J. Hazard. Mater.* **2022**, *424*, 127550. [CrossRef]

54. Banerjee, S.S.; Hait, S.; Natarajan, T.S.; Wießner, S.; Stöckelhuber, K.W.; Jehnichen, D.; Janke, A.; Fischer, D.; Heinrich, G.; Busfield, J.J.; et al. Water-responsive and mechanically adaptive natural rubber composites by in situ modification of mineral filler structures. *J. Phys. Chem. B* **2019**, *123*, 5168–5175. [CrossRef] [PubMed]

55. Wang, X.D.; Summers, C.J.; Wang, Z.L. Mesoporous single-crystal ZnO nanowires epitaxially sheathed with Zn_2SiO_4. *Adv. Mater.* **2004**, *16*, 1215–1218. [CrossRef]

56. Yang, H.P.; Chen, H.P.; Ju, F.D.; Yan, R.; Zhang, S.H. Influence of pressure on coal pyrolysis and char gasification. *Energy Fuels* **2007**, *21*, 3165–3170. [CrossRef]

57. Chen, H.P.; Luo, Z.W.; Yang, H.P.; Ju, F.D.; Zhang, S.H. Pressurized pyrolysis and gasification of Chinese typical coal samples. *Energy Fuels* **2008**, *22*, 1136–1141. [CrossRef]

58. Cheng, J.; Zhang, Y.S.; Wang, T.; Norris, P.; Chen, W.Y.; Pan, W.P. Thermogravimetric-Fourier transform infrared spectroscopy-gas chromatography/mass spectrometry study of volatile organic compounds from coal pyrolysis. *Energy Fuels* **2017**, *31*, 7042–7051. [CrossRef]

59. Seidelt, S.; Müller-Hagedorn, M.; Bockhorn, H. Description of tire pyrolysis by thermal degradation behavior of main components. *J. Anal. Appl. Pyrolysis* **2006**, *75*, 11–18. [CrossRef]

60. Yu, D.; Ma, Z.; Wang, R. Efficient smart grid load balancing via fog and cloud computing. *Math. Probl. Eng.* **2022**, 3151249. [CrossRef]

61. Osipov, A.A.; Ul'yanov, V.V.; Gulevskii, V.A.; Mel'nikov, V.P.; Kharchuk, S.E. Thermodynamics of processes in the liquid-metal pyrolysis of waste car tires. *Theor. Found. Chem. Eng.* **2019**, *53*, 1035–1047. [CrossRef]

62. Selbes, M.; Yilmaz, O.; Khan, A.A.; Karanfil, T. Leaching of DOC, DN, and inorganic constituents from scrap tires. *Chemosphere* **2015**, *139*, 617–623. [CrossRef] [PubMed]

63. Larionov, K.B.; Slyusarskiy, K.V.; Ivanov, A.A.; Mishakov, I.V.; Pak, A.Y.; Jankovsky, S.A.; Stoyanovskii, V.O.; Vedyagin, A.A.; Gubin, V.E. Comparative analysis of the characteristics of carbonaceous material obtained via single-staged steam pyrolysis of waste tires. *J. Air Waste Manag.* **2022**, *72*, 161–175. [CrossRef] [PubMed]

Article

Multiphysics Numerical Simulation Model and Hydraulic Model Experiments in the Argon-Stirred Ladle

Chengjian Hua, Yanping Bao * and Min Wang

State Key Laboratory of Advanced Metallurgy, University of Science and Technology Beijing, Beijing 100083, China
* Correspondence: baoyp@ustb.edu.cn

Abstract: The argon-stirred ladle is a standard piece of steelmaking refining equipment. The molten steel quality will improve when a good argon-stirred process is applied. In this paper, a Multiphysics model that contained fluid flow, bubble transport, alloy transport, bubble heat flux, alloy heat flux, alloy melting, and an alloy concentration species transport model was established. The fluid model and bubble transport model that were used to calculate the fluid velocity were verified by the hydraulic model of the ladle that was combined with particle image velocimetry measurement results. The numerical simulation results of the temperature fields and steel–slag interface shape were verified by a ladle that contained 25 t of molten steel in a steel plant. The velocity difference between the hydraulic model and numerical model decreased when the C_L (integral time-scale constant) increased from 0 to 0.3; then, the difference increased when the C_L increased from 0.3 to 0.45. The results showed that a C_L of 0.3 approached the experiment results more. The bubble heat flux model was examined by the industrial practice, and the temperature decrease rate was 0.0144 K/s. The simulation results of the temperature decrease rate increased when the initial bubble temperature decreased. When the initial bubble temperature was 800 °C, the numerical simulation results showed that the temperature decrease rate was 0.0147 K/s, and the initial bubble temperature set at 800 °C was more appropriate. The average melting time of the alloy was 12.49 s and 12.71 s, and the mixture time was approximately the same when the alloy was added to two slag eyes individually. The alloy concentration had fewer changes after the alloy was added in the ladle after 100 s.

Keywords: argon-stirred ladle; particle image velocimetry; numerical simulation; fluid flow; bubble

Citation: Hua, C.; Bao, Y.; Wang, M. Multiphysics Numerical Simulation Model and Hydraulic Model Experiments in the Argon-Stirred Ladle. *Processes* **2022**, *10*, 1563. https://doi.org/10.3390/pr10081563

Academic Editors: Elio Santacesaria, Riccardo Tesser and Vincenzo Russo

Received: 18 July 2022
Accepted: 3 August 2022
Published: 10 August 2022

Publisher's Note: MDPI stays neutral with regard to jurisdictional claims in published maps and institutional affiliations.

1. Introduction

In metallurgy, the argon-stirred ladle is a low-cost, highly efficient, and common refining method for the improvement of the molten steel temperature, the homogeneity of molten steel concentration [1–9], inclusion floatation [10–16], and increase in the kinetic reaction [17–20] in the ladle.

Studies on the argon-stirred ladle numerical simulation have focused on the following topics: (1) fluid-flow simulation [21], (2) bubble transport modeling, (3) temperature simulation, (4) alloy melting, (5) alloy concentration or other species transport [22], and (6) kinetic reaction in the ladle. In the fluid-flow simulation, the Euler–Euler approach [23,24], the single-phase methods by exerting an additional force on the fluid flow [25–27], the multiphase volume of fluid (VOF) method [2,28–30], and the Euler–Lagrange method [31–33] considered the bubble as a discrete phase and the molten steel and molten slag as the continuum phase. In the bubble transport modeling, Xue et al. [34] studied the effect of small bubbles on inclusion removal, and the results showed that the small bubble is beneficial to inclusion removal. Zhu et al. [35] studied the effect of an argon-stirred ladle on the inclusion removal, and the results showed that a larger size of inclusion will be removed more quickly and the difficulty of the inclusion removal will increase when the inclusion diameter decreases. In the molten steel temperature simulation, Urióstegui-Hernández [36] considered the heat transfer between the fluid and the bubble using the

Euler–Euler methods. Cheng et al. [37] studied the effect of the slot plug on the flow in the ladle and temperature exchange with the argon. In the alloy melting simulation, Bhattacharjee et al. [38] studied the alloy melting in the argon-stirred ladle by numerical simulation, and the results showed that dissolution rates are a function of gas flow rate and vessel dimensions and depend on the diffusivity of the alloy and initial size. In the alloy concentration or other species, Liu et al. [28] studied the species transport in the argon-stirred ladle, and monitoring the species concentration varies with solution time. In the kinetic reaction in the ladle studies, Kwon et al. [39] studied the Al-deoxidation in the argon-stirred ladle to predict the inclusion formation. Lachmund et al. [40] established a desulfurization model and predicted the desulphurization process by numerical simulation. Singh et al. [41] studied the desulfurization with kinetic data acquired from Thermo-Calc software, and the results showed that dual plugs are better than a single plug when the flow is the same.

Recently, Mantripragada [36] studied the argon plugs' position and flow rate on the argon-stirred ladle refined by numerical simulation. They pointed out a dimensional variation in the optimized ladle. Wondrak et al. [42] used Sn-Bi alloy in industrial experiments, and the results showed the 'slag eye' size and velocity magnitude in the molten alloy surface. The experimental results will benefit the numerical simulation optimization. Uriostegui-Hernandez et al. [43] studied the fluid flow and mass transfer of sulfur; the results showed that the desulfurization rates increase as the argon gas flow rate increases. Wang et al. [44] studied the effect of the argon-stirred process on the ladle refractory erosion by numerical simulation and industrial practice. The results showed that the refractory erosion will increase near the slag layer and near the bottom gas inlet. Joubert [45] studied the mass transfer of the species in the steel–slag phase in the argon-stirred ladle by a numerical simulation and a hydraulic model simulation. The results showed that the numerical simulation model of species transport showed good consistency with the hydraulic simulation. Guo et al. [16] studied the effect of plug position, argon flow rate, and inclusion size on the inclusion removal. The results showed that the inclusion removal rate will increase when the argon flow rate grows. Riabov et al. [46] studied the bubble diameter, plume area, and bubble velocity using a hydraulic model combined with particle image velocimetry technology. The results showed that the increase in the diameter of the porous plug will lead to a small bubble diameter and poorer mixing conditions. Li et al. [47] studied the bubble transport and fluid flow in an argon-stirred ladle by the volume of the fluid (VOF) model with a finer scale of mesh coupling with a sub-grid-scale large eddy simulation. The results showed that the slag eye size varies with time, and the bubble detachment frequency has a direct effect on the slag eyes. A new correlation of the slag eye and the modified Froude number was established.

In the previous work, the argon-stirred ladle that contained molten steel was 80~300 t. The tundish was applied in the argon-stirred ladle during the casting process. The tundish is helpful in the molten steel temperature, species homogeneity, and inclusion removal. However, there is no tundish in the casting process; when the argon-stirred steel that contained molten steel is about 25~30 t, the effect of molten steel temperature, species homogeneity, and fluid flow in the argon-stirred ladle is essential. In this paper, a numerical simulation model was established that contained a fluid flow simulation, bubble transport, alloy transport, alloy melting, alloy concentration species transport, and a bubble and alloy heat transfer simulation. In this paper, a random walk model was introduced to simulate the bubble transport and bubble drive flow novelty. The effect of random transport parameters on the bubble drive flow simulation was examined by particle image velocimetry results of the hydraulic model. The heat transfer model of the bubble and the molten steel was introduced for the first time, and the bubble heat transfer simulation was examined by industrial practice where the ladle contained 25 t of molten steel in a steel plant. The molten steel temperature distribution in the ladle was revealed for the first time. The effect of the alloy-added situation on the alloy concentration diffusion was discussed. The numerical simulation model was helpful in the argon-stirred ladle design and process modification.

2. Experimental Methods

2.1. The measurement of the Fluid Flow in the Ladle Water Model

A hydraulic model of the ladle was established by the principle of hydraulic similarity. The ratio was 1:2.4 between the water model and the prototype. The dimensions of the argon-stirred ladle in the prototype and water model are shown in Table 1. The angle between the two plugs was 135° in the prototype. The fluid velocity in the hydraulic model was measured by particle image velocimetry (PIV) technology. The prototype ladle can contain 25 t of molten steel.

Table 1. The dimensions of argon-stirred ladle in prototype and water model.

Item	Parameters
The bottom diameter of the hydraulic model/mm	661.75
Top diameter of the hydraulic model/mm	752
Height of the hydraulic model/mm	910.5
Height of water/mm	716.5
Position of argon plug in the hydraulic model	$0.6\,R_\text{m}$
The bottom diameter of the prototype ladle/mm	1612
Top diameter of prototype ladle/mm	1848
Height of the prototype ladle/mm	2374
Hight of molten steel/mm	1574
Position of argon plugs in prototype	$0.6\,R_\text{p}$ and $0.67\,R_\text{p}$

Where R represents the radius, subscript m represents the hydraulic model, and p represents the prototype.

Figure 1 shows the schematic of the velocity measurements in the hydraulic ladle. The plug was placed at $0.6R_\text{m}$, and the laser sheet emission was from the right side and passed through the plug. The camera was vertical to the plane where the laser sheet was at. The flow rate was 6 L·min^{-1}, controlled by the valve, and the argon flowed through the tube and out from the plug. The PIV equipment contained: laser (Litron lasers LPY704-100 PIV, Rugby, UK), guiding arm (LaVision Gmbh, Göttingen, Germany), sheet optics, Nikon AF-S 50 mm f/1.4 G lens (Nikon Corporation, Tokyo, Japan), calibration plate (LaVision Gmbh, Germany), camera (LaVision Gmbh, Germany), programmable timing unit X (PTU X), workstation (Intel(R), Xeon(R), CPU E5-2650 v3 @ 2.30 GHz, two processors), laser power equipment, and laser cooling equipment. The measurement and postprocessing were accomplished by the DaVis 8.4.0 (LaVision GmbH, Germany). The seeding particle was vestosint and the seeding particle diameter was 55 μm; it had a good flow following. The laser shutter opened twice in a very short time, and the camera exposed twice following the laser shutter opening. The time interval of the laser shutter opening twice was 0.01428 s, the camera acquisition rate was 70 Hz, and the acquisition time lasted two seconds. During the measurement, the sampling count was 100 and the time-averaged results were calculated by DaVis 8.4.0. The postprocessing was a cross-correlation model, the interrogation window size was 96×96 window size, and the interrogation window was decreased to 64×64 window size.

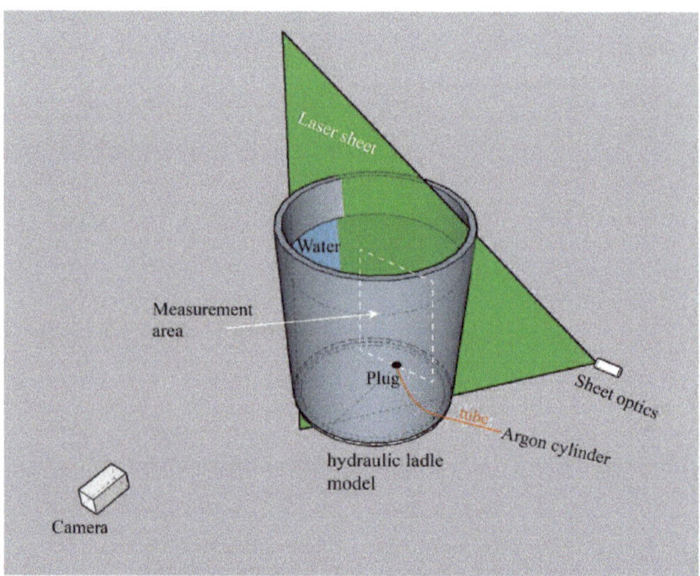

Figure 1. The schematic of the velocity measurement setup.

2.2. Numerical Simulation Models

Figure 2 shows the schematic of the numerical simulation model included in the paper. The numerical simulation model had the following part: the fluid flow model, the bubble and alloy transport model, the bubble and fluid heat exchange model, the alloy melting model, the VOF model, and the alloy concentration diffusion model.

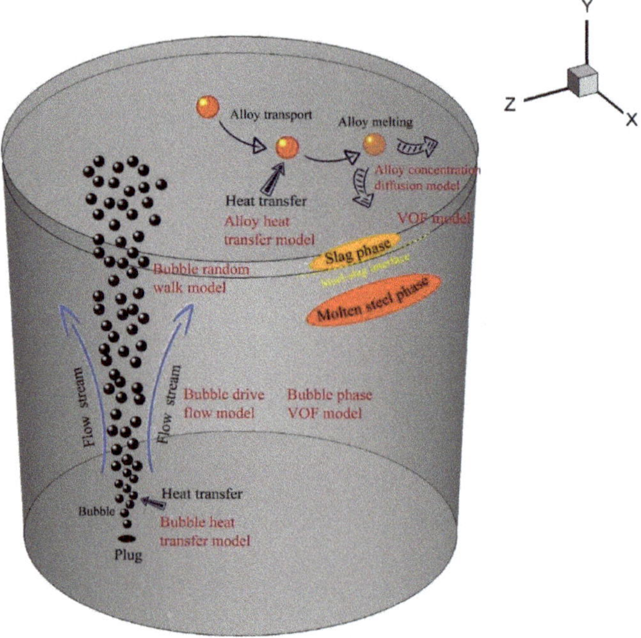

Figure 2. Schematic of the simulation model in the argon-stirred ladle.

2.2.1. Fluid Flow and VOF Model

In this paper, the fluid flow was driven by bubble transport. The governing equations are shown in Equations (1)–(7), and a source term was added to simulate the flow driven by the bubble. The drag force was considered in this paper as the bubble-driven flow, and the source term was added. In this paper, three kinds of fluid phases were considered: molten steel phase, molten slag phase, and argon bubble phase. The continuum surface stress and surface tension models were used in the VOF model. The volume fraction of the bubble in the fluid domain was adjusted by the User-Defined Function (UDF) module. The volume fraction of the bubble was updated according to the bubble's position.

$$\frac{1}{\rho_q}\frac{\partial}{\partial t}(\alpha_q \rho_q) + \nabla \cdot (\alpha_q \rho_q \vec{v}) = 0 \tag{1}$$

$$\sum_{q=1}^{n} \alpha_q = 1 \tag{2}$$

$$\frac{\partial}{\partial t}(\rho\vec{v}) + \nabla(\rho\vec{v}\vec{v}) = -\nabla p + \nabla(\mu(\nabla\vec{v} + \nabla\vec{v}^{\mathrm{T}}) - \frac{2}{3}\nabla\vec{v}I) + \rho\vec{g} + \vec{F} \tag{3}$$

$$\frac{\partial}{\partial t}(\rho k) + \frac{\partial}{\partial x_i}(\rho k u_i) = \frac{\partial}{\partial x_j}(\alpha_k \mu_{eff}\frac{\partial k}{\partial x_j}) + G_k + G_b - \rho\varepsilon \tag{4}$$

$$\frac{\partial}{\partial t}(\rho\varepsilon) + \frac{\partial}{\partial x_i}(\rho\varepsilon u_i) = \frac{\partial}{\partial x_j}(\alpha_\varepsilon \mu_{eff}\frac{\partial\varepsilon}{\partial x_j}) + C_{1\varepsilon}\frac{\varepsilon}{k}(G_k + C_{2\varepsilon}G_b) + C_{2\varepsilon}\rho\frac{\varepsilon^2}{k} - R_\varepsilon + S_\varepsilon \tag{5}$$

$$\vec{F} = \frac{\alpha_l}{\Delta V_{\text{cell}}}\sum_i^N (\vec{F}_{D,i})Q\,\Delta t \tag{6}$$

$$\vec{F}_{D,i} = \frac{18\mu C_{D,i}\mathrm{Re}_i}{24d_{b,i}^2\rho_{b,i}}(\vec{u}_l - \vec{u}_{b,i}) \tag{7}$$

where α_q represents the volume fraction of the qth phase, ρ_q represents the density in the qth phase, v represents the velocity, n represents the phase count in the simulation, ρ represents the density, μ represents the molecular viscosity, I is the unit tensor, $\vec{F}$ is the source term of the bubble-driven flow, k represents the turbulent kinetic energy, G_k represents the generation of turbulence kinetic energy due to the mean velocity gradient, G_b is the generation of turbulence kinetic energy due to buoyancy, ε is turbulent kinetic energy dissipation, α_l represents the volume fraction of the liquid phase, ΔV_{cell} represents the volume of the cell, N represents the bubble count in the cell, Q represents the flow rate of the argon, Δt represents the time step, $C_{D,i}$ represents the coefficient of drag force, Rei represents the Reynolds number of bubble, $d_{b,i}$ represents the diameter of the bubble, $\rho_{b,i}$ represents the density of the bubble, u_l is the velocity of the fluid, and $u_{b,i}$ is the velocity of the bubble.

2.2.2. Bubble and Alloy Transport Model

The discrete phase model (DPM) was applied in the bubble and inclusion transport modeling. The bubble and alloy transport equations are shown in Equations (8)–(10). The discrete random walk model was applied in the bubble transport simulation. The randomness of the inclusion transport will increase if the C_L is increased. The bubble will be removed when the bubble reaches the steel–slag interface. The steel–slag interface is when the volume fraction of slag is 0.5. The bubble diameter will increase if the bubble floats. The bubble diameter variation mechanism is shown in Equation (11).

$$\frac{d\vec{u}_p}{dt} = \frac{\vec{u}_l - \vec{u}_p}{\tau_r} + \frac{\vec{g}(\rho_p - \rho_f)}{\rho_p} \tag{8}$$

$$T_E = -T_L \ln(r) \tag{9}$$

$$T_L = C_L \frac{k}{\varepsilon} \tag{10}$$

$$d_{b,i} = \sqrt[3]{\frac{p_0 + \rho_l g h}{p_0 + \rho_l g (H - y_i)}} d_{\text{bottom}} \tag{11}$$

where $\vec{u}_p$ and $\vec{u}_l$ are, respectively, the particle vector and the fluid vector; ρ_p is the particle density; ρ_f is the fluid density; T_E is the eddy lifetime; r is the random Gaussian number; T_L is the integral time scale; C_L (integral time-scale constant) is a constant.

2.2.3. Energy Conservation and Bubble Heat Exchange Model

The governing equation of the temperature is shown in Equations (12) and (13). In this paper, a heat transport model of molten steel and the bubble was applied. The equations of the bubble heat transport are shown in Equations (14)–(19). The bubble temperature is lower than the molten steel temperature, and the heat will transfer from the molten steel to the bubble. The heat transfer energy is based on the energy conservation law. The bubble will stop heat transfer if the bubble temperature is the same as that of the molten steel. The effect of bubble initial injection temperature on the molten steel temperature simulation is discussed in Section 3.2.2.

$$\frac{\partial}{\partial t}(\rho E) + \nabla(\vec{v}(\rho E + p)) = \nabla(k_{eff}\nabla T - \sum_j h_j \vec{J}_j + (\overline{\overline{\tau}}_{eff} \cdot \vec{v})) \tag{12}$$

$$E = h - \frac{p}{\rho} + \frac{v^2}{2} \tag{13}$$

$$E_T = C_{b,i} \cdot m_{b,i} \cdot (T_{\text{cell}} - T_{b,init}) \tag{14}$$

$$E_{step} = Nu \cdot k_m \cdot \pi \cdot d_{b,i} \cdot (T_{\text{cell}} - T_{b,pre.temp}) \cdot \triangle t \tag{15}$$

$$Nu = 2 + 0.6 \cdot \text{Re}^{1/2} \cdot \text{Pr}^{1/3} \tag{16}$$

$$T_{b,cur.temp} = T_{b,pre.temp} + \frac{E_{step}}{(C_{b,i} \cdot m_{b,i})} \tag{17}$$

$$\text{Pr} = \frac{\mu_f \cdot C_p}{k_m} \tag{18}$$

$$\text{Re}_p = \frac{\rho_{b,i} d_{b,i}(\vec{u}_l - \vec{u}_{b,i})}{\mu_l} \tag{19}$$

where h is sensible enthalpy, E_T is the total energy of the bubble, $C_{b,i}$ is the heat capacity of the bubble, $m_{b,i}$ is the bubble mass, T_{cell} is the cell's temperature where the bubble is injected into the fluid domain, $T_{b,init}$ is the initial bubble temperature, E_{step} is the energy exchange in each timestep, Nu is the Nusselt number, k_m is the fluid thermal conductivity, $T_{b,pre.temp}$ is the bubble temperature in the previous timestep, and C_p is the heat capacity of the fluid.

2.2.4. Alloy Melting and Alloy Concentration Diffusion Model

In this paper, a solid crust will form when the alloy is added to the high-temperature molten steel. At this time, the alloy began melting inside the solid crust. The alloy concentration was released until the solid crust melted at a specific time. The alloy concentration numerical simulation governing equation is shown in Equations (20)–(24). The source term was added in a cell where the alloy melted.

$$t_{melt} = \frac{\rho_A C_{P,A} d_A}{2\pi h_c} \frac{T_s - T_0}{T_M - T_s} \tag{20}$$

$$Nu = \frac{d_A h_c}{k_m} \tag{21}$$

$$Nu = 2 + (0.4\mathrm{Re}_A^{1/2} + 0.06\mathrm{Re}_A^{2/3}) \cdot \mathrm{Pr}^{2/5} \tag{22}$$

$$\frac{\partial \rho C}{\partial t} + \frac{\partial}{\partial x_i}(\rho u_i C - \Gamma_k \frac{\partial C}{\partial x_i}) = S_C \tag{23}$$

$$S_C = \frac{m_{alloy}}{m_{cell}} \tag{24}$$

where t_{melt} is the alloy solid crust melting time, ρ_A is the alloy density, $C_{P,A}$ is the heat capacity of the alloy, d_A is the diameter of the alloy, h_c is the coefficient of the heat transfer, T_s is the solidification temperature of molten steel, T_M is the temperature of molten steel, T_0 is the initial alloy temperature, C is the mass concentration of alloy, S_c is the source term of the alloy melting, m_{alloy} is the mass of the melted alloy, and m_{cell} is the mass of the cell where the alloy is melting.

2.3. Boundary Condition and Solution Strategy

Table 2 shows the relevant parameters that are required in the numerical simulation. The initial bubble diameter and the initial bubble velocity were calculated by Equations (23) and (24). The numerical simulation was calculated by ANSYS Fluent software (ANSYS, Inc., Canonsburg, PA, USA). The inlet boundary condition was set as follows: the bubbles were injected from the plugs, and the inject position was defined by the User-Defined Function (UDF) that was built in the ANSYS Fluent software (ANSYS, Inc., Canonsburg, PA, USA). The boundary name, position, and condition are shown in Figure 3. The bubble flow rate was 5 L·min^{-1}. The bubble injection velocity was 2.63 m·s^{-1} in this paper. The initial bubble diameter was 3.924 mm. The position of the alloys thrown was above the slag eye by 2.5 m; based on the law of free fall, the velocity of the alloy contacting with the molten steel was 7 m·s^{-1}. Figure 3 shows the hexahedral mesh of the fluid zone in the prototype and hydraulic model. The argon bubble was injected from the bottom of the ladle, and the injection position is shown in Figure 3. In the prototype, argon was injected at two places, and the angle between them was 135°. The solution initialization used standard initialization and computing from all zones. The solution step was 0.001 s, the injection rate of the bubble was one bubble per solution step, and the N_{per} was 1000. The free surface was used on the top of the slag (in prototype modeling) or the top of the air (in the hydraulic modeling). The rest boundary was taken as the wall boundary. The SIMPLE scheme was used in the pressure-velocity coupling. The residual in the governing equations was 0.001. The heat flux in the ladle wall, the molten slag surf, and the ladle bottom was 0.006 W^2·mm^{-2}.

$$d_{b,init} = (0.35 \cdot (\frac{Q^2}{9.8})^{0.2}) \tag{25}$$

$$v_{b.init} = \frac{Q}{(N_{per} \cdot \frac{1}{6} \cdot \pi \cdot d_{b,init}^3)} \tag{26}$$

where $v_{b,init}$ is the initial velocity when the bubble is injected into the fluid zone and N_{per} is the bubble injection count in 1 s.

Table 2. Relevant parameters in the numerical simulation.

Item	Parameters
Water density (25 °C)/kg·m^{-3}	997.04
Water viscosity (25 °C)/Pa·s	0.8949×10^{-3}
Air phase density (25 °C)/kg·m^{-3}	1.185
Air phase viscosity (25 °C)/Pa·s	1.8315×10^{-5}
Water surface tension/N·m^{-1}	0.07197
Argon bubble flow rate in hydraulic model/L·min^{-1}	6
Molten steel density/kg·m^{-3}	7000
Molten steel viscosity/Pa·s	0.0067
Molten slag density/kg·m^{-3}	3500
Molten slag viscosity/Pa·s	0.06
Argon density/kg·m^{-3}	1.784
Argon viscosity/Pa·s	2.2624×10^{-5}
Molten steel–slag interface surface tension/N·m^{-1}	1.6
Molten steel specific heat/J·kg^{-1}·K^{-1}	680
Molten slag specific heat/J·kg^{-1}·K^{-1}	1100
Argon specific heat/J·kg^{-1}·K^{-1}	1006.43
Molten steel thermal conductivity/W·m^{-1}·K^{-1}	40.3
Molten slag thermal conductivity/W·m^{-1}·K^{-1}	34
Argon thermal conductivity/W·m^{-1}·K^{-1}	0.0242
Alloy density/kg·m^{-3}	5780
Alloy specific heat/J·kg^{-1}·K^{-1}	364
Alloy diameter/mm	50
Alloy thermal conductivity/W·m^{-1}·K^{-1}	26.32
Initial alloy temperature/°C	25
Initial bubble temperature/°C	25, 400, 800
Alloy concentration diffusion coefficient/m^2·s^{-1}	1.566×10^{-14}
Argon bubble flow rate in prototype/L·min^{-1}	5

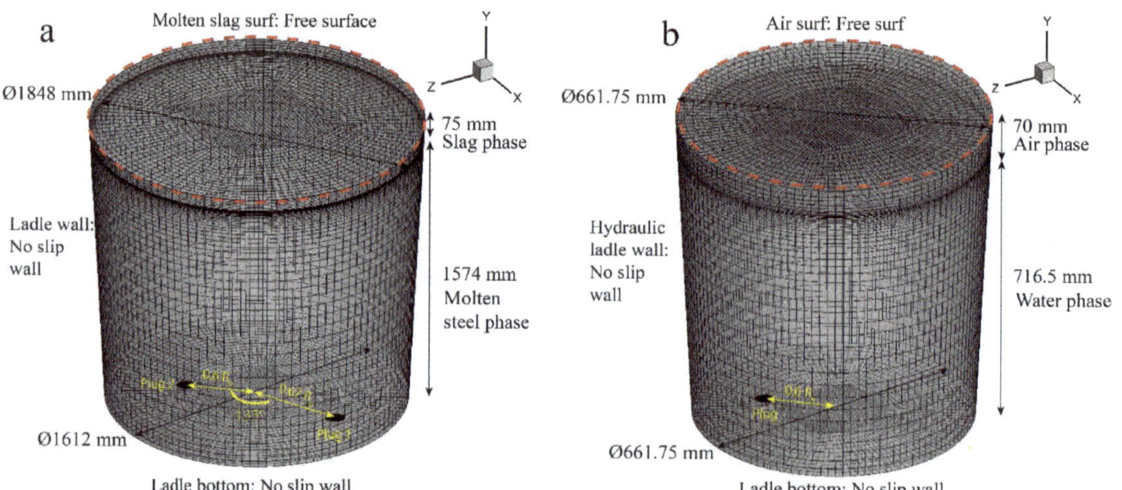

Figure 3. Hexahedral mesh of the fluid zone, boundary name, boundary condition, and the schematic of the argon injection position: (**a**) prototype; (**b**) hydraulic model.

3. Results and Discussion

3.1. Model Validation

3.1.1. Effect of Discrete Random walk Model on the Simulation Results

Figure 4 shows the simulation results of the water model simulation and the PIV measurements results of the water model. Figure 4a shows the slice position of the numerical simulation results. The PIV measurements area is identical to the slice position. Figure 4b shows the velocity contour of the slice position, the velocity is higher in the center, and the bubbles are transported in the center, which accelerates the fluid velocity when the bubble random transport model is deactivated. The bubble is upward in a straight line, and the velocity contour is symmetrical. Figure 4c shows the velocity contour when the bubble random transport model is activated, and the C_L is 0.15. The area of the velocity at a higher speed is in a reverse cone shape. The maximum velocity is lower than when the random walk model is deactivated. The bubbles are transported in a random way; the degree of dispersion of the bubble position is higher, which results in a lower maximum velocity. Figure 4d shows the PIV measurement result, and the measurement area is plotted in Figure 4c in a red scatter line. Figure 4d shows that the velocity is higher at the top of the ladle model, the velocity in a higher-speed area is in a reverse cone shape, and the velocity contour distribution of the numerical simulation is similar to the PIV measurement results.

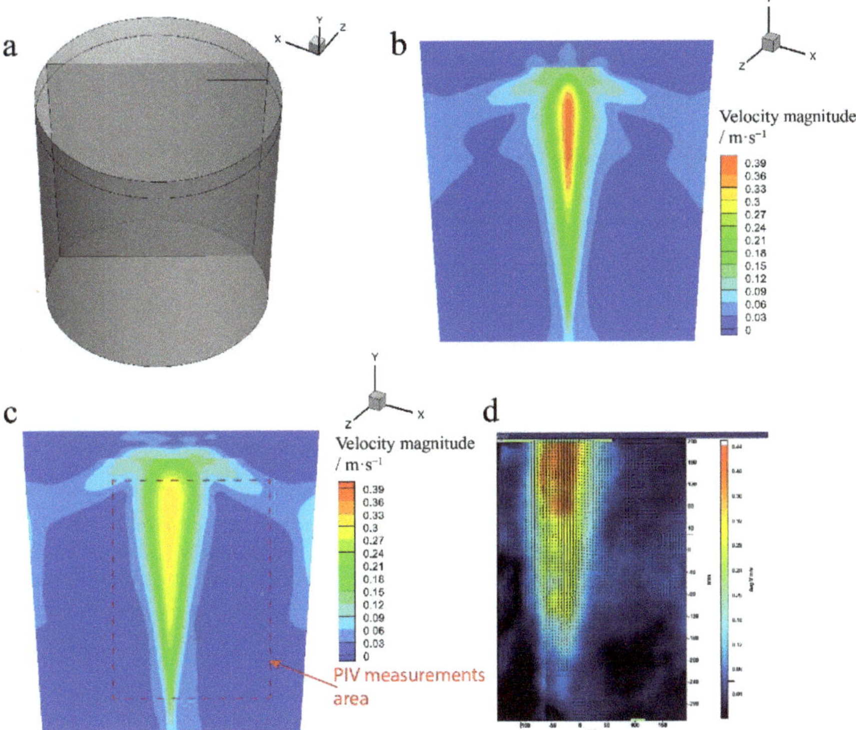

Figure 4. (**a**) The slice position of the simulation results of the water model, (**b**) the velocity magnitude contour in the slice when the random walk model was deactivated, (**c**) the velocity magnitude contour in the slice when the random walk model was activated and the C_L was 0.15, and (**d**) the time-averaged results of the PIV measurements in the water model.

Figure 5 shows the quantitative data of the velocity magnitude of the numerical simulation and PIV measurement results. Figure 5a–c show the simulation results of the bubble random walk model being deactivated and the PIV measurement results. The results show that the velocity magnitude distribution is sharp, and the velocity peak is narrow and higher compared to the PIV results. The flow of the fluid in the water model is driven by the bubbles. When the bubble random walk model is deactivated, the flow of the bubbles in the fluid presents a vertical upward movement, so the velocity peaks appear high and narrow. Figure 5d–f show the simulation results when the bubble random walk model is activated, the C_L is 0.15, and the PIV measurements results. The velocity magnitude peak is smaller and broader than when the bubble random walk model is deactivated.

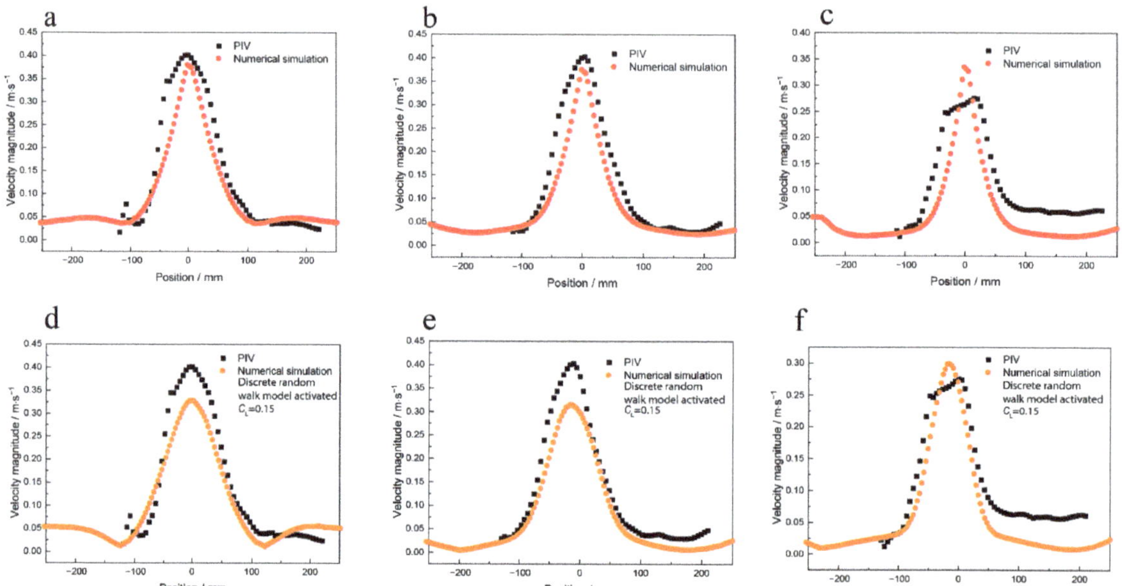

Figure 5. The quantitative velocity magnitude data of numerical simulation of water model and PIV measurements results, when the random walk model was deactivated (**a**) near the water surface, (**b**) under the water surface by 100 mm, and (**c**) under the water surface by 200 mm; when the random walk model was activated and the C_L was 0.15 (**d**) near the water surface, (**e**) under the water surface by 100 mm, and (**f**) under the water surface by 200 mm.

3.1.2. Effect of Time Scale Factor on the Simulation Results

Figure 6 shows the velocity magnitude of the numerical simulation results and the PIV measurements. The numerical simulation result under the case of C_L is 0.3 or 0.45. Figure 7 shows the variance in the velocity difference between the numerical simulation and the PIV measurement results. The calculation range is the velocity data from −100 mm to 100 mm.

In Figure 7, D_L represents the variance in the velocity magnitude difference between the numerical simulation and the PIV measurements, where a C_L of 0 illustrates the case where the bubble random transport model is deactivated. The results show that the variance is decreased when the C_L increases from 0 to 0.3, then the variance is increased when the C_L increases from 0.3 to 0.45. The C_L has a significant influence on the surf velocity. The randomness of the bubble transport is higher at the top of the ladle, then the variation in C_L has a significant impact on the surf velocity. The velocity difference is less when the C_L is 0.3.

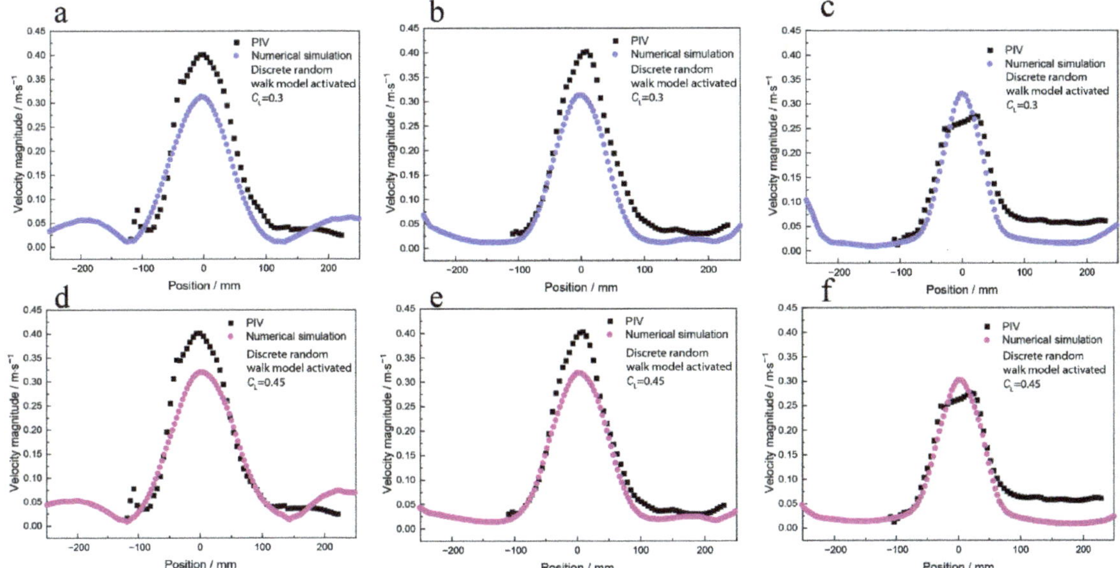

Figure 6. The quantitative velocity magnitude data of numerical simulation of water model and PIV measurements results, when the random walk model was activated and C_L was 0.3 (**a**) near the water surface, (**b**) under the water surface by 100 mm, and (**c**) under the water surface by 200 mm; when the random walk model was activated and the C_L was 0.45 (**d**) near the water surface, (**e**) under the water surface by 100 mm, and (**f**) under the water surface by 200 mm.

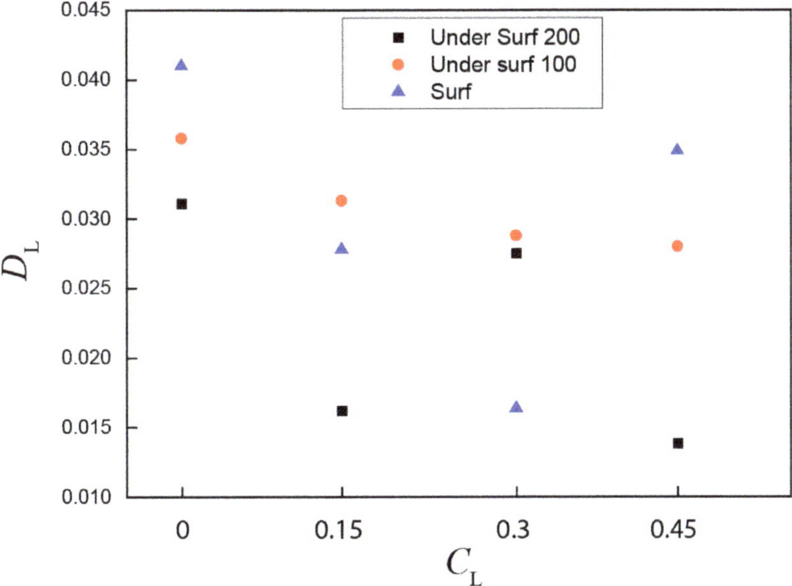

Figure 7. The variance in the velocity difference between PIV measurements and numerical simulation at −100 mm to 100 mm in Figures 5 and 6.

3.2. Model Application of the Prototype Argon-Stirred Ladle

3.2.1. The Fluid Flows

Figure 8 shows the streamline, the slice position, and the quantitative data of the velocity magnitude in the prototype. Figure 8a shows the streamline and the sliced contour of the velocity. Figure 8a shows that the streamline flows from the ladle bottom where the plug is situated, the streamline is forward to the top of the ladle, and the streamline is reverse to the ladle bottom. Figure 8b shows that the velocity is higher in the bubble transport route, and the velocity magnitude is decreased near the surf of the ladle. The flow is divided into two streams when the stream flows to the surf of the ladle. Figure 8c shows the velocity magnitude in the three lines shown in Figure 8b. The results show that the stream velocity peak is broader in Line 1 compared to Lines 2 and 3. The maximum velocity in the stream is 0.172 m·s^{-1}.

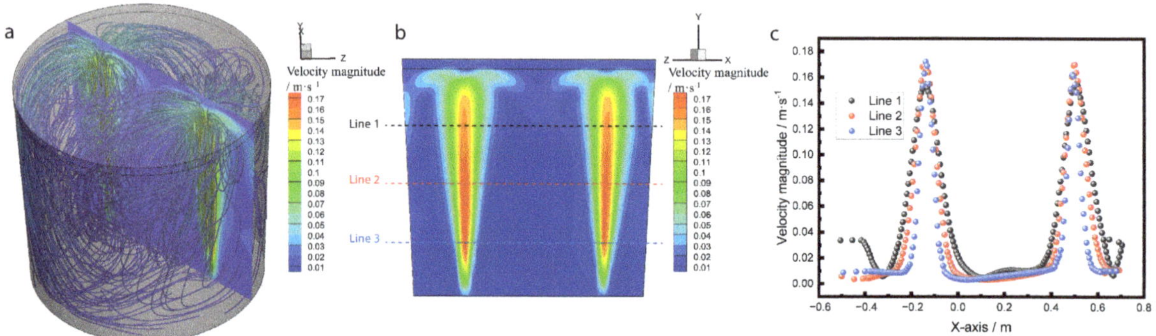

Figure 8. The prototype numerical simulation results: (**a**) streamline results and slice contour position, (**b**) velocity contour at the slice position, and (**c**) velocity magnitude results at three lines.

3.2.2. The Temperature Fields

Effect of Initial Bubble Temperature on the Temperature Filed

Figure 9 shows the temperature measurement position in industrial practice, and the numerical simulation of the temperature varies with the solution time at the temperature measurement position. In industrial practice, the temperature is 1759 K before the argon blow, and the temperature is 1740 K when the argon blow lasts for 22 min. The temperature drop rate is 0.0144 K/s. The bubble is injected from the ladle bottom, and then the bubble temperature is lower than that of the molten steel, which will decrease the molten steel temperature. Figure 9b shows the temperature variation with the solution time under different initial bubble temperatures. The temperature decrease rate is 0.0147 K/s, 0.023 K/s, and 0.0313 K/s when the initial bubble temperature is 800 °C, 400 °C, and 25 °C, respectively. The temperature decrease rate is increased when the initial bubble temperature decreases. The temperature decrease rate is similar to the industrial practice when the initial bubble injection temperature is 800 °C.

Figure 10 shows the iso-surface of the temperature under the different initial bubble temperatures when the argon blow simulation time lasts 150 s. The molten steel temperature is lower where the bubble is placed. The low-temperature area is diffused along the fluid flow direction.

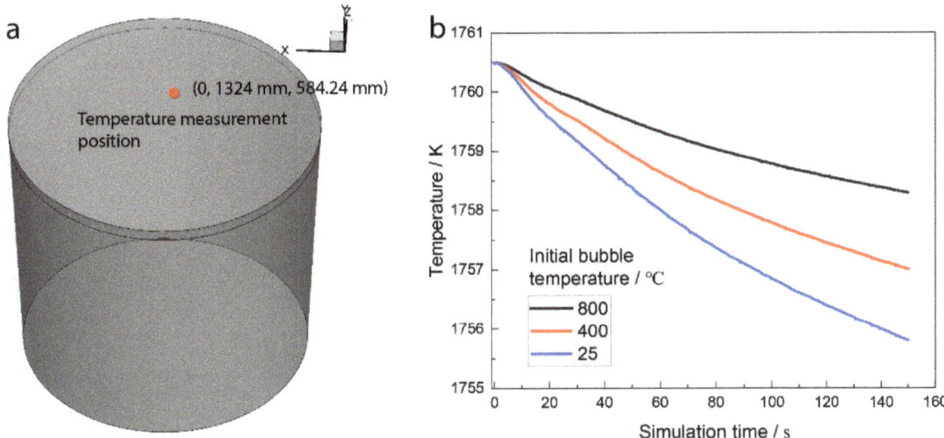

Figure 9. (**a**) Temperature measurements position in the industrial practice; (**b**) temperature varies with solution time at the temperature measurements position.

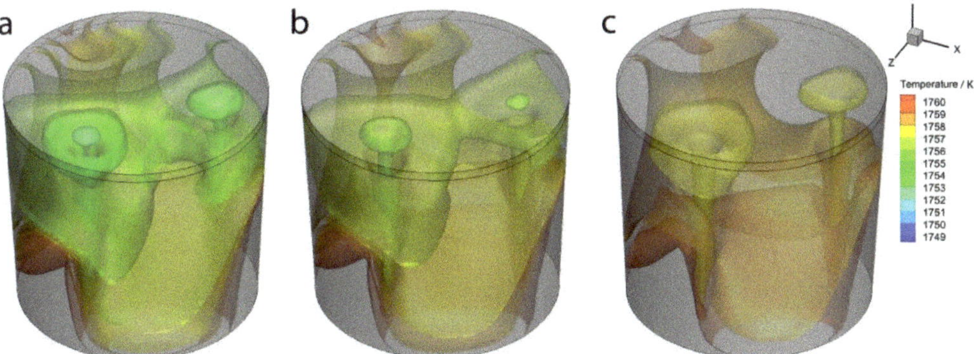

Figure 10. The temperature iso-surface under different initial bubble temperatures: (**a**) 25 °C, (**b**) 400 °C, and (**c**) 800 °C.

Figure 11 shows the sliced contour of the temperature when the initial bubble temperature is 800 °C and the solution time is 150 s. The results show that the temperature in the ladle bottom is higher than those in the other slices. The temperature in the bubble plume area is lower than those in the different regions. The temperature distribution is asymmetric because the plugs distribution is asymmetric, which results in the flow being asymmetric. The high-temperature area is diffused to the low-temperature area.

Figure 12a shows the streamline in the top view of the ladle. The stream in the ladle is divided into two main circulations. Plug 1 is set at $0.67R_m$ and plug 2 is set at $0.6R_m$. Figure 12b shows the schematic of the circulation, and the circulation is larger in plug 1. The circulation area is asymmetric, which results in the temperature field asymmetry. The temperature fields are diffused in the flow direction. The circulation area of plug 1 is larger than that of plug 2, then the molten steel temperature above plug 1 is higher, as shown in Figure 11f.

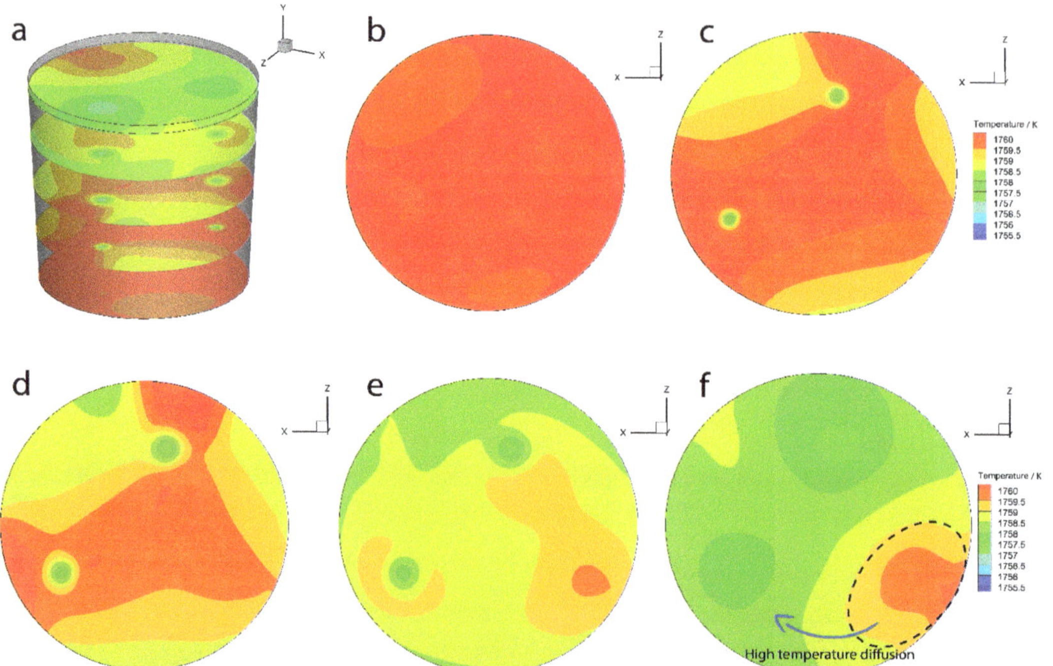

Figure 11. The temperature slice contour when the initial bubble temperature is 800 °C: (**a**) slice position, (**b**) slice position near the bottom, (**c**) slice position at Y = 0.3935, (**d**) slice position at Y = 0.787, (**e**) slice position at Y = 1.18, and (**f**) slice position at near the surface.

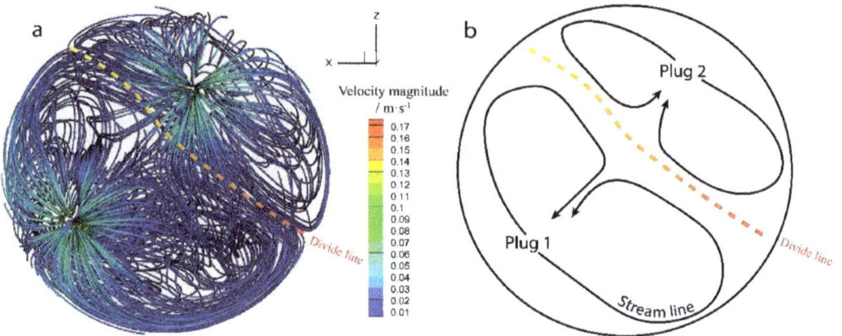

Figure 12. (**a**) The streamline in the top view; (**b**) the schematic of the streamline in the different plugs.

3.2.3. The Steel–Slag Interface Shape

Figure 13a shows the argon-stirred process in a steel plant. The ladle contains 25 t of molten steel. There are two plugs used in the argon-stirred process, and there are two places called 'slag eyes' plotted by blue dash circles. Figure 13b shows the simulation results of the steel–slag interface shape during the argon-stirred process, and there are two 'peaks' formed above the two plugs. The argon bubbles drive the fluid upward, then push the slag away, which results in the two 'slag eyes' formed. Figure 13b shows that the maximum height of the steel–slag interface is 7.95 mm, and the simulation results of the position of the 'slag eyes' are similar to industrial practice.

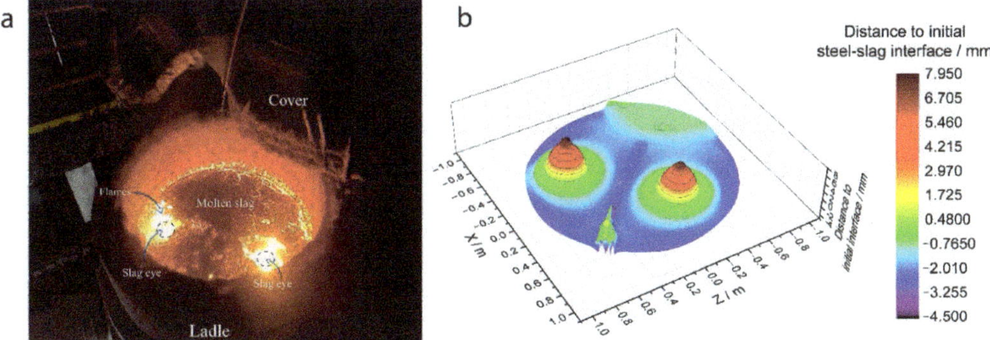

Figure 13. (**a**) The molten steel and slag surface during the argon-stirred process, and (**b**) the numerical simulation results of the steel–slag interface position.

3.2.4. Alloy Melting and Alloy Species Diffusion

Figure 14 shows the sliced contour of the dimensional concentration. Figure 14a–c show the alloy added in the 'slag eye' above plug 1. The results show that the alloy melts and releases the species of the alloy, then the species is diffused along the fluid flow direction. Figure 14d–f show that the alloy added above the 'slag eye' is above plug 2.

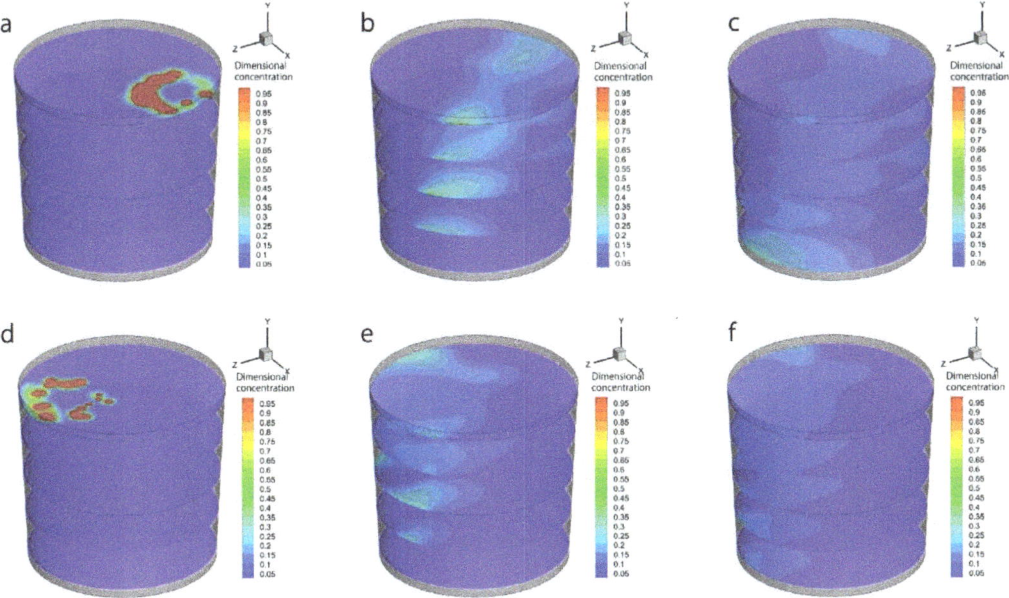

Figure 14. The contour of the slice of the dimensional concentration of the alloy species in the ladle when the alloy is added on the 'slag eye' above plug 1, after (**a**) 10 s, (**b**) 65 s, and (**c**) 120 s; when the alloy is added on the 'slag eye' above plug 2, after (**d**) 10 s, (**e**) 65 s, and (**f**) 120 s.

Figure 15a,b show the alloy melting position when the alloy is added in the 'slag eye' above plug 1 and plug 2. Figure 15a,b show that the alloy melting position is in a circle shape, near the 'slag eye'. Figure 15c shows the statistical results of the melting time, and the results show that the average melting time is 12.49 s and 12.71 s when the alloy added in the 'slag eye' is above plug 1 and plug 2, respectively. The melting time when the alloy is added in the 'slag eye' above plug 1 is less than that of plug 2. Figures 11f and 12 show that the circulation area in plug 1 is higher than that of plug 2, the molten steel temperature in the 'slag eye' above plug 1 is higher than that of plug 2, and the melting time is decreased when the molten steel temperature increases. Figure 15d shows the average dimensional concentration varying with the solution time, and the average dimensional concentration is increased and reaches a maximum after the alloy is added after 20 s. The maximum of the average dimensional concentration is higher when the alloy is added on the 'slag eye' above plug 1, because the melting time is less, and the average dimensional concentration has fewer changes after the alloy is added after 100 s.

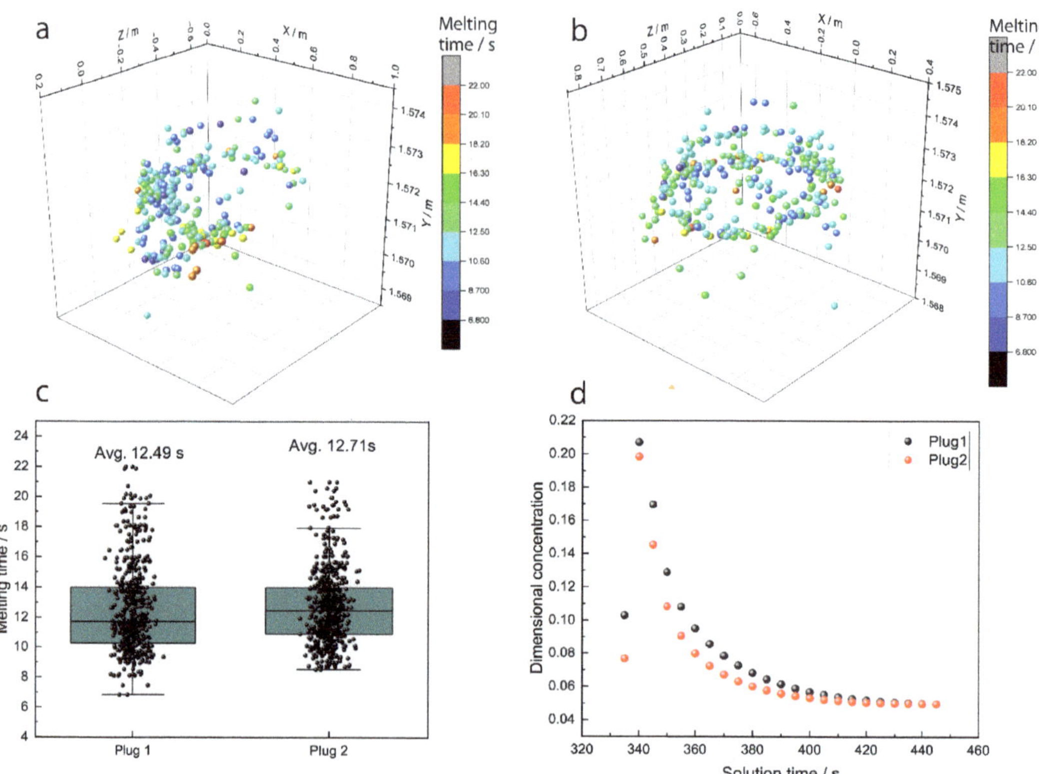

Figure 15. The alloy melting position when the alloy is added on the 'slag eye' above (**a**) plug 1 and (**b**) plug 2; (**c**) the alloy melting time and (**d**) the average dimensional concentration under different alloy add-in positions.

4. Conclusions

In this paper, a multi-physics numerical simulation model in an argon-stirred ladle was established, the random walk model was used in the bubble transport model, and the effect of C_L on the fluid flow simulation was studied for the first time. In the temperature simulation, a heat flux transfer model in molten steel and bubbles was introduced for the first time. The effect of initial bubble temperature on the molten steel temperature simulation was discussed. The alloy melting and alloy concentration diffusion were simulated. The multi-physics simulation model was beneficial in the ladle design and the modification of the parameters.

(1) The random walk model needs to be applied to the bubble transport model. The velocity difference between the numerical simulation and the hydraulic model decreases when the C_L decreases from 0 to 0.3 and increases when the C_L increases from 0.3

to 0.45. The velocity difference between the numerical simulation and the hydraulic model is minimum when the C_L is 0.3.

(2) There are two circulations of fluid flow in the prototype. The velocity is higher in the center of the plume. The velocity is smaller near the steel–slag interface. The maximum velocity is 0.172 m·s^{-1}.

(3) The molten steel temperature will decrease when the argon bubbles are injected into the molten steel from the plugs. The temperature decrease rate will increase when the initial bubble temperature decreases. The temperature decrease rate in industrial practice is 0.0144 K/s. The numerical simulation results of the temperature decrease rate are 0.0147 K/s when the initial bubble temperature is 800 °C.

(4) The steel–slag interface position is higher above the plugs. The maximum height of the steel–slag interface is 7.95 mm.

(5) The average alloy melting time is 12.49 s or 12.71 s when the alloy is added on the two slag eyes separately. The alloy melting time has a little difference because the molten steel temperature has little difference in the two slag eyes. The average alloy concentration in the ladle is increased when the alloy is added to the molten steel in 20 s, and the average alloy concentration decreases during the alloy species diffusion. The alloy concentration has fewer changes when the alloy is added to the molten steel after 100 s.

Author Contributions: Conceptualization, C.H. and Y.B.; Methodology, C.H.; Software, C.H.; Validation, C.H. and Y.B.; Formal Analysis, C.H.; Investigation, C.H.; Resources, Y.B.; Data Curation, C.H.; Writing—Original Draft Preparation, C.H.; Writing—Review and Editing, C.H., Y.B. and M.W.; Visualization, C.H.; Supervision, Y.B.; Project Administration, Y.B.; Funding Acquisition, Y.B. All authors have read and agreed to the published version of the manuscript.

Funding: This work was supported by the Open Project of State Key Laboratory of Advanced Special Steel, Shanghai Key Laboratory of Advanced Ferrometallurgy, Shanghai University (SKLASS 2020-02), and the Science and Technology Commission of Shanghai Municipality (No. 19DZ2270200).

Institutional Review Board Statement: Not applicable.

Informed Consent Statement: Not applicable.

Data Availability Statement: All the data are given in the manuscript.

Conflicts of Interest: The authors have declared no conflict of interest.

Nomenclature

$\vec{v}$	velocity
$\vec{u}_l$	fluid vector
$\vec{g}$	Gravitational acceleration
$\vec{u}_p$	particle vector
$\vec{F}$	source term of the bubble-driven flow
C	mass concentration of alloy
$C_{b,i}$	heat capacity of the bubble
$C_{D,i}$	coefficient of drag force
C_L	integral time-scale constant
C_p	heat capacity of the fluid
$C_{P,A}$	heat capacity of the alloy
d_A	diameter of the alloy
$d_{b,i}$	diameter of the bubble
D_L	variance in the velocity magnitude difference between the numerical simulation and the PIV measurements
E_{step}	energy exchange in each timestep
E_T	total energy of the bubble

G_b	generation of turbulence kinetic energy due to buoyancy
G_k	generation of turbulence kinetic energy due to the mean velocity gradient
h	sensible enthalpy
h_c	coefficient of the heat transfer
I	unit tensor
k	turbulent kinetic energy
k_m	fluid thermal conductivity
m_{alloy}	mass of the melted alloy
$m_{b.i}$	bubble mass
m_{cell}	mass of the cell where the alloy melting
N	bubble count in the cell
n	phase count
N_{per}	bubble injection count in 1 s
Nu	Nusselt number
Q	flow rate of the argon
r	random Gaussian number
Rei	Reynolds number of bubble
R_m	the radius in the hydraulic model
R_p	the radius in the prototype(ladle)
S_c	source term of the alloy melting
T_0	initial alloy temperature
$T_{b,\,pre.temp}$	bubble temperature in the previous timestep
$T_{b.init}$	initial bubble temperature
T_{cell}	cell's temperature where the bubble is injected into the fluid domain
T_E	eddy lifetime
T_L	integral time scale
T_M	temperature of molten steel
t_{melt}	alloy solid crust melting time
T_s	solidification temperature of molten steel
$u_{b,i}$	velocity of the bubble
u_l	velocity of the fluid
$v_{b,init}$	initial velocity when the bubble is injected into the fluid zone
α_l	volume fraction of liquid phase
α_q	volume fraction of qth phase in the fluid
Δt	time step
ΔV_{cell}	volume of the cell
ε	turbulent kinetic energy dissipation
μ	molecular viscosity
ρ	density of the fluid
ρ_A	alloy density
$\rho_{b,i}$	density of the bubble
ρ_f	fluid density
ρ_p	particle density
ρ_q	density of qth phase in the fluid

References

1. Conejo, A.N.; Kitamura, S.; Maruoka, N.; Kim, S.J. Effects of Top Layer, Nozzle Arrangement, and Gas Flow Rate on Mixing Time in Agitated Ladles by Bottom Gas Injection. *Metall. Mater. Trans. B* **2013**, *44*, 914–923. [CrossRef]
2. Liu, Z.; Li, L.; Li, B. Modeling of Gas-Steel-Slag Three-Phase Flow in Ladle Metallurgy: Part I. Physical Modeling. *ISIJ Int.* **2017**, *57*, 1971–1979. [CrossRef]
3. Manuel Amaro-Villeda, A.; Aurelio Ramirez-Argaez, M.; Conejo, A.N. Effect of Slag Properties on Mixing Phenomena in Gas-stirred Ladles by Physical Modeling. *ISIJ Int.* **2014**, *54*, 1–8. [CrossRef]
4. Mazumdar, D.; Dhandapani, P.; Sarvanakumar, R. Modeling and Optimisation of Gas Stirred Ladle Systems. *ISIJ Int.* **2017**, *57*, 286–295. [CrossRef]
5. Szekely, J.; Wang, H.J.; Kiser, K.M. Flow Pattern Velocity and Turbulence Energy Measurements and Predictions in a Water Model of an Argon-Stirred Ladle. *Metall. Trans. B* **1976**, *7*, 287–295. [CrossRef]
6. Li, Y.; Zhu, H.; Wang, R.; Ren, Z.; He, Y. Bubble behavior and evolution characteristics in the RH riser tube-vacuum chamber. *Int. J. Chem. React. Eng.* **2022**. [CrossRef]

7. Li, Y.; Zhu, H.; Wang, R.; Ren, Z.; Lin, L. Prediction of two phase flow behavior and mixing degree of liquid steel under reduced pressure. *Vacuum* **2021**, *192*, 110480. [CrossRef]
8. Pan, Q.; Johansen, S.T.; Olsen, J.E.; Reed, M.; Sætran, L.R. On the turbulence modelling of bubble plumes. *Chem. Eng. Sci.* **2021**, *229*, 116059. [CrossRef]
9. Mantripragada, V.T.; Sahu, S.; Sarkar, S. Morphology and flow behavior of buoyant bubble plumes. *Chem. Eng. Sci.* **2021**, *229*, 116098. [CrossRef]
10. Xu, Y.; Ersson, M.; Jonsson, P.G. A Numerical Study about the Influence of a Bubble Wake Flow on the Removal of Inclusions. *ISIJ Int.* **2016**, *56*, 1982–1988. [CrossRef]
11. Li, Y.-H.; Bao, Y.-P.; Wang, R.; Ma, L.-F.; Liu, J.-S. Modeling study on the flow patterns of gas-liquid flow for fast decarburization during the RH process. *Int. J. Miner. Metall. Mater.* **2018**, *25*, 153–163. [CrossRef]
12. Yuan, F.; Xu, A.-J.; Gu, M.-Q. Development of an improved CBR model for predicting steel temperature in ladle furnace refining. *Int. J. Miner. Metall. Mater.* **2021**, *28*, 1321–1331. [CrossRef]
13. Gu, C.; Liu, W.-Q.; Lian, J.-H.; Bao, Y.-P. In-depth analysis of the fatigue mechanism induced by inclusions for high-strength bearing steels. *Int. J. Miner. Metall. Mater.* **2021**, *28*, 826–834. [CrossRef]
14. Xiao, W.; Bao, Y.-Q.; Gu, C.; Wang, M.; Liu, Y.; Huang, Y.-S.; Sun, G.-T. Ultrahigh cycle fatigue fracture mechanism of high-quality bearing steel obtained through different deoxidation methods. *Int. J. Miner. Metall. Mater.* **2021**, *28*, 804–815. [CrossRef]
15. Li, Y.; He, Y.; Ren, Z.; Bao, Y.; Wang, R. Comparative Study of the Cleanliness of Interstitial-Free Steel with Low and High Phosphorus Contents. *Steel Res. Int.* **2021**, *92*, 2000581. [CrossRef]
16. Guo, X.; Godinez, J.; Walla, N.J.; Silaen, A.K.; Oltmann, H.; Thapliyal, V.; Bhansali, A.; Pretorius, E.; Zhou, C.Q. Computational Investigation of Inclusion Removal in the Steel-Refining Ladle Process. *Processes* **2021**, *9*, 1048. [CrossRef]
17. Krishnapisharody, K.; Irons, G.A. A Model for Slag Eyes in Steel Refining Ladles Covered with Thick Slag. *Metall. Mater. Trans. B* **2015**, *46*, 191–198. [CrossRef]
18. Patil, S.P.; Satish, D.; Peranandhanathan, M.; Mazumdar, D. Mixing Models for Slag Covered, Argon Stirred Ladles. *ISIJ Int.* **2010**, *50*, 1117–1124. [CrossRef]
19. Peranandhanthan, M.; Mazumdar, D. Modeling of Slag Eye Area in Argon Stirred Ladles. *ISIJ Int.* **2010**, *50*, 1622–1631. [CrossRef]
20. Conejo, A.N. Physical and Mathematical Modelling of Mass Transfer in Ladles due to Bottom Gas Stirring: A Review. *Processes* **2020**, *8*, 750. [CrossRef]
21. Hua, J.; Wang, C.-H. Numerical simulation of bubble-driven liquid flows. *Chem. Eng. Sci.* **2000**, *55*, 4159–4173. [CrossRef]
22. Chen, C.; Jonsson, L.T.I.; Tilliander, A.; Cheng, G.; Jönsson, P.G. A mathematical modeling study of the influence of small amounts of KCl solution tracers on mixing in water and residence time distribution of tracers in a continuous flow reactor-metallurgical tundish. *Chem. Eng. Sci.* **2015**, *137*, 914–937. [CrossRef]
23. Geng, D.-Q.; Lei, H.; He, J.-C. Optimization of mixing time in a ladle with dual plugs. *Int. J. Miner. Metall. Mater.* **2010**, *17*, 709–714. [CrossRef]
24. Li, L.; Liu, Z.; Li, B.; Matsuura, H.; Tsukihashi, F. Water Model and CFD-PBM Coupled Model of Gas-Liquid-Slag Three-Phase Flow in Ladle Metallurgy. *ISIJ Int.* **2015**, *55*, 1337–1346. [CrossRef]
25. Uriostegui-Hernandez, A.; Garnica-Gonzalez, P.; Angel Ramos-Banderas, J.; Alberto Hernandez-Bocanegra, C.; Solorio-Diaz, G. Multiphasic Study of Fluid-Dynamics and the Thermal Behavior of a Steel Ladle during Bottom Gas Injection Using the Eulerian Model. *Metals* **2021**, *11*, 1082. [CrossRef]
26. Mazumdar, D.; Guthrie, R.I.L. Modeling Energy Dissipation in Slag-Covered Steel Baths in Steelmaking Ladles. *Metall. Mater. Trans. B* **2010**, *41*, 976–989. [CrossRef]
27. Zhu, M.Y.; Inomoto, T.; Sawada, I.; Hsiao, T.C. Fluid-Flow and Mixing Phenomena in the Ladle Stirred by Argon through Multi-Tuyere. *ISIJ Int.* **1995**, *35*, 472–479. [CrossRef]
28. Liu, H.; Qi, Z.; Xu, M. Numerical Simulation of Fluid Flow and Interfacial Behavior in Three-phase Argon-Stirred Ladles with One Plug and Dual Plugs. *Steel Res. Int.* **2011**, *82*, 440–458. [CrossRef]
29. Llanos, C.A.; Garcia-Hernandez, S.; Ramos-Banderas, J.A.; Barret, J.D.J.; Solorio-Diaz, G. Multiphase Modeling of the Fluidynamics of Bottom Argon Bubbling during Ladle Operations. *ISIJ Int.* **2010**, *50*, 396–402. [CrossRef]
30. Tang, H.; Guo, X.; Wu, G.; Wang, Y. Effect of Gas Blown Modes on Mixing Phenomena in a Bottom Stirring Ladle with Dual Plugs. *ISIJ Int.* **2016**, *56*, 2161–2170.
31. Huang, A.; Gu, H.; Zhang, M.; Wang, N.; Wang, T.; Zou, Y. Mathematical Modeling on Erosion Characteristics of Refining Ladle Lining with Application of Purging Plug. *Metall. Mater. Trans. B* **2013**, *44*, 744–749. [CrossRef]
32. Li, L.; Li, B. Investigation of Bubble-Slag Layer Behaviors with Hybrid Eulerian-Lagrangian Modeling and Large Eddy Simulation. *JOM* **2016**, *68*, 2160–2169. [CrossRef]
33. Liu, W.; Tang, H.; Yang, S.; Wang, M.; Li, J.; Liu, Q.; Liu, J. Numerical Simulation of Slag Eye Formation and Slag Entrapment in a Bottom-Blown Argon-Stirred Ladle. *Metall. Mater. Trans. B* **2018**, *49*, 2681–2691. [CrossRef]
34. Xue, Z.L.; Wang, Y.F.; Wang, L.T.; Li, Z.B.; Zhang, J.W. Inclusion removal from molten steel by attachment small bubbles. *Acta Metall. Sin.* **2003**, *39*, 431–434.
35. Zhu, M.Y.; Zheng, S.G.; Huang, Z.Z.; Gu, W.P. Numerical simulation of nonmetallic inclusions behaviour in gas-stirred ladies. *Steel Res. Int.* **2005**, *76*, 718–722. [CrossRef]

36. Mantripragada, V.T.; Sarkar, S. Multi-Objective Optimization of Bottom Purged Steelmaking Ladles. *Trans. Indian Inst. Met.* **2022**, 1–10. [CrossRef]
37. Cheng, R.; Zhang, L.; Yin, Y.; Ma, H.; Zhang, J. Influence of Argon Gas Flow Parameters in the Slot Plug on the Flow Behavior of Molten Steel in a Gas-Stirred Ladle. *Trans. Indian Inst. Met.* **2021**, *74*, 1827–1837. [CrossRef]
38. Bhattacharjee, A.; Mazumdar, D. Mathematical-Modeling of Fluid-Flow, Alloy Dissolution and Mixing in Industrial Argon Stirred Ladles. *Trans. Indian Inst. Met.* **1992**, *45*, 153–161.
39. Kwon, Y.-J.; Zhang, J.; Lee, H.-G. A CFD-based nucleation-growth-removal model for inclusion behavior in a gas-agitated ladle during molten steel deoxidation. *ISIJ Int.* **2008**, *48*, 891–900. [CrossRef]
40. Lachmund, H.; Xie, Y.K.; Buhles, T.; Pluschkell, W. Slag emulsification during liquid steel desulphurisation by gas injection into the ladle. *Steel Res. Int.* **2003**, *74*, 77–85. [CrossRef]
41. Singh, U.; Anapagaddi, R.; Mangal, S.; Padmanabhan, K.A.; Singh, A.K. Multiphase Modeling of Bottom-Stirred Ladle for Prediction of Slag-Steel Interface and Estimation of Desulfurization Behavior. *Metall. Mater. Trans. B* **2016**, *47*, 1804–1816. [CrossRef]
42. Wondrak, T.; Timmel, K.; Bruch, C.; Gardin, P.; Hackl, G.; Lachmund, H.; Lungen, H.B.; Odenthal, H.-J.; Eckert, S. Large-Scale Test Facility for Modeling Bubble Behavior and Liquid Metal Two-Phase Flows in a Steel Ladle. *Metall. Mater. Trans. B* **2022**, *53*, 1703–1720. [CrossRef]
43. Uriostegui-Hernandez, A.; Garnica-Gonzalez, P.; Hernandez-Bocanegra, C.-A.; Ramos-Banderas, J.-A.; Montes-Rodriguez, J.-J.; Ortiz-Castillo, J.-R. Study of fluid dynamics and sulfar mass transfer between steel and slag in a ladle furnace considering drag and non-drag forces. *Dyna* **2022**, *97*, 176–183. [CrossRef]
44. Wang, Q.; Liu, C.; Pan, L.; He, Z.; Li, G. Numerical Understanding on Refractory Flow-Induced Erosion and Reaction-Induced Corrosion Patterns in Ladle Refining Process. *Metall. Mater. Trans. B* **2022**, *53*, 1617–1630. [CrossRef]
45. Joubert, N.; Gardin, P.; Popinet, S.; Zaleski, S. Experimental and numerical modelling of mass transfer in a refining ladle. *Metall. Res. Technol.* **2022**, *119*, 109. [CrossRef]
46. Riabov, D.; Gain, M.M.; Kargul, T.; Volkova, O. Influence of Gas Density and Plug Diameter on Plume Characteristics by Ladle Stirring. *Crystals* **2021**, *11*, 475. [CrossRef]
47. Li, Q.; Pistorius, P.C. Interface-Resolved Simulation of Bubbles-Metal-Slag Multiphase System in a Gas-Stirred Ladle. *Metall. Mater. Trans. B* **2021**, *52*, 1532–1549. [CrossRef]

Article

A Detailed Process and Techno-Economic Analysis of Methanol Synthesis from H_2 and CO_2 with Intermediate Condensation Steps

Bruno Lacerda de Oliveira Campos *, Kelechi John, Philipp Beeskow, Karla Herrera Delgado *, Stephan Pitter, Nicolaus Dahmen and Jörg Sauer

Institute of Catalysis Research and Technology (IKFT), Karlsruhe Institute of Technology (KIT), Hermann-von Helmholtz-Platz 1, D-76344 Eggenstein-Leopoldshafen, Germany
* Correspondence: bruno.campos@kit.edu (B.L.d.O.C.); karla.herrera@kit.edu (K.H.D.)

Citation: Lacerda de Oliveira Campos, B.; John, K.; Beeskow, P.; Herrera Delgado, K.; Pitter, S.; Dahmen, N.; Sauer, J. A Detailed Process and Techno-Economic Analysis of Methanol Synthesis from H_2 and CO_2 with Intermediate Condensation Steps. *Processes* **2022**, *10*, 1535. https://doi.org/10.3390/pr10081535

Academic Editors: Elio Santacesaria, Riccardo Tesser and Vincenzo Russo

Received: 13 July 2022
Accepted: 3 August 2022
Published: 5 August 2022

Publisher's Note: MDPI stays neutral with regard to jurisdictional claims in published maps and institutional affiliations.

Abstract: In order to increase the typically low equilibrium CO_2 conversion to methanol using commercially proven technology, the addition of two intermediate condensation units between reaction steps is evaluated in this work. Detailed process simulations with heat integration and techno-economic analyses of methanol synthesis from green H_2 and captured CO_2 are presented here, comparing the proposed process with condensation steps with the conventional approach. In the new process, a CO_2 single-pass conversion of 53.9% was achieved, which is significantly higher than the conversion of the conventional process (28.5%) and its equilibrium conversion (30.4%). Consequently, the total recycle stream flow was halved, which reduced reactant losses in the purge stream and the compression work of the recycle streams, lowering operating costs by 4.8% (61.2 M€·a^{-1}). In spite of the additional number of heat exchangers and flash drums related to the intermediate condensation units, the fixed investment costs of the improved process decreased by 22.7% (94.5 M€). This was a consequence of the increased reaction rates and lower recycle flows, reducing the required size of the main equipment. Therefore, intermediate condensation steps are beneficial for methanol synthesis from H_2/CO_2, significantly boosting CO_2 single-pass conversion, which consequently reduces both the investment and operating costs.

Keywords: methanol synthesis; CO_2 utilization; power-to-X; intermediate condensation steps; product removal; techno-economic analysis; heat integration; plant simulation

1. Introduction

Sustainable solutions are required to reduce the greenhouse gas emissions of the transportation and industrial sectors, shrinking the dependency on fossil fuels. With the continuously increasing installed capacity of wind and solar power plants [1], adequate energy storage solutions have to be implemented in order to deal with their fluctuating nature. Therefore, the conversion of electricity into valuable chemicals and fuels, a concept often called power-to-fuels or power-to-X, can make an important contribution to the future energy system [2]. In this context, key process steps are hydrogen generation via electrolysis (primary conversion) [3,4] and methanol synthesis (secondary conversion) [5].

Methanol can be used as a fuel substitute or additive, either in fuel cells or via combustion [2], as a feedstock in the production of base chemicals (e.g., formaldehyde) and liquid fuels (e.g., gasoline, oxymethylene ethers, jetfuel) [6–8], and also as a solvent. Methanol fuel has recently attracted significant interest, especially in China, where the consumption of methanol for thermal applications (e.g., boilers, kilns, cooking stoves) and in the transportation section amounted to 5.7 Mton·a^{-1} (year 2019) [9].

The current global capacity of methanol production is 164 Mton·a^{-1} (year 2021), with an annual increase of 10% projected for the next decade [10]. Traditionally, methanol synthesis is fed by fossil-based syngas, which comes either from steam reforming of natural gas or from coal gasification [11]. However, sustainable syngas production is gaining importance, such as from renewable electricity and captured CO_2, and also from biomass. In Figure 1, a scheme is presented showing the intermediate position of methanol in the conversion of both fossil-based and sustainable syngas to added-value chemicals and fuels, as well as methanol end-use applications.

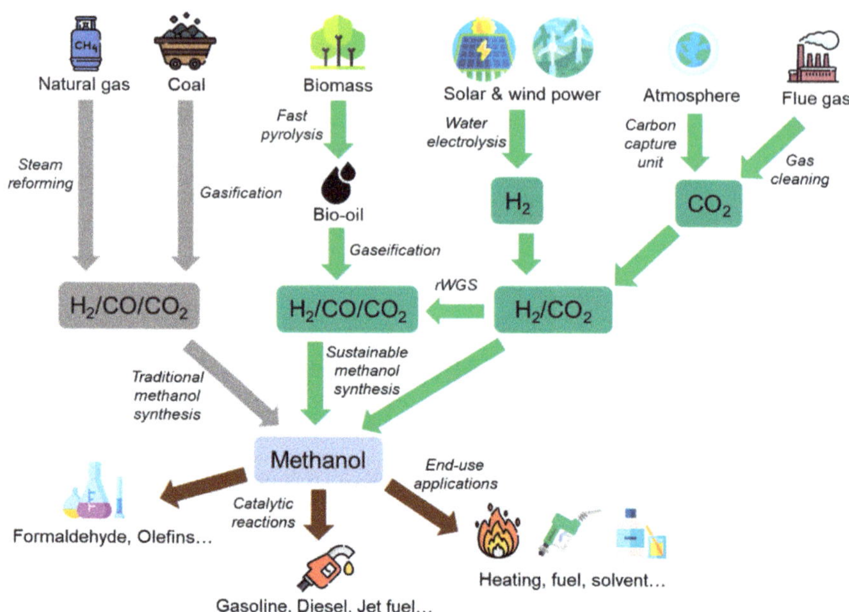

Figure 1. The key position of methanol to convert syngas sources into added-value chemicals and fuels. Icons from: Freepik, Flaticon [12].

The current world-scale technology for methanol synthesis is mostly based on the application of $Cu/ZnO/Al_2O_3$ (CZA) catalysts in either multi-tube reactors with boiling water as the cooling fluid, normally called isothermal reactors (e.g., the Lurgi process, the Linde process), or adiabatic reactors with intermediate cold syngas quenching, generally named quench reactors (e.g., ICI and the Casale process, the Haldor Topsoe process) [11,13]. Less common but also industrially applied are the adiabatic reactors with intermediate cooling (e.g., the Kellogg process, the Toyo process) [11,14]. Normally, temperatures between 200 and 300 °C and pressures between 50 and 100 bar are applied [13].

Methanol can be produced from either CO (Equation (1)) or CO_2 (Equation (2)), with the reverse water–gas shift reaction (rWGSR, Equation (3)) also occurring. If the feed gas contains both CO and CO_2, there is a prevailing opinion that direct CO hydrogenation (Equation (1)) on Cu/Zn-based catalysts is significantly slower than CO_2 hydrogenation [15,16], and kinetic models not considering this reaction can adequately simulate experimental data [17–19].

$$CO_{(g)} + 2 \cdot H_{2(g)} \rightleftarrows CH_3OH_{(g)} \qquad \Delta H^0_{25\,°C} = -90.6\ \text{kJ} \cdot \text{mol}^{-1} \tag{1}$$

$$CO_{2(g)} + 3 \cdot H_{2(g)} \rightleftarrows CH_3OH_{(g)} + H_2O_{(g)} \quad \Delta H^0_{25\,°C} = -49.4\ \text{kJ} \cdot \text{mol}^{-1} \tag{2}$$

$$CO_{2(g)} + H_{2(g)} \rightleftarrows CO_{(g)} + H_2O_{(g)} \qquad \Delta H^0_{25\,°C} = 41.2\ \text{kJ} \cdot \text{mol}^{-1} \tag{3}$$

With regard to the general use of CO_2 as a carbon source, several process simulations and techno-economic analyses of methanol synthesis from green hydrogen and captured CO_2 have been reported, with the CO_2 source being either a cleaned industrial flue gas or concentrated atmospheric CO_2 (i.e., through carbon capture units, CCUs) [20–22]. Pérez-Fortes et al. [20] and Szima et al. [21] simulated a methanol synthesis plant with CZA, having a total production of 440 and 100 kton MeOH·a^{-1}, respectively. Heat integration was considered in both studies, with the plant being energetically self-sufficient. Cordero-Lanzac et al. [22] simulated the production of 275 kton MeOH·a^{-1} with an In$_2$O$_3$/Co catalyst. In all studies, it was concluded that economic viability can be achieved if reactant prices significantly decrease or if CO$_2$ taxation is enforced. Nonetheless, it is expected that the costs of electrolysis and CCUs might considerably decrease in the foreseeable future [23,24], and the first industrial-scale plants to produce e-methanol and e-gasoline are expected to start operating in 2024–2025 [25,26].

A general argument is the thermodynamic restrictions of CO$_2$ conversion to methanol compared to CO conversion (see Figure 2), whereby only limited methanol yields are achievable even at elevated pressures and lower temperatures. Consequently, a low CO$_2$ single-pass conversion ($X_{CO_2,SP}$) is obtained independently of the reactor size, leading to large recycle streams, which increase operating costs and cause higher reactant losses in purge streams.

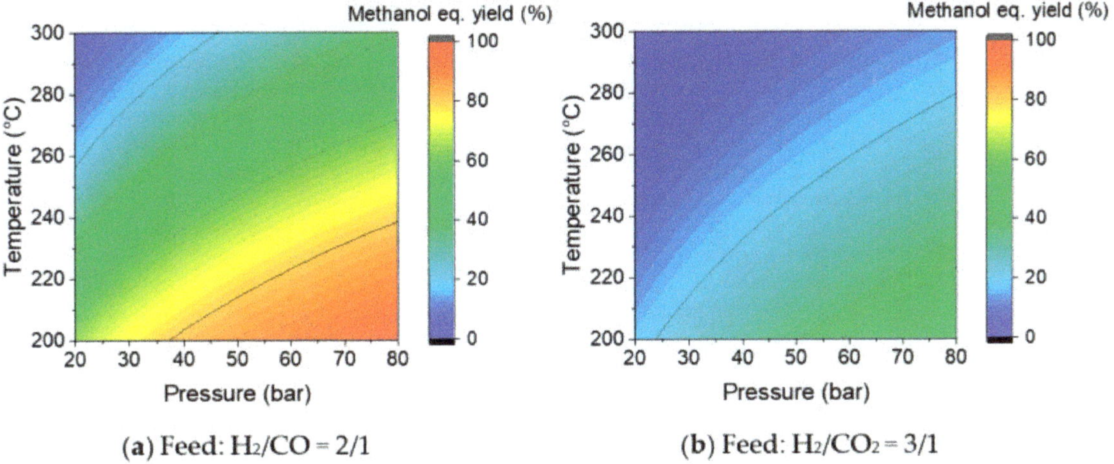

Figure 2. Methanol equilibrium yield as a function of temperature and pressure. Data generated with Aspen Plus. (**a**) H$_2$/CO feed in a 2:1 ratio. (**b**) H$_2$/CO$_2$ feed in a 3:1 ratio.

If the products (i.e., methanol and water) are removed from the reacting system, the thermodynamic equilibrium is shifted towards a higher methanol yield. This strategy has been studied using alternative reactor designs with in situ condensation [27,28] or membrane reactors [29], but these technologies are not yet ready for commercialization. A feasible approach using commercially proven technology is the implementation of intermediate condensation steps between reactor units displaced in series. In the Davy series loop methanol process, two reactors with an intermediate condensation unit are proposed for large scale methanol production from CO-rich syngas [13,30]. Although the implementation of intermediate condensation steps is a promising strategy to increase methanol yield from H$_2$/CO$_2$ syngas, such an approach has still not received particular attention, and plant simulations with heat integration and techno-economic analyses are not available in the literature yet, to the best of our knowledge.

In this work, the conventional approach (named here the 'one-step process') is compared with a new alternative approach including two condensation steps (named here the 'three-step process'). Using our recently developed kinetic model for methanol synthesis [19], both processes were implemented in Matlab in order to critically analyze and select key process parameters (i.e., cooling fluid temperature, number of reactor modules, and purge fraction). With the optimized parameters, detailed methanol synthesis plants with heat integration were implemented in Aspen Plus, and techno-economic analyses were performed.

2. Methodology

2.1. Process Overview

In the present work, a methanol synthesis plant from H_2/CO_2 with a production of 145 ton·h^{-1} is considered. This value is based on an ongoing power-to-gasoline project via H_2/CO_2 conversion to methanol [26], whose final goal is a gasoline production of $5.5 \cdot 10^8$ L·a^{-1}, which corresponds to a methanol production of 1.16 Mton·a^{-1} or 145 ton·h^{-1} (assuming a yield of 80% in the methanol-to-gasoline process and plant operating hours of 8000 h·a^{-1}).

In our simulations, feed carbon dioxide comes from the cleaned flue gas of nearby industries (e.g., a cement industry) at 25 °C and 1 bar, with a purity of 99.5% mol/mol (the rest was water). Feed hydrogen comes from water electrolysis at 25 °C and 30 bar, with a purity of 99.5% mol/mol (the rest was nitrogen). Although it is possible to obtain these feedstocks in an extremely high purity (e.g., 99.99% mol/mol) [31,32], we chose a more conservative scenario, which also allows a proper simulation of inert material accumulation in the plant.

As pressure has a significant influence on the thermodynamic equilibrium of methanol synthesis (see Figure 2), the reactor operating pressure was set to 70 bar. Although higher pressures are reported to have potential in methanol synthesis [33,34], they were out of the scope of this work, since considerable extrapolations in the kinetic model would be necessary, and condensation inside the reactor might have to be taken into account. Besides, higher pressures increase compression costs and might also require more expensive materials to build the equipment.

The dimensions of the reactor modules were chosen to be close to the upper size limits that are currently commercially available. That is, each reactor module consisted of a shell containing 33,000 tubes with 12.5 m length and an inner diameter of 3.75 cm. Since the heat generation in CO_2 hydrogenation is lower than in CO hydrogenation (Equations (1) and (2)), less heat transfer area is necessary. Because of that, the tube inner diameter chosen in this work (3.75 cm) was larger than the size typically used for CO conversion to methanol (2.5 cm). Considering 1050 kg·m^{-3} as the apparent catalyst bed density [35], the total CZA catalyst loading of each reactor module was 478.13 ton. A total pressure loss of 0.75 bar was considered for each reactor module [36].

2.1.1. One-Step Approach—Process Description

In Figure 3, a detailed flowsheet of the one-step process is presented. This is an adapted version from a concept reported in the literature [37–39]. Feed CO_2 is mixed with a low-pressure recycle stream, and then compressed from 1 to 70 bar in a three-stage process, including intermediate cooling (reducing compression work) and intermediate phase separation (to remove condensed methanol and water from the recycle stream). The resulting compressed stream is mixed with feed H_2 (compressed from 30 to 70 bar in one stage) and with a high pressure recycle stream. The mixed stream is preheated with the product gas and enters the inner tubes of parallel reactor modules, with the temperature being controlled by boiling water on the shell side.

The product stream is cooled down to 30 °C in four heat exchangers, condensing water, methanol, and some CO_2, which are separated from the light gases in a flash drum. A fraction of the gas stream is purged, and the remaining stream is recompressed to 70 bar and recycled. The liquid stream from the flash drum is depressurized to 1 bar and heated to 30 °C, vaporizing most remaining CO_2. A liquid–gas separation is performed in another flash drum. A fraction of the gas stream from the low-pressure flash drum is purged, and the rest is recycled by mixing with feed CO_2. The liquid stream from the low-pressure flash drum is preheated and fed to a packed column, where methanol in high purity (>99.5% m/m) is recovered in the liquid distillate, water is recovered in the bottom, and most of the remaining CO_2 is recovered in the gas distillate.

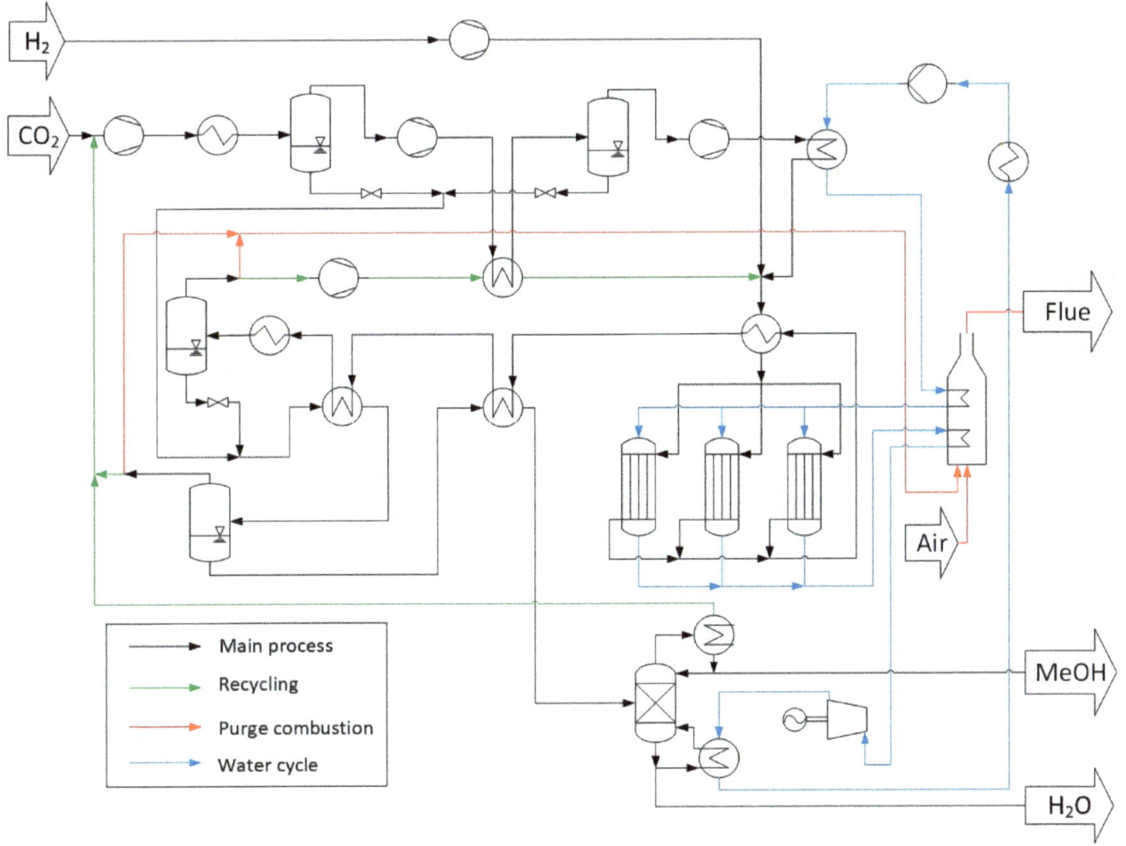

Figure 3. One-step process—detailed flowsheet with a total of three reactor modules. Cooling water streams are omitted.

The purge streams are burned with 15% air excess in a fired heater [40]. The heat of reaction of both the purge combustion and the methanol synthesis are used in a water Rankine cycle to produce electricity. The cycle starts with liquid water at 1 bar and 99.6 °C being pumped to a certain pressure, whose boiling temperature corresponds to the desired reactor temperature. Pressurized water reaches its boiling temperature in two steps (heat exchanger and fired heater) and vaporizes inside the reactor modules. The produced saturated steam is further heated in the fired heater and then performs work in a turbine, with a discharge pressure of 1.43 bar ($T_{boiling}$ = 110 °C). The resulting low-pressure steam condenses partially in the column reboiler, and total condensation is completed in a heat exchanger, closing the water cycle.

2.1.2. Three-Step Approach—Process Description

In Figure 4, a detailed flowsheet of the three-step process is presented. In this approach, the feed compression and recycling of non-converted reactants occurs similarly to the one-step process. The mixed feed stream is preheated and enters the first reactor module. The product gas is cooled down to 45 °C in three steps, and the condensed stream (mostly water, methanol, and some CO_2) is separated from the light gases in a flash drum. The gas stream is preheated and enters the second reactor module. The second product gas is cooled down to 30 °C in three steps, and the condensed stream (mostly water, methanol, and some CO_2) is separated from the gas stream in another flash drum.

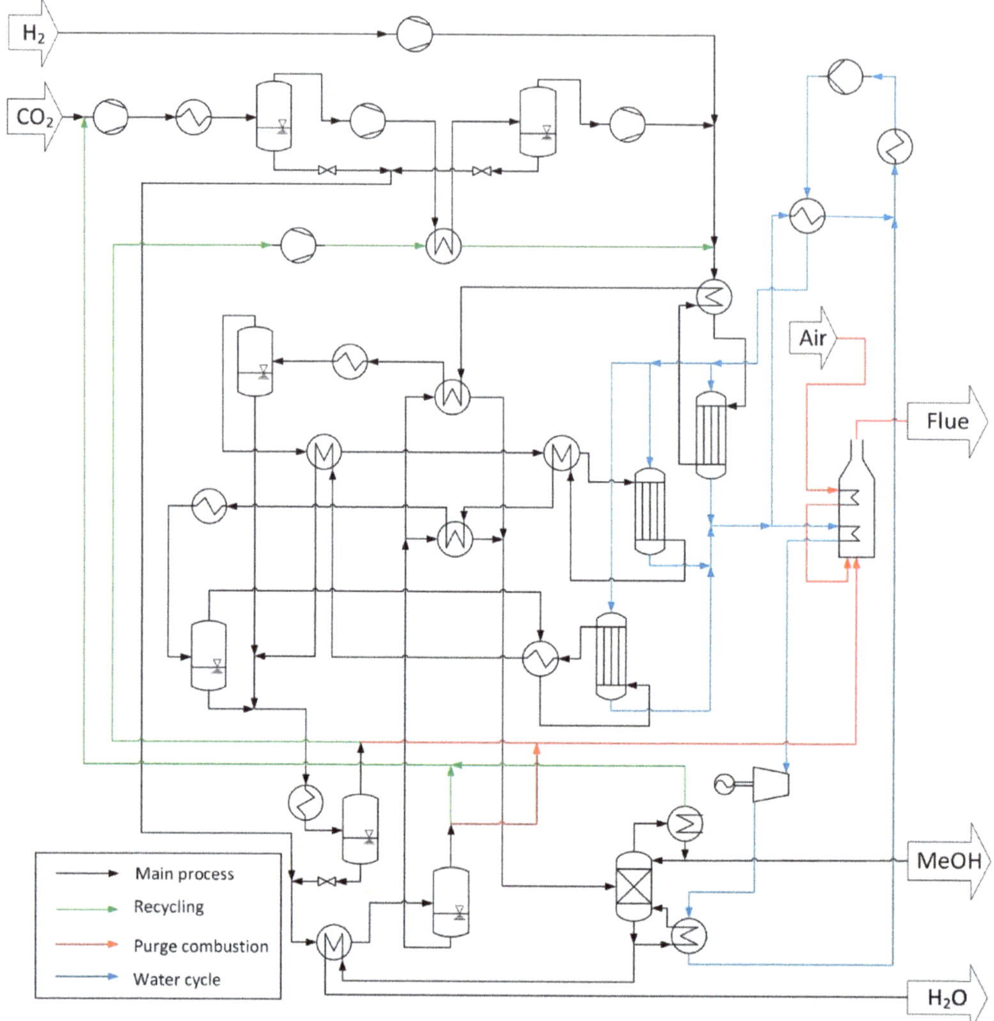

Figure 4. Three-step process—detailed flowsheet with a total of three reactor modules. Cooling water streams are omitted.

The gas phase is preheated and enters the third reactor module. The third product gas is cooled down, mixed with the condensed streams from the first and second reaction stages, and further cooled down to 30 °C.

Similar to the one-step process, component separation of the product stream is performed with one flash drum at high pressure, one flash drum at ambient pressure, and one distillation column.

The purge stream is burned in a fired heater with preheated air. In the water cycle, pressurized water is preheated and distributed to the reactor modules. A fraction of the produced saturated steam is split and used to preheat the water while the remaining steam is further heated in the fired heater. Supersaturated steam performs work in a turbine, with a discharge pressure of 1.43 bar ($T_{boiling}$ = 110 °C). The resulting low-pressure steam is partially condensed in the column reboiler, and total condensation is completed in a heat exchanger, closing the water cycle.

2.2. Process Simulation in Matlab

Before implementing the final version of each plant in Aspen Plus, different scenarios were investigated in Matlab. Therefore, optimal key parameters were selected, such as the total number of reactor modules, the purge fraction, and the temperature of the cooling fluid in the reactor.

In order to simulate the reactor, the following considerations were made: there are only variations along the length of the reactor (1D assumption), the influence of back-mixing is neglected (plug flow assumption), and the cooling fluid temperature (T_w, in K) is constant. Starting from mass and energy balances, the differential equations of the total mole flow of a single tube ($\dot{n}$, in mol·s^{-1}), the mole fraction of each component j (y_j), and the temperature (T, in K) in the axial direction z are shown as follows:

$$\frac{\mathrm{d}\dot{n}}{\mathrm{d}z} = \frac{m_{Cat}}{L} \cdot \sum_{j=1}^{6} \sum_{k=1}^{2} (v_{jk} \cdot r_k) \tag{4}$$

$$\frac{\mathrm{d}y_j}{\mathrm{d}z} = \frac{1}{\dot{n}} \cdot \left\{ \frac{m_{Cat}}{L} \cdot \sum_{k=1}^{2} (v_{jk} \cdot r_k) - y_j \cdot \frac{\mathrm{d}\dot{n}}{\mathrm{d}z} \right\} \tag{5}$$

$$\frac{\mathrm{d}T}{\mathrm{d}z} = \frac{1}{\left(\dot{n} \cdot C_{P,f}\right)} \cdot \left[-\frac{\mathrm{d}\dot{n}}{\mathrm{d}z} \cdot h_f - \dot{n} \cdot \sum_{j=1}^{6} \left(h_j \cdot \frac{\mathrm{d}y_j}{\mathrm{d}z} \right) + U \cdot \pi \cdot D_i \cdot (T_w - T) \right] \tag{6}$$

where m_{Cat} is the catalyst mass (kg), L is the reactor length (m), y_{jk} is the stoichiometric coefficient of component j in reaction k, r_k is the rate of reaction k (mol·kg$_{cat}$·s^{-1}), $C_{P,f}$ is the heat capacity of the fluid (J·mol·K^{-1}), h_f is the specific enthalpy of the fluid (J·mol^{-1}), h_j is the specific enthalpy of component j, U is the global heat transfer coefficient (W·m^{-2}·K^{-1}), and D_i is the inner diameter of a single tube (m).

The temperature-dependent parameters ($C_{P,f}$, h_f, h_j, U) were updated in each integration point in the axial direction. Heat capacity and enthalpy were calculated with the thermodynamic functions provided by Goos et al. [41], which are detailed in the Supplementary Material (Section A) along with the derivation of the differential equations. The global heat transfer coefficient was estimated (U) by summing the heat transfer resistances in the axial direction, according to the methodology described in the literature [42,43] (see Section B of the Supplementary Material).

The methodology to calculate the reaction rates (r_k) is described in Section 2.3. The system of differential equations was solved with the Matlab function ode45, with absolute and relative tolerances set to 10^{-10}.

In order to simplify the simulation of the separation steps in Matlab, the following procedure was applied. Both processes were implemented in Aspen Plus, considering a total of six reactor modules, a purge fraction of 2%, and T_w = 235 °C. The values of the split ratio of each component in the liquid and gas phase of each flash drum and the distillation column were extracted. For example, in the column of the one-step process, the methanol distribution in the outlet streams was: 3.80% in the gas distillate, 96.13% in the liquid distillate, and 0.06% in the bottom. The split ratio of all the components were taken from Aspen Plus and were considered constant for the different scenarios investigated in Matlab (i.e., variations in the number of reactor modules, purge fraction, and T_w). These split ratios are provided in the Supplementary Material (Section C).

Flowsheet convergence was achieved in Matlab by an iterative method, as there were two cycles of streams due to recycling unconverted reactants. First, educated initial guesses of the composition and total mole flow of each recycle stream were given. In each iteration, the recycle stream mole flow and its composition were calculated and used in the next iteration until the tolerance criterion was fulfilled:

$$\frac{\left(\dot{n}_{Ref,k+1} - \dot{n}_{Ref,k} \right)^2}{\left(\dot{n}_{Ref,k+1} \right)^2} \leq Tolerance \tag{7}$$

where $\dot{n}_{Ref,k}$ is the total mole flow of the recycle stream at iteration k. The tolerance of the inner cycle and the outer cycle were set to 10^{-9} and 10^{-8}, respectively.

2.3. Kinetic Modeling of the Methanol Synthesis

The kinetic simulation of the methanol synthesis was performed with our previously published six-parameter model (Model-6p) [19], whose considerations regarding the reaction mechanism, the assumption of the rate determining steps, and the most abundant surface species were based on our detailed microkinetic model [15]. This six-parameter model was validated with 496 experimental points from different laboratory plants [15,18,44], which contemplated temperatures between 210 and 260 °C, pressures between 20 and 60 bar, gas hourly space velocities (GHSV) between 1.8 and 40 $L_S \cdot h^{-1} \cdot g_{cat}^{-1}$, and a variety of syngas ($H_2/CO/CO_2/N_2$) feed compositions, including 126 points with only $H_2/CO_2/N_2$ in feed.

In this model, two main reactions are considered: CO_2 hydrogenation (Equation (2)) and the rWGSR (Equation (3)). The reaction rates ($r_{CO_2\ hyd.}$, r_{rWGSR}) in $mol \cdot kg_{cat}^{-1} \cdot s^{-1}$ are described as follows:

$$r_{CO_2\ hyd.} = \exp\left(A_2 - \frac{E_{A,2}}{R \cdot T} \right) \cdot \phi_{Zn} \cdot \theta_b \cdot \theta_c \cdot f_{H_2}^{1.5} \cdot f_{CO_2} \cdot \left(1 - \frac{f_{CH_3OH} \cdot f_{H_2O}}{f_{H_2}^3 \cdot f_{CO_2} \cdot K_{P,CO_2\ hyd.}^0} \right) \tag{8}$$

$$r_{rWGSR} = \exp\left(A_3 - \frac{E_{A,3}}{R \cdot T} \right) \cdot \phi_{Zn} \cdot \theta_b \cdot \theta_c \cdot f_{CO_2} \cdot f_{H_2O} \cdot \left(1 - \frac{f_{CO} \cdot f_{H_2O}}{f_{H_2} \cdot f_{CO_2} \cdot K_{P,rWGSR}^0} \right) \tag{9}$$

Here, A_{2-3} and $E_{A,2-3}$ are kinetic parameters, R is the universal gas constant, ϕ_{Zn} is the zinc coverage on the surface, θ_b and θ_c are the free active sites b and c, f_j is the fugacity of gas component j (bar), and $K_{P,k}^0$ is the global equilibrium constant of reaction k.

The Peng–Robinson equation of state is used to calculate the fugacities [45], considering the binary interaction parameters reported by Meng et al. [46] and Meng and Duan [47], and an effective hydrogen acentric factor of -0.05 proposed by Deiters et al. [48].

The zinc coverage is dependent on temperature and gas concentration [49], and its exact quantification under reaction conditions is difficult to predict. The zinc coverage is then considered to be constant and equal to $\phi_{Zn} = 0.50$ for a general case, while it is set to $\phi_{Zn} = 0.10$ for CO_2-rich feed ($CO_2/CO_x > 0.90$). The free active sites are calculated with the following equations:

$$\theta_b = \left(\overline{K_2} \cdot f_{H_2}^{0.5} \cdot f_{CO_2} + 1 \right)^{-1} \tag{10}$$

$$\theta_c = \left(\overline{K_3} \cdot f_{H_2}^{-0.5} \cdot f_{H_2O} + 1 \right)^{-1} \tag{11}$$

where $\overline{K_{2-3}}$ are adsorption parameters. In Table 1, the equilibrium constants as well as the previously estimated kinetic and adsorption parameters are summarized [19].

The side products of methanol synthesis on Cu/Zn-based catalysts (e.g., hydrocarbons or dimethyl ether) are typically at low concentrations [13,50]. Several studies reported that syngas conversion to hydrocarbons or dimethyl ether on commercial CZA at moderate temperatures (T $\leq$ 260 °C) is significantly low or even below detection range [15,18,44], while Condero-Lanzac et al. [22] reported low methane production from H_2/CO_2 on CZA at high temperatures (T $\geq$ 275 °C). Saito et al. [51] observed that side product formation is further reduced by increasing CO_2/CO_x feed concentration. Therefore, the generation of side products is not considered in this work.

Table 1. Equilibrium constants, kinetic and adsorption parameters of Model-6p [19]. Reprinted with permission from [19]. Copyright 2021 American Chemical Society.

Parameter	Value | Equation	Unit
A_2	14.41 ± 0.99	–
A_3	29.13 ± 1.74	–
$E_{A,2}$	94.73 ± 4.18	$kJ \cdot mol^{-1}$
$E_{A,3}$	132.79 ± 7.46	$kJ \cdot mol^{-1}$
K_2	0.1441 ± 0.0289	$bar^{-1.5}$
K_3	49.44 ± 11.08	$bar^{-0.5}$
$K^0_{P,CO_2\ hyd.}$	$T^{-4.481} \cdot \exp\left(\dfrac{4755.7}{T} + 8.369 \right)$	bar^{-2}
$K^0_{P,rWGSR}$	$T^{-1.097} \cdot \exp\left(\dfrac{-5337.4}{T} + 12.569 \right)$	–

2.4. Process Analysis and Optimization

Considering the fixed methanol production of 145 ton·h^{-1} or 1257.1 mol·s^{-1} and the 99.5% mol/mol purity of the reactants, the minimum required feed is 1263.4 mol·s^{-1} of CO_2 and 3790.3 mol·s^{-1} of H_2, totalizing $\dot{n}_{feed,min}$ = 5053.7 mol·s^{-1}. Since there are reactant losses in the purge and product streams, an excess of feed is required. With a fix feed ratio H_2:CO_2 of 3:1, the excess of feed (*Exc*) is defined here as:

$$Exc = \frac{\left(\dot{n}_{feed} - \dot{n}_{feed,min} \right)}{\dot{n}_{feed,min}} \cdot 100\% \tag{12}$$

It is, of course, of interest to minimize feed consumption, due to its high costs. Feed consumption is affected by key variables, such as reactor temperature and pressure, the number of reactor modules (which defines the total catalyst mass), and purge fraction. Avoiding large recycle streams is also important, as compressor work is required to get the pressure back to 70 bar, and larger equipment (i.e., heat exchangers, flash drums, compressors) are required to process higher flows.

Simulations were performed for a different number of reactor modules (from 3 to 12) and different purge fractions (from 0.5 to 5%). For each case, an initial guess for the feed excess was given (*Exc* = 5%), and a fix feed ratio H_2:CO_2 of 3:1 (a stoichiometric ratio) was applied. Then, an optimization problem was solved in Matlab with the function fminsearch (function tolerance = 0.1 mol·s^{-1}, step tolerance = 0.1 °C), whose objective was to maximize methanol production by varying the reactor coolant temperature (T_w).

With the optimum T_w, the required excess of feed was calculated to meet the methanol demand (1257.1 mol·s^{-1}) with Newton's method (function tolerance: 0.1 mol·s^{-1}). The steps of the temperature optimization and *Exc* calculations were repeated until the temperature update was lower than 0.25 °C.

2.5. Detailed Plant Simulation in Aspen Plus

After analyzing the results of the Matlab simulations, optimum parameters were selected for each approach (i.e., the number of reactor modules, purge fraction, cooling fluid temperature) and a detailed plant simulation including heat integration was implemented in Aspen Plus V10.

The Peng–Robinson property method was selected for the reactor modules. All other equipment were simulated with the Non-Random Two-Liquid Model with a second set of binary parameters (NRTL2) as the property method.

The methanol synthesis reactor was simulated with the rigorous plug flow reactor model (RPLUG unit) and the kinetics described in Section 2.3 were implemented as a Langmuir–Hinshelwood–Hougen–Watson (LHHW) reaction model. The rearrangement of the model parameters to follow the software's specific input format is detailed in the Supplementary Material (Section D). Since the reactor cooling fluid is at a constant temperature due to water evaporation, both co-current and counter-current operations give the same results. Therefore, the co-current operation was selected in order to simplify the mathematical calculations.

The combustion of the purge streams in a fired heater was simulated with the RGIBBS unit, which considers that chemical equilibrium is achieved when the free Gibbs energy of the system is minimized.

The heat exchangers were simulated in counter-current flow with the HeatX unit, with a minimum temperature approach of 25 °C for the heat exchangers located inside the fired heater and a minimum temperature approach of 10 °C for all the other heat exchangers.

The compressors were modeled using the ASME method, assuming a mechanical efficiency of 0.95 and an isentropic efficiency of 0.80 [22]. The pump was simulated assuming an efficiency of 0.70. The turbine was simulated with the ASME method, assuming a mechanical efficiency of 0.95 and an isentropic efficiency of 0.90 [52].

The distillation column was simulated with the rigorous RadFrac model, considering a kettle reboiler and a partial condenser at 53 °C with liquid and vapor distillate. A Murphree efficiency of 0.75 was set to all intermediate stages [53,54]. In both processes, the column had 30 stages and a reflux ratio of 2, with the feed entering above the 24th stage.

The relative tolerance of all equipment calculations was set to 10^{-5}. Flowsheet convergence was achieved using the Broyden method, with a relative tolerance of 10^{-4}, which corresponds to a mass balance closure of 99.99%.

2.6. Efficiency Evaluation

The chemical conversion efficiency (η_{CCE}) accounts for how much fuel energy remains in the final product in relation to the reactants. For methanol synthesis from H_2/CO_2, it is calculated as follows: [55]

$$\eta_{CCE} = \frac{\dot{m}_{MeOH} \cdot LHV_{MeOH}}{\dot{m}_{H_2} \cdot LHV_{H_2}} \tag{13}$$

where $\dot{m}_{MeOH}$ is the methanol mass production, $\dot{m}_{H_2}$ is the hydrogen feed demand, and LHV is the low heating value. The maximum possible efficiency ($\eta_{CCE,max}$) occurs at 100% overall H_2 conversion to methanol (stoichiometric conversion):

$$\eta_{CCE,max} = \frac{M_{MeOH} \cdot LHV_{MeOH}}{3 \cdot M_{H_2} \cdot LHV_{H_2}} = 0.876 \tag{14}$$

Here, M_j is the molar mass of component j. In order to also account for heat and the work input, the exergy efficiency (η_{Ex}) is calculated: [39]

$$\eta_{Ex} = \frac{\dot{m}_{MeOH} \cdot e_{MeOH}}{\dot{m}_{H_2} \cdot e_{H_2} + \dot{m}_{CO_2} \cdot e_{CO_2} + P_{el} + E_Q} \tag{15}$$

where e_j is the specific exergy of component j, P_{el} is the total required electric power, and E_Q is the total exergy input associated with heat demand.

The specific exergy of a component (e_j) is divided between thermal and chemical exergy: [39]

$$e_j(T,p) = \left[e_{j,therm}\right] + e_{j,chem} = \left[H_j - S_j \cdot T_0 - H_j^0 + S_j^0 \cdot T_0\right] + HHV_j \tag{16}$$

Here, $e_{j,therm}$ and $e_{j,chem}$ are the thermal and chemical exergies, H_j is enthalpy, S_j is entropy, H_j^0 and S_j^0 are the enthalpy and entropy at reference conditions (298.15 K and 1 bar), T_0 is the reference temperature, and HHV_j is the high heating value. In the exergy efficiency calculation, the HHV is used instead of the LHV, as water is liquid at reference conditions.

2.7. Techno-Economic Evaluation

In order to calculate the production costs, the standardized methodology from Albrecht et al. [56] was considered, which is a further development based on the work of Peters et al. [57].

The main equipment costs (EC) were estimated based on reference equipment costs [57,58]. The scale up to the required capacity was performed with specific equipment scaling factors, and price inflation was corrected to 2020 with the Chemical Engineering Plant Cost Indexes ($CEPCI$). In Equation (23), the costs of equipment j (EC_j) is described:

$$EC_j = EC_{j,ref} \cdot \left(\frac{C_j}{C_{j,ref}} \right)^M \cdot \left(\frac{CEPCI_{2020}}{CEPCI_{ref}} \right) \tag{17}$$

Here, the subscription ref relates to the reference equipment, C is the characteristic capacity, and M is the equipment scaling factor. The equipment is constructed with carbon steel. When the reference price is in US dollars (USD), a conversion to euros (EUR) of 1.13 USD·EUR^{-1} is applied (February 2022) [59].

The dimensions of the flash drums and the packed distillation column were calculated with the methodology reported by Towler and Sinnott [60]. The required heat transfer area of the heat exchangers, column condenser, and reboiler were estimated by assuming the typical global heat transfer coefficients reported by the VDI Atlas [43], according to each specific situation. Equipment dimensioning is detailed in the Supplementary Material (Section G).

The fixed capital investment (FCI) was estimated by multiplying the total EC with the Lang Factor (LF), which accounts for all direct and indirect costs related to the plant construction. In this work, LF was assumed to be 4.86 (details are provided in the Supplementary Material, Section H) [56,57]. A working capital (WC) of 10% of the total capital expenses ($CAPEX$) was considered [56]. Summarizing the equations:

$$FCI = LF \cdot \sum EC_j \tag{18}$$

$$CAPEX = FCI + WC \tag{19}$$

$$WC = 0.10 \cdot CAPEX \tag{20}$$

The equivalent annual capital costs (ACC) were estimated by applying the annuity method on the FCI, assuming an annual interest rate (IR) of 10%, a plant operating life (t_P) of 20 years, and no salvage value [61]. The working capital does not depreciate in value, and only its interest has to be taken into account [56].

$$ACC_{FCI} = \frac{FCI \cdot IR \cdot (1 + IR)^{t_P}}{[(1 + IR)^{t_P} - 1]} \tag{21}$$

$$ACC_{WC} = WC \cdot IR \tag{22}$$

$$ACC = ACC_{FCI} + ACC_{WC} = \frac{FCI \cdot IR \cdot (1 + IR)^{t_P}}{[(1 + IR)^{t_P} - 1]} + WC \cdot IR \tag{23}$$

The operating expenses ($OPEX$) were divided between direct and indirect costs. The costs related to the direct OPEX ($OPEX_{dir}$) are presented in Table 2, which include raw materials, catalysts, process water treating, and electricity. A catalyst lifetime of three years was considered. In the Rankine water cycle, a clean water replacement of 1% of the total flow was considered [62].

The indirect $OPEX$ consisted of operating labor (OL), operating supervision, mainte-nance, operating supplies, laboratory charges, taxes on property, insurance, plant overhead, administration, distribution, marketing, research, and development. The estimation of each of these items was based on typical values, which are dependent on OL, FCI, and the net production costs (NPC) (see Section H of the Supplementary Material) [56,57]. The total indirect $OPEX$ ($OPEX_{ind}$) is calculated as follows:

$$OPEX_{ind} = 2.2125 \cdot OL + 0.081 \cdot FCI + 0.10 \cdot NPC \tag{24}$$

Table 2. Costs of feedstock, catalyst, water treating, and electricity.

Item	Costs	Ref.
Hydrogen	3097.4 €·ton^{-1}	[22]
Carbon dioxide	44.3 €·ton^{-1}	[22]
Cooling water	0.00125 €·ton^{-1}	[56]
Clean water	2 €·ton^{-1}	[56]
Total organic carbon (TOC) abatement of process water	1938 €·(ton C)$^{-1}$	[63]
Electricity	90 €·MWh^{-1}	[53]
Catalyst (Cu/ZnO/Al$_2$O$_3$)	18,100 €·ton^{-1}	[64]

The required number of operators in a shift (n_{OP}) was estimated with the following equation: [65,66]

$$n_{OP} = (6.29 + 0.23 \cdot N_{np})^{0.5} \tag{25}$$

where N_{np} is the number of non-particulate main processing units. Considering daily working shifts, resting periods and vacations, the number of operators to fulfill each position in a continuous operation is approximately $F_{OP} = 4.5$. Therefore, the total number of operators (N_{OP}) is: [65,66]

$$N_{OP} = F_{OP} \cdot n_{OP} \tag{26}$$

The total costs of operating labor (OL) is then calculated as follows:

$$OL = W_{OP} \cdot N_{OP} \tag{27}$$

where W_{OP} is the wage rate of each operator ($W_{OP} = 72{,}000$ €·a^{-1}) [53].

The net production costs (NPC) are calculated in terms of average annual costs and in terms of average costs per kg of methanol:

$$NPC \left[\frac{€}{a} \right] = ACC + OPEX_{dir} + OPEX_{ind} \tag{28}$$

$$NPC \left[\frac{€}{kg} \right] = \frac{(ACC + OPEX_{dir} + OPEX_{ind})}{\dot{m}_{MEOH}} \tag{29}$$

3. Results and Discussion

In this section, process simulation and analysis are presented separately for the one-step and the three-step approaches. Finally, the techno-economic analysis of both approaches is presented and discussed jointly.

3.1. One-Step Process

3.1.1. One-Step Process—Selecting Key Parameters

The one-step process was successfully implemented in Matlab. Different scenarios were simulated by varying the number of reactor modules and the purge fraction, with the optimal temperature for a fixed methanol production (145 ton·h^{-1}) being estimated in each case. In Figure 5, several contour plots are shown, where CO$_2$ single-pass conversion ($X_{CO_2,SP}$) (Figure 5a), the required feed excess (Figure 5b), the optimal temperature (Figure 5c), and the total recycle stream (Figure 5d) are plotted against the number of reactor modules and the purge fraction.

CO$_2$ single-pass conversion (Figure 5a) was considerably enhanced by increasing the number of reactor modules. This was not only because the gas hourly space velocity (GHSV) decreased, but also because the optimal temperature had lower values (Figure 5c), shifting the thermodynamic equilibrium towards higher methanol concentrations. In contrast, reducing the purge fraction had little effect on $X_{CO_2,SP}$. This should be the result of two competing effects: on one hand, a lower purge fraction means higher recycle streams (Figure 5d), which increases the GHSV, reducing $X_{CO_2,SP}$. On the other hand, the recycle stream has a H$_2$:CO$_2$ ratio greater than three due to a limited rWGSR extension.

By increasing the recycle stream, the H_2:CO_2 ratio of the reactor feed stream is enhanced, positively contributing to $X_{CO_2,SP}$.

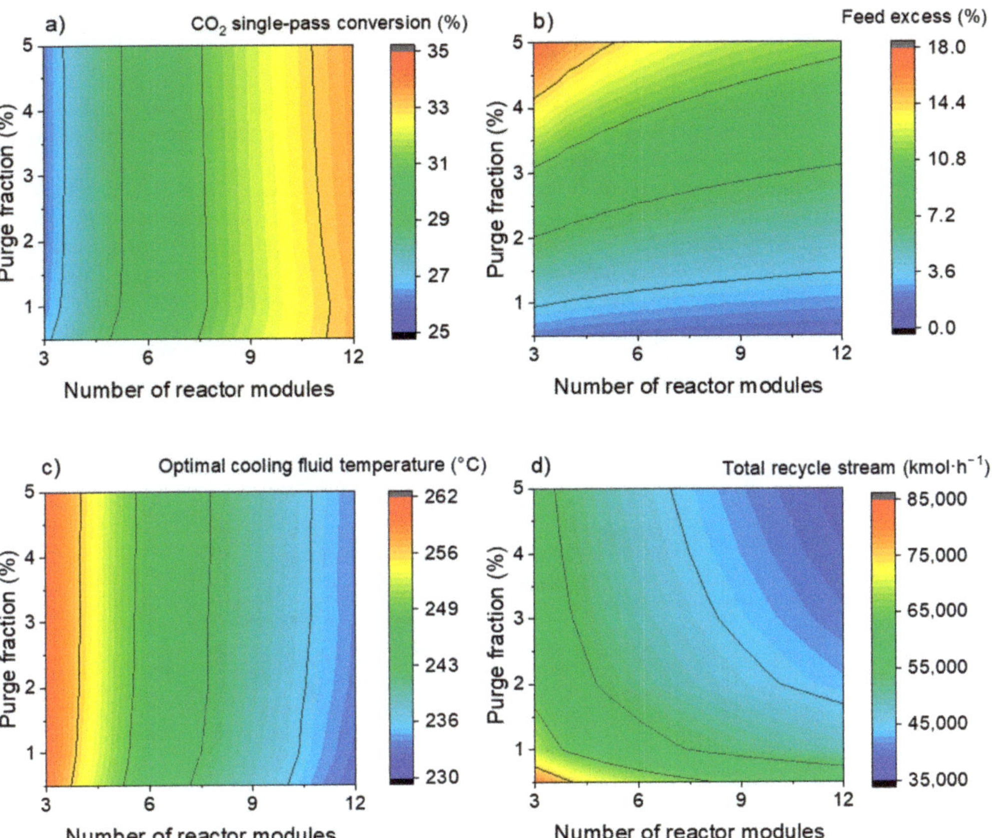

Figure 5. One-step process—CO_2 single-pass conversion (**a**), required feed excess (**b**), optimal temperature (**c**), and total recycle stream (**d**) as a function of the number of reactor modules and the purge fraction.

The required feed excess was significantly decreased both by increasing the number of reactor modules and by reducing the purge fraction. This occurred because the former procedure increased $X_{CO_2,SP}$ and the latter maintained $X_{CO_2,SP}$ roughly constant while increasing the gas flow inside the reactor modules.

Since the reactants (H_2, CO_2) represent the highest costs of the plant, it is important to minimize the required feed excess, which according to Figure 5b, occurred at 0.5% purge fraction. However, with such a low purge fraction, the total recycle stream was considerably high, demanding larger heat exchangers, flash drums, and compressors, as well as higher power consumption. Therefore, an intermediate value of 2% as the purge fraction was selected for the detailed simulation in Aspen Plus, agreeing with other studies and typical industrial values [22,39,58].

With the purge fraction fixed at 2%, six reactor modules were used in the detailed study, because further increasing the number of reactor modules only slightly reduced the required excess feed, not justifying further expenses in equipment and catalyst.

The value of the global heat transfer coefficient was updated point by point within mathematical integration along the reactor length. For the selected condition, $U_{z=0} = 160 \, \text{W} \cdot \text{m}^{-2} \cdot \text{K}^{-1}$

and $U_{z=12.5} = 150$ W·m^{-2}·K^{-1}. Since Aspen Plus requires a constant value, the average value was used ($U_{avg} = 155$ W·m^{-2}·K^{-1}).

3.1.2. One-Step Process—Detailed Plant Simulation and Process Analysis

A detailed flowsheet of the one-step process presented in Figure 3 was implemented in Aspen Plus, considering 2% purge fraction, six reactor modules working in parallel, and the optimized temperature of the reactor cooling fluid ($T_w = 247.5$ °C). A picture of the flowsheet in Aspen Plus, the properties of the streams, and a detailed plant description are provided in the Supplementary Material (Section E).

In Figure 6, the concentration of the products along the reactor length is shown. The methanol and water feed concentrations were close to zero, and their outlet concentrations were 7.4 and 7.2% mol/mol, respectively. The nitrogen concentration remained relatively low (inlet: 4.95% mol/mol, outlet: 5.65% mol/mol). Due to the recycle streams, CO entered the reactor modules at 1.50% mol/mol, although it was not a feedstock in the plant. CO was produced through the rWGSR until the length of 1.5 m, where its concentration reached 3.3% mol/mol. Then, due to the high water concentration (4.30% mol/mol), the WGSR became faster than its reverse reaction and started to consume CO, which exited the reactor at 1.76% mol/mol and a marginal selectivity (0.5%). This virtually stabilized CO content in the plant led to a high methanol selectivity (99.5%).

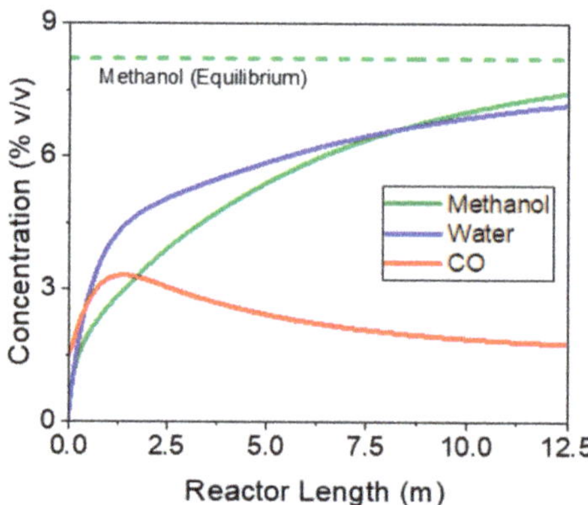

Figure 6. One-step process—product concentration along the reactor length. Reactor feed concentration: $H_2/CO/CO_2/CH_3OH/H_2O/N_2 = 71.3/1.5/21.9/0.3/0.0/5.0\%$ mol/mol.

The CO_2 single-pass conversion was 28.5%, close to the equilibrium conversion (30.4%), while the feed excess was 6.05%, which corresponded to an overall CO_2 conversion to methanol of 94.3%. These values are in agreement with the Matlab simulations ($X_{CO_2,SP} = 29.7\%$, feed excess = 5.75%, overall CO_2 conversion to MeOH = 94.6%).

The chemical conversion efficiency (η_{CCE}) of the process was 82.6%, which was close to the maximum possible value ($\eta_{CCE,max} = 87.6\%$). With the heat integration, the one-step process was not only self-sufficient, but had a heat excess that could be supplied to other processes, in agreement with the literature [21,39]. In our case, as is commonly performed in industrial methanol synthesis plants, the heat excess was used to generate electricity via a water Rankine cycle, reducing the electricity consumption from 47.4 to 17.6 MW.

In Figure 7, a global exergy balance and an exergy loss distribution are provided. No distinction was made between exergy destruction and exergy losses via side output streams (i.e., cooling water, process water, and flue gas). The exergy efficiency (η_{Ex}) of the process was 76.4%, with a total exergy loss of 281.9 MW. The main losses occurred due to the exothermic chemical reactions with heat recovery at low temperatures (reactor modules: 58.1%, fired heater: 14.8%). Additionally, exergy losses in the heat exchangers (11.1%) and in the column (9.0%) were also significant, mainly due to heat transfer to cooling water.

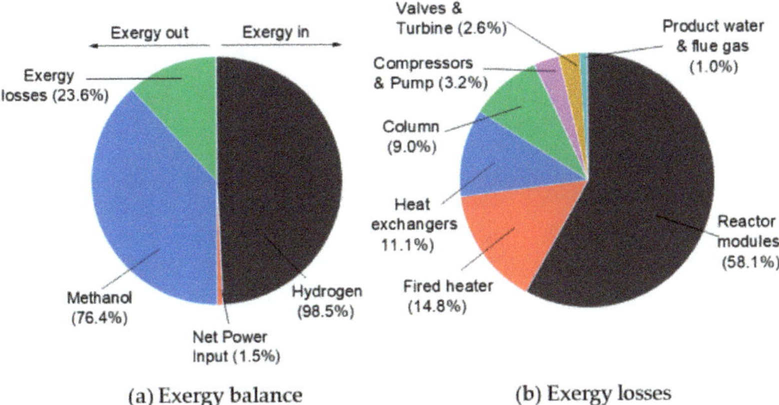

(**a**) Exergy balance   (**b**) Exergy losses

Figure 7. One-step process—exergy analysis. (**a**) Global exergy balance. Total exergy input: 1194.5 MW. (**b**) Distribution of exergy losses (total = 281.9 MW).

3.2. Three-Step Process

3.2.1. Three-Step Process—Selecting Key Parameters

The three-step process was successfully implemented in Matlab. In Figure 8, CO_2 single-pass conversion ($X_{CO_2,SP}$) (Figure 8a), the required feed excess (Figure 8b), the optimal temperature (Figure 8c), and the total recycle stream (Figure 8d) are described as a function of the number of reactor modules and the purge fraction. Since this process considers three reaction steps with intermediate cooling, the simulations were limited to multiples of three as the total number of reactor modules.

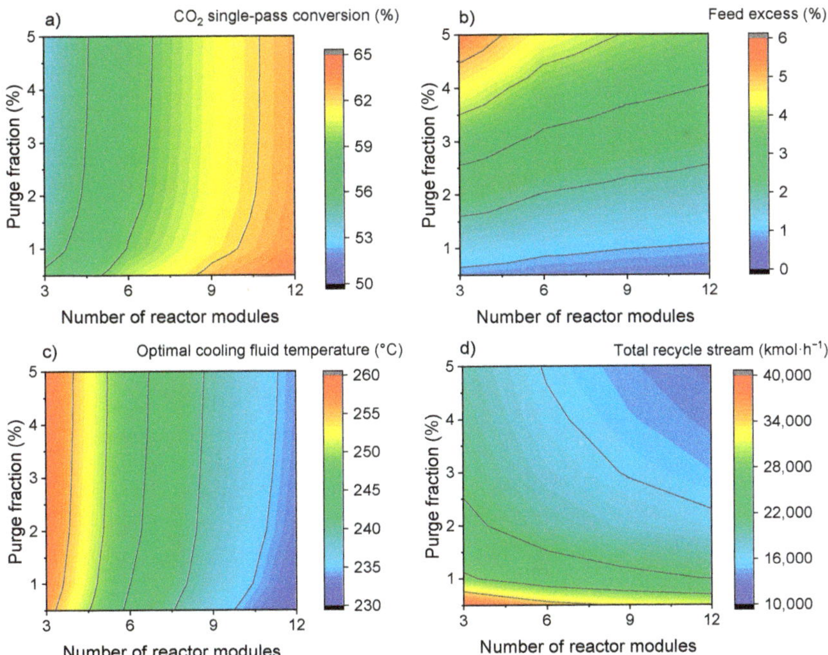

Figure 8. Three-step process—CO_2 single-pass conversion (**a**), required feed excess (**b**), optimal temperature (**c**), and total recycle stream (**d**) as a function of the number of reactor modules and the purge fraction.

A significant improvement was seen in the three-step process in relation to the one-step approach. For similar conditions (i.e., the same total number of reactor modules and purge stream fraction), CO_2 single-pass conversion had approximately doubled, the required feed excess decreased by 60–70%, and the total recycle stream decreased by 50–70%. The optimal values for the reactor cooling fluid remained close to the ones of the first approach (between 230 and 260 °C).

Similarly to the one-step process, a purge fraction equal to 2% was chosen here, having a good compromise between minimizing the feed requirements and minimizing the total recycle stream. With this fixed purge fraction, a number of reactor modules equal to three was selected, as further increasing this amount gave limited improvement in the required feed excess and the total recycle stream, while considerably increasing equipment and catalyst costs.

When analyzing different scenarios in Matlab, the same cooling fluid temperature (T_w) was considered for all reactors. A further optimization was possible by allowing this temperature to be independently operated in each reactor. This possibility was checked for the chosen condition (2% purge fraction, three reactor modules), but only a marginal improvement was obtained (see Table 3), probably not justifying the increase in plant complexity. Therefore, in the detailed analysis, the cooling fluid temperature of all the reactors was set to 258.5 °C.

Table 3. Three-step synthesis—Performance indicators for two process approaches: same cooling fluid temperature in all reactors, and independent optimization of the cooling fluid temperature in each reactor.

Approach	$T_{w,R1}$ (°C)	$T_{w,R2}$ (°C)	$T_{w,R3}$ (°C)	$X_{CO_2,SP}$ (%)	Feed Excess (%)	Total Recycle Stream (kmol·h^{-1})
Same T_w	258.5	258.5	258.5	54.1	2.42	23,038
Varying T_w	264.6	259.9	249.4	54.6	2.35	22,464

Similarly to the one-step process, the average heat transfer coefficients were obtained for each reactor and given to Aspen Plus: U_1 = 327 W·m^{-2}·K^{-1}, U_2 = 285 W·m^{-2}·K^{-1}, U_3 = 246 W·m^{-2}·K^{-1}. The decrease in the coefficient values is associated with a decrease in total flow, due to intermediate product removal. Still, the heat transfer coefficients were higher than in the one-step process (155 W·m^{-2}·K^{-1}), which had lower flows for each reactor module because of parallel operation.

3.2.2. Three-Step Process—Detailed Plant Simulation and Process Analysis

A detailed flowsheet of the three-step process presented in Figure 4 was implemented in Aspen Plus, considering a 2% purge fraction, three reactor modules working in series with intermediate product condensation, and the previously optimized temperature of the reactor cooling fluid (T_w = 258.5 °C). A detailed plant description, stream properties, and a picture of the flowsheet in Aspen Plus are provided in the Supplementary Material (Section F).

In Figure 9, the concentration of the products along the length of the three reactors is shown, as well as the product removal through the intermediate condensation steps. In Reactor 1, CO entered at a low concentration (1.3% mol/mol), peaked at z = 2.5 m, and left the reactor with a higher concentration (2.7% mol/mol). This CO production through the rWGSR increased the water concentration ($y_{H_2O}^{R_1,out}$ = 5.6% mol/mol) and slowed down methanol production ($y_{MeOH}^{R_1,out}$ = 4.7% mol/mol).

In Reactors 2 and 3, the CO inlet concentration was significantly higher (3.0% mol/mol for both cases), causing its concentration peak to come much sooner (at 1.8 m and 1.25 m, respectively). After that, the WGSR was faster than its reverse reaction and the CO concentration decreased, leaving both reactors with an overall positive CO consumption. Therefore, the water concentration in Reactors 2 and 3 was maintained at lower levels ($y_{H_2O}^{R_2,out}$ = 4.7% mol/mol, $y_{H_2O}^{R_3,out}$ = 4.5% mol/mol), enhancing the final methanol concentration ($y_{MeOH}^{R_2,out}$ = 5.6% mol/mol, $y_{MeOH}^{R_3,out}$ = 5.8% mol/mol).

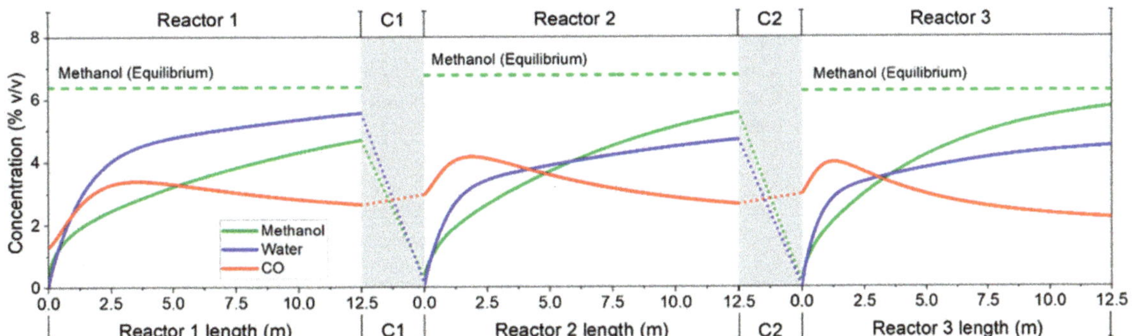

Figure 9. Three-step process—methanol, water, and CO concentration along the length of each reactor, as well as in the intermediate condensation steps (C1 and C2).

Water is known to accelerate the deactivation of Cu-based catalysts [67]. Therefore, the lower water concentration of the three-step process in relation to the one-step process ($y_{H_2O}^{1s,out} = 7.2\%$) should not only benefit the reaction rates, but also the catalyst lifetime.

In Table 4, the operating conditions and split ratios of the intermediate condensation steps are provided, while the reactor information is summarized in Table 5. Methanol and water were almost fully removed from the gas phase, but at the cost of ca. 9–13% CO_2 condensation. The split ratios of CO_2 and methanol were strongly dependent on temperature, with the chosen values ($T_1 = 45\,°C$, $T_2 = 30\,°C$) being derived from a sensitivity analysis.

Table 4. Three-step process—operating conditions and split ratios of the intermediate condensation steps.

Cond. Step	Temp. (°C)	Pres. (Bar)	Phase	Split Ratio CO_2 (%)	MeOH (%)	H_2O (%)
#1	45	69.25	Gas	90.66	5.31	1.17
			Liquid	9.34	94.69	98.83
#2	30	68.50	Gas	87.36	2.46	0.52
			Liquid	12.64	97.54	99.48

Table 5. Three-step process—heat transfer, inlet mole flow, mole fractions, and methanol production in the reactor modules.

Reactor	#1	#2	#3
$\dot{Q}$ (MW)	−18.5	−25.7	−22.7
$\dot{n}_{in}$ (kmol·h^{-1})	40,833	33,209	26,366
$y_{H_2,in}$ (% mol/mol)	69.1	69.0	69.5
$y_{CO,in}$ (% mol/mol)	1.3	3.0	3.0
$y_{CO_2,in}$ (% mol/mol)	20.6	17.3	14.4
$y_{MeOH,out}$ (% mol/mol)	4.7	5.6	5.8
$y_{H_2O,out}$ (% mol/mol)	5.6	4.7	4.5
$\Delta\dot{n}_{MeOH}$ (kmol·h^{-1})	1616	1589	1325

The methanol production was similar in Reactors 1 and 2 (1616 and 1589 kmol·h^{-1}, respectively), while it was 18% lower in Reactor 3 (1325 kmol·h^{-1}). This shows the positive effect of a higher CO concentration in the reactor feed, despite the lower total feed flow and CO_2 inlet concentration of Reactors 2 and 3 in relation to Reactor 1.

The CO_2 single-pass conversion ($X_{CO_2,SP}$) was 53.9%, with a selectivity to methanol of 99.8% and a selectivity to CO of 0.2%. The feed excess was 2.35%, leading to an overall conversion of CO_2 to methanol of 97.7%. These values are in agreement with the Matlab simulations ($X_{CO_2,SP} = 54.1\%$, $Exc = 2.42\%$, overall CO_2 conversion to MeOH = 97.6%).

The three-step approach was significantly superior to the one-step process, even using only half the number of reactor modules (three vs. six). This superiority is clear when looking at the CO_2 single-pass conversion (53.9% vs. 28.5%), leading to a considerably higher overall conversion to methanol (97.7% vs. 94.3%).

With the heat integration, the three-step process was also self-sufficient in heat, while electricity was produced through a water Rankine cycle, reducing the total power consumption from 42.7 to 21.8 MW. The chemical conversion efficiency was $\eta_{CCE}^{3s} = 85.6\%$, which was higher than the value of the one-step process ($\eta_{CCE}^{1s} = 82.3\%$) and, therefore, even closer to the maximum possible value ($\eta_{CCE,max} = 87.6\%$).

In Figure 10, an exergy analysis of the process is presented. The exergy efficiency was $\eta_{Ex}^{3s} = 78.8\%$, an improvement from the previous approach ($\eta_{Ex}^{1s} = 76.4\%$), with the total exergy losses decreasing in 13% (245.3 vs. 281.9 MW). Although the total power consumption decreased (42.7 vs. 47.4 MW), the net power consumption increased slightly (21.8 vs. 17.6 MW). This occurred because power generation was significantly lower in the three-step approach (20.8 vs. 29.8 MW) due to the much lower heat duty of the fired heater, as less reactant was lost in the purge streams.

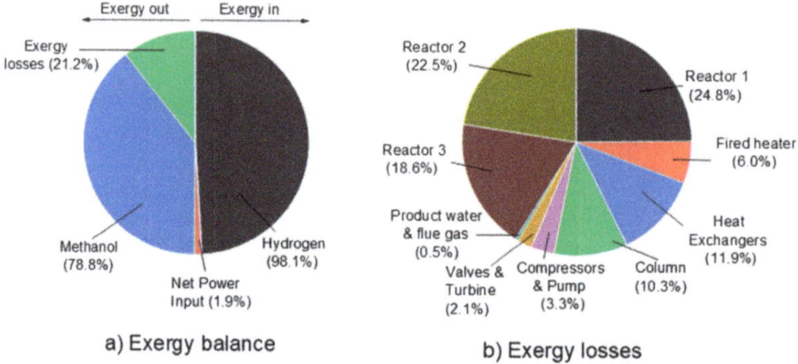

Figure 10. Three-step process—exergy analysis. (**a**) Global exergy balance (total exergy input = 1157.5 MW). (**b**) Distribution of exergy losses (total = 245.3 MW).

Chemical reactions with heat recovery at low temperatures was also the main cause of exergy losses in the three-step approach (reactor modules: 66.0%, fired heater: 6.0%). Both processes lost approximately the same exergy in the reactor modules and the distillation column. The main improvement in relation to the one-step process was a much lower exergy loss in the fired heater (14.7 vs. 41.4 MW), as the total purge stream flow decreased by 59% (455 against 1100 kmol·h^{-1}). Despite the higher number of cooling and warming operations and the higher total heat transfer duty in the three-step process (357.1 vs. 310.2 MW), the exergy losses in the heat exchangers were slightly lower for the three-step process (29.4 vs. 31.4 MW). Finally, moderate improvements were also seen in the compressors and pump (8.2 vs. 9.1 MW) and in the valves and turbine (5.0 vs. 7.4 MW).

In Table 6, the data comparing both processes is summarized, once again emphasizing the superior performance of the three-step approach.

Table 6. Data comparison between the one-step and the three-step approach.

Item	One-Step	Three-Step
Total methanol production (kmol·h^{-1})	4527	4525
CO_2 single-pass conversion (%)	28.5	53.9
Overall CO_2 conversion to methanol (%)	94.3	97.7
Feed excess (%)	6.05	2.35
Methanol selectivity (%)	99.5	99.8
Total recycle stream flow (kmol·h^{-1})	54,290	22,581
Maximum water concentration (% mol/mol)	7.2	5.6
Total exergy loss (MW)	281.9	245.3
Exergy efficiency (%)	76.4	78.8

3.3. Techno-Economic Analysis

In Figure 11a, the distribution of the equipment costs (EC) is presented, with the reactor modules and the compressors representing the majority of the costs (>75%). The total EC was 85.5 and 66.1 M€ for the one-step and the three-step approach, respectively. This significant improvement of the three-step process was a consequence of the intermediate condensation steps, requiring a lower total reactor volume (due to an enhanced reaction velocity), lower compressor size (due to a lower recycle flow), and lower furnace, turbine, and generator size (due to a lower purge flow). The cost reduction in the aforementioned equipment was significantly higher than the additional costs of the heat exchangers and flash drums from the intermediate condensation units. The total fixed capital investment (FCI) was 415.9 and 321.4 M€ for the one-step and three-step approach, respectively. The detailed estimated capacity and price of each equipment is presented in the Supplementary Material (Section H).

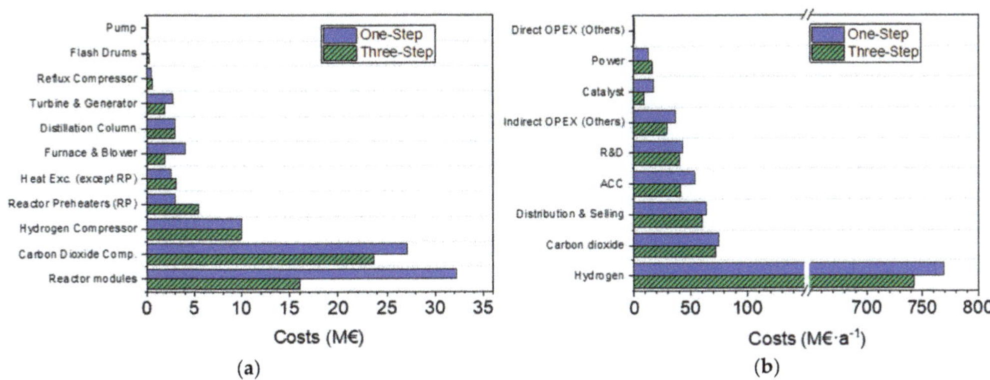

Figure 11. Distribution of the costs. (**a**) Equipment Costs (EC). (**b**) Net Production Costs (NPC).

In Figure 11b, the distribution of the net production costs (NPC) is detailed. The main operating costs were the reactant expenses (78–80% of NPC), with ACC contributing with only 4–5%, while the catalysts and electricity consisted of less than 3% of the NPC. Due to the higher overall CO_2 conversion to methanol, the NPC of the three-step process was 5.7% lower than the one-step approach. The detailed $OPEX$ costs are presented in the Supplementary Material (Section H).

In Table 7, a summary of the overall costs is presented. The NPC was 920 and 868 €·ton^{-1} for the one-step and the three-step process, respectively, corresponding to an improvement of 5.7% for the new process. Besides the hydrogen and carbon dioxide costs, the fixed capital investment (FCI) and the discount rate were the most sensitive parameters to the methanol selling price, as shown in the tornado analysis (see Figure 12).

Table 7. Summary of the costs of the one-step and the three-step process.

Item	Costs		Decrease (%)
	One-Step	**Three-Step**	
Equipment Costs (EC)	85.5 M€	66.1 M€	22.7
Fixed capital investment (FCI)	415.9 M€	321.4 M€	22.7
Working capital (WC)	46.2 M€	35.7 M€	22.7
Total $CAPEX$	462.1 M€	357.1 M€	22.7
Annual Capital Costs (ACC)	53.5 M€·a^{-1}	41.3 M€·a^{-1}	22.7
Direct $OPEX$	874.9 M€·a^{-1}	839.6 M€·a^{-1}	4.0
Indirect $OPEX$	143.4 M€·a^{-1}	129.6 M€·a^{-1}	9.6
Total $OPEX$	1018.3 M€·a^{-1}	969.3 M€·a^{-1}	4.8
Net Production Costs (NPC)	1071.8 M€·a^{-1}	1010.6 M€·a^{-1}	5.7

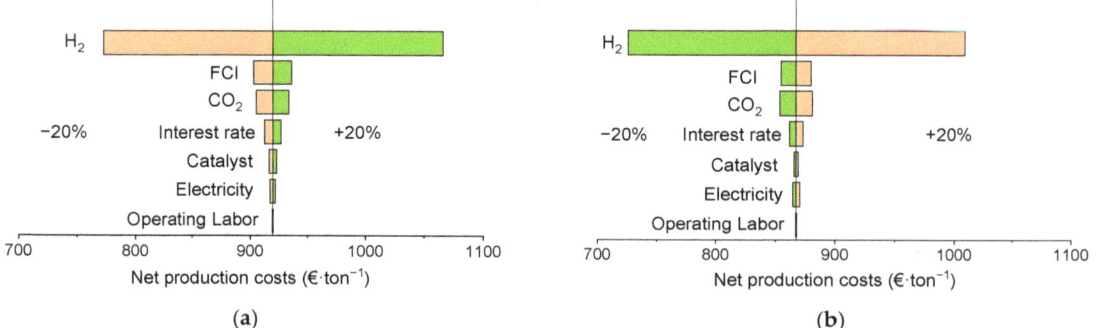

Figure 12. Sensitivity analysis of the main cost factors in relation to the net production costs (NPC). Variation of $\pm$ 20% in each factor. (**a**) One-step process. (**b**) Three-step process.

In Figure 13, the net production costs are plotted against the hydrogen price. Although the methanol market price in Europe was still significantly below the values (495 €·ton^{-1} in February 2022), [68,69] the green methanol produced from the proposed process would become economically competitive if the green hydrogen price reached 1468 €·ton^{-1}.

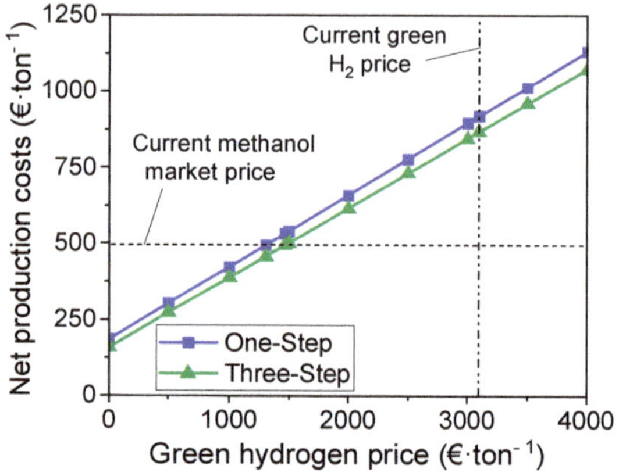

Figure 13. Net production costs of methanol as a function of green hydrogen price.

4. Conclusions

A detailed study of a methanol synthesis plant from H$_2$ and CO$_2$ with intermediate condensation units (the three-step process) is presented and compared with the conventional approach (the one-step process). The total production was fixed at 1.16 Mton MeOH·a^{-1}. The processes were first implemented in Matlab in order to critically analyze the number of reactor modules, the purge fraction, and the reactor operating temperature. Using the most suitable process parameters, detailed plants of both approaches were implemented in Aspen Plus, including heat integration and a water Rankine cycle to make use of the reaction enthalpy. Finally, techno-economic analyses were applied. Both processes offered an excess of heat, which was used to generate electricity in our work, but could alternatively supply other plants (e.g., CCU, OME synthesis) in a larger process integration.

It was demonstrated that CO$_2$ single-pass conversion almost doubled when including intermediate condensation steps (53.9 vs. 28.5%), resulting in a significantly higher overall conversion to methanol (97.7 vs. 94.3%) and in a higher exergy efficiency (78.8 vs. 76.4%). Because of the enhanced conversion, the new process required lower recycle and feed streams, decreasing net production costs by 61.2 M€·a^{-1} (5.7%). Although additional equipment (i.e., heat exchangers and gas–liquid separators) is necessary, the improved process was significantly more efficient than the conventional approach, requiring lower sizes of the main equipment (e.g., compressors, reactors, fired heater). Consequently, according to our analysis, the total investment costs were 94.5 M€ (22.7%) lower than for the conventional process.

Intermediate condensation steps are therefore highly recommended for methanol production from H_2/CO_2, reducing costs by improving CO_2 equilibrium conversion to methanol while using commercially proven technology. Besides, since water accelerates the deactivation of Cu-based catalysts, product intermediate removal should increase catalyst lifetime, as the average water concentration in the reactor is significantly lower than in the conventional process.

With our proposed process, the methanol net production costs amounted to 868 €·ton^{-1}, which are still significantly higher than the current market price (495 €·ton^{-1}) but is believed to become economically viable with an effective reduction in the price of green hydrogen.

Supplementary Materials: The following supporting information can be downloaded at: https://www.mdpi.com/article/10.3390/pr10081535/s1, Figure S1: One-step process—Aspen Plus flowsheet; Figure S2: Three-step process—Aspen Plus flowsheet; Table S1: Parameters for the estimation of the specific heat capacity and specific enthalpy of selected components in the gas phase; Table S2: Liquid and gas fractions (% mol/mol) of the phase separation via flash drums and the separation via the distillation column in the one-step process. Values taken from Aspen Plus calculations and used for the Matlab simulations; Table S3: Liquid and gas fractions (% mol/mol) of the phase separation via flash drums and the separation via the distillation column in the three-step process. Values taken from Aspen Plus calculations and used for the Matlab simulations; Table S4: Aspen kinetic factor and Model-6p corresponding expressions; Table S5: Coefficients of the driving force constant and the corresponding expressions from Model-6p; Table S6: Concentration exponents (v_j) of the driving force expression; Table S7: Adsorption constants and the corresponding expression of Model-6p; Table S8: Concentration exponents and the corresponding values of Model-6p; Table S9: Properties of the streams from the one-step process; Table S10: Molar composition (% mol/mol) of the streams from the one-step process; Table S11: Properties of the streams from the three-step process; Table S12: Molar composition (% mol/mol) of the streams from the three-step process; Table S13: Dimension of the flash drums of the one-step and the three-step processes; Table S14: Global heat transfer coefficients, heat transfer duty, and estimated surface area of the heat exchangers of the one-step and the three-step process; Table S15: Calculation of the Capital Expenses (CAPEX) depending on the total equipment costs (EC); Table S16: Estimation of indirect operating expenses ($OPEX_{ind}$); Table S17: Equipment characteristic dimensions and equipment costs (EC) of the one-step approach. All equipment was built with carbon steel. All equipment reference prices were taken from Peters et al., except for the power generator, whose ref. price was taken from Henning and Haase; Table S18: Equipment characteristic dimensions and equipment costs (EC) of the three-step approach. All equipment was built with carbon steel, and the costs included 10% delivery costs. All equipment reference prices were taken from Peters et al., except for the power generator, whose ref. price was taken from Henning and Haase; Table S19: Detailed operating expenditures (OPEX) of the one-step and the three-step approach.

Author Contributions: Conceptualization, B.L.d.O.C. and J.S.; methodology, B.L.d.O.C., K.J. and P.B.; software, B.L.d.O.C., K.J. and P.B.; formal analysis, B.L.d.O.C.; writing—original draft preparation, B.L.d.O.C.; writing—review and editing, K.J., P.B., K.H.D., S.P., N.D. and J.S.; visualization, B.L.d.O.C. and P.B.; supervision, K.H.D., S.P., N.D. and J.S.; funding acquisition, K.H.D., S.P., and J.S. All authors have read and agreed to the published version of the manuscript.

Funding: This research was funded by Coordenação de Aperfeiçoamento de Pessoal de Nível Superior (CAPES) (process number: 88881.174609/2018-01) and by the Helmholtz Research Program "Materials and Technologies for the Energy Transition (MTET), Topic 3: Chemical Energy Carriers". We also acknowledge the support from the KIT-Publication Fund of the Karlsruhe Institute of Technology.

Conflicts of Interest: The authors declare no conflict of interest. The funders had no role in the design of the study; in the collection, analyses, or interpretation of data; in the writing of the manuscript; or in the decision to publish the results.

References

1. Our World in Data. Renewable Energy. Available online: https://ourworldindata.org/renewable-energy (accessed on 4 April 2022).
2. Daiyan, R.; MacGill, I.; Amal, R. Opportunities and challenges for renewable power-to-X. *ACS Energy Lett.* **2020**, *5*, 3843–3847. [CrossRef]
3. Younas, M.; Shafique, S.; Hafeez, A.; Javed, F.; Rehman, F. An overview of hydrogen production: Current status, potential, and challenges. *Fuel* **2022**, *316*, 123317. [CrossRef]
4. Chisholm, G.; Zhao, T.; Cronin, L. 24—Hydrogen from water electrolysis. In *Storing Energy*, 2nd ed.; Letcher, T.M., Ed.; Elsevier: Amsterdam, The Netherlands, 2022; pp. 559–591. [CrossRef]
5. Artz, J.; Müller, T.E.; Thenert, K.; Kleinekorte, J.; Meys, R.; Sternberg, A.; Bardow, A.; Leitner, W. Sustainable conversion of carbon dioxide: An integrated review of catalysis and life cycle assessment. *Chem. Rev.* **2018**, *118*, 434–504. [CrossRef] [PubMed]
6. Wang, L.; Chen, M.; Küngas, R.; Lin, T.-E.; Diethelm, S.; Maréchal, F.; Van Herle, J. Power-to-fuels via solid-oxide electrolyzer: Operating window and techno-economics. *Renew. Sustain. Energy Rev.* **2019**, *110*, 174–187. [CrossRef]
7. Bongartz, D.; Burre, J.; Ziegler, A.L.; Mitsos, A. Power-to-OME1 via direct oxidation of methanol: Process design and global flowsheet optimization. In *Computer Aided Chemical Engineering*; Türkay, M., Gani, R., Eds.; Elsevier: Amsterdam, The Netherlands, 2021; Volume 50, pp. 273–278.
8. Schmidt, P.; Batteiger, V.; Roth, A.; Weindorf, W.; Raksha, T. Power-to-liquids as renewable fuel option for aviation: A review. *Chem. Ing. Tech.* **2018**, *90*, 127–140. [CrossRef]
9. Methanol Institute—Methanol Fuel in China 2020. Available online: https://www.methanol.org/methanol-fuel-in-china-full-report/ (accessed on 5 July 2022).
10. Global Data—Methanol Market Capacity 2021. Available online: https://www.globaldata.com/store/report/methanol-market-analysis/ (accessed on 27 June 2022).
11. Bozzano, G.; Manenti, F. Efficient methanol synthesis: Perspectives, technologies and optimization strategies. *Prog. Energy Combust. Sci.* **2016**, *56*, 71–105. [CrossRef]
12. Freepik. Flaticon Icons. Available online: https://www.flaticon.com/authors/freepik (accessed on 27 June 2022).
13. Ott, J.; Gronemann, V.; Pontzen, F.; Fiedler, E.; Grossmann, G.; Kersebohm, D.B.; Weiss, G.; Witte, C. Methanol. In *Ullmann's Encyclopedia of Industrial Chemistry*; Wiley: New York, NY, USA, 2012. [CrossRef]
14. Toyo Engineering. G-Methanol. Available online: https://www.toyo-eng.com/jp/en/solution/g-methanol/ (accessed on 9 May 2022).
15. Lacerda de Oliveira Campos, B.; Herrera Delgado, K.; Wild, S.; Studt, F.; Pitter, S.; Sauer, J. Surface reaction kinetics of the methanol synthesis and the water gas shift reaction on $Cu/ZnO/Al_2O_3$. *React. Chem. Eng.* **2021**, *6*, 868–887. [CrossRef]
16. Studt, F.; Behrens, M.; Kunkes, E.L.; Thomas, N.; Zander, S.; Tarasov, A.; Schumann, J.; Frei, E.; Varley, J.B.; Abild-Pedersen, F.; et al. The mechanism of CO and CO_2 hydrogenation to methanol over Cu-based catalysts. *ChemCatChem* **2015**, *7*, 1105–1111. [CrossRef]
17. Bussche, K.M.V.; Froment, G.F. A steady-state kinetic model for methanol synthesis and the water gas shift reaction on a commercial $Cu/ZnO/Al_2O_3$ Catalyst. *J. Catal.* **1996**, *161*, 1–10. [CrossRef]
18. Slotboom, Y.; Bos, M.J.; Pieper, J.; Vrieswijk, V.; Likozar, B.; Kersten, S.R.A.; Brilman, D.W.F. Critical assessment of steady-state kinetic models for the synthesis of methanol over an industrial $Cu/ZnO/Al_2O_3$ catalyst. *Chem. Eng. J.* **2020**, *389*, 124181. [CrossRef]
19. Lacerda de Oliveira Campos, B.; Herrera Delgado, K.; Pitter, S.; Sauer, J. Development of consistent kinetic models derived from a microkinetic model of the methanol synthesis. *Ind. Eng. Chem. Res.* **2021**, *60*, 15074–15086. [CrossRef]
20. Pérez-Fortes, M.; Schöneberger, J.C.; Boulamanti, A.; Tzimas, E. Methanol synthesis using captured CO_2 as raw material: Techno-economic and environmental assessment. *Appl. Energy* **2016**, *161*, 718–732. [CrossRef]
21. Szima, S.; Cormos, C.-C. Improving methanol synthesis from carbon-free H_2 and captured CO_2: A techno-economic and environmental evaluation. *J. CO_2 Util.* **2018**, *24*, 555–563. [CrossRef]
22. Cordero-Lanzac, T.; Ramirez, A.; Navajas, A.; Gevers, L.; Brunialti, S.; Gandía, L.M.; Aguayo, A.T.; Mani Sarathy, S.; Gascon, J. A techno-economic and life cycle assessment for the production of green methanol from CO_2: Catalyst and process bottlenecks. *J. Energy Chem.* **2022**, *68*, 255–266. [CrossRef]
23. FFE München. Electrolysis, the Key Technology for Power-to-X. Available online: https://www.ffe.de/veroeffentlichungen/elektrolyse-die-schluesseltechnologie-fuer-power-to-x/ (accessed on 4 April 2022).
24. Fasihi, M.; Efimova, O.; Breyer, C. Techno-economic assessment of CO_2 direct air capture plants. *J. Clean. Prod.* **2019**, *224*, 957–980. [CrossRef]
25. Carbon Recycling International. Renewable Methanol from Carbon Dioxide. Available online: https://www.carbonrecycling.is/ (accessed on 4 April 2022).
26. Siemens Energy. The Haru Oni Hydrogen Plant. Available online: https://www.siemens-energy.com/global/en/news/magazine/2021/haru-oni.html (accessed on 1 February 2022).
27. van Bennekom, J.G.; Venderbosch, R.H.; Winkelman, J.G.M.; Wilbers, E.; Assink, D.; Lemmens, K.P.J.; Heeres, H.J. Methanol synthesis beyond chemical equilibrium. *Chem. Eng. Sci.* **2013**, *87*, 204–208. [CrossRef]
28. Bos, M.J.; Brilman, D.W.F. A novel condensation reactor for efficient CO_2 to methanol conversion for storage of renewable electric energy. *Chem. Eng. J.* **2015**, *278*, 527–532. [CrossRef]

29. Seshimo, M.; Liu, B.; Lee, H.R.; Yogo, K.; Yamaguchi, Y.; Shigaki, N.; Mogi, Y.; Kita, H.; Nakao, S.-i. Membrane reactor for methanol synthesis using Si-Rich LTA Zeolite Membrane. *Membranes* **2021**, *11*, 505. [CrossRef]

30. Johnson Mattey. Methanol Synthesis Technology. Available online: https://matthey.com/en/products-and-services/chemical-processes/core-technologies/synthesis (accessed on 9 May 2022).

31. Guo, Y.; Li, G.; Zhou, J.; Liu, Y. Comparison between hydrogen production by alkaline water electrolysis and hydrogen production by PEM electrolysis. *IOP Conf. Ser. Earth Environ. Sci.* **2019**, *371*, 042022. [CrossRef]

32. Porter, R.T.J.; Fairweather, M.; Kolster, C.; Mac Dowell, N.; Shah, N.; Woolley, R.M. Cost and performance of some carbon capture technology options for producing different quality CO_2 product streams. *Int. J. Greenh. Gas Control.* **2017**, *57*, 185–195. [CrossRef]

33. Bennekom, J.G.v.; Winkelman, J.G.M.; Venderbosch, R.H.; Nieland, S.D.G.B.; Heeres, H.J. Modeling and experimental studies on phase and chemical equilibria in high-pressure methanol synthesis. *Ind. Eng. Chem. Res.* **2012**, *51*, 12233–12243. [CrossRef]

34. Gaikwad, R.; Bansode, A.; Urakawa, A. High-pressure advantages in stoichiometric hydrogenation of carbon dioxide to methanol. *J. Catal.* **2016**, *343*, 127–132. [CrossRef]

35. Süd-Chemie AG. Safety Data Sheet—MegaMax 700. 2006.

36. Zhu, J.; Araya, S.S.; Cui, X.; Sahlin, S.L.; Kær, S.K. Modeling and design of a multi-tubular packed-bed reactor for methanol steam reforming over a $Cu/ZnO/Al_2O_3$ Catalyst. *Energies* **2020**, *13*, 610. [CrossRef]

37. Van-Dal, É.S.; Bouallou, C. Design and simulation of a methanol production plant from CO_2 hydrogenation. *J. Clean. Prod.* **2013**, *57*, 38–45. [CrossRef]

38. Pontzen, F.; Liebner, W.; Gronemann, V.; Rothaemel, M.; Ahlers, B. CO_2-based methanol and DME—Efficient technologies for industrial scale production. *Catal. Today* **2011**, *171*, 242–250. [CrossRef]

39. Bongartz, D.; Burre, J.; Mitsos, A. Production of oxymethylene dimethyl ethers from hydrogen and carbon dioxide—Part I: Modeling and analysis for OME1. *Ind. Eng. Chem. Res.* **2019**, *58*, 4881–4889. [CrossRef]

40. Burners for Fired Heaters in General Refinery Services. In *Reference Practice (RP) 535*, 3rd ed.; American Petroleum Institute (API): Washington, DC, USA, 2014.

41. Goos, E.; Burcat, A.; Ruscic, B. New NASA Thermodynamic Polynomials Database. Available online: http://garfield.chem.elte.hu/Burcat/THERM.DAT (accessed on 4 March 2022).

42. Gruber, M. Detaillierte Untersuchung des Wärme und Stofftransports in einem Festbett-Methanisierungsreaktor für Power-to-Gas Anwendungen. Ph.D. Thesis, Karlsruhe Institute of Technology (KIT), Karlsruhe, Germany, 2019; 275p.

43. *VDI Heat Atlas*, 2nd ed.; Springer: Berlin-Heidelberg, Germany, 2010; p. 1585.

44. Seidel, C.; Jörke, A.; Vollbrecht, B.; Seidel-Morgenstern, A.; Kienle, A. Kinetic modeling of methanol synthesis from renewable resources. *Chem. Eng. Sci.* **2018**, *175*, 130–138. [CrossRef]

45. Peng, D.-Y.; Robinson, D.B. A new two-constant equation of State. *Ind. Eng. Chem. Fundam.* **1976**, *15*, 59–64. [CrossRef]

46. Meng, L.; Duan, Y.-Y.; Wang, X.-D. Binary interaction parameter kij for calculating the second cross-virial coefficients of mixtures. *Fluid Phase Equilibria* **2007**, *260*, 354–358. [CrossRef]

47. Meng, L.; Duan, Y.-Y. Prediction of the second cross virial coefficients of nonpolar binary mixtures. *Fluid Phase Equilibria* **2005**, *238*, 229–238. [CrossRef]

48. Deiters, U.K. Comments on the modeling of hydrogen and hydrogen-containing mixtures with cubic equations of state. *Fluid Phase Equilibria* **2013**, *352*, 93–96. [CrossRef]

49. Kuld, S.; Thorhauge, M.; Falsig, H.; Elkjær, C.F.; Helveg, S.; Chorkendorff, I.; Sehested, J. Quantifying the promotion of Cu catalysts by ZnO for methanol synthesis. *Science* **2016**, *352*, 969–974. [CrossRef] [PubMed]

50. CLARIANT. Catalysts for Methanol Synthesis, Product Data Sheet. 2017. Available online: https://www.clariant.com/-/media/Files/Solutions/Products/Additional-Files/M/18/Clariant-Brochure-Methanol-Synthesis-201711-EN.pdf (accessed on 4 April 2022).

51. Saito, M.; Murata, K. Development of high performance Cu/ZnO-based catalysts for methanol synthesis and the water-gas shift reaction. *Catal. Surv. Asia* **2004**, *8*, 285–294. [CrossRef]

52. Campos Fraga, M.M.; Lacerda de Oliveira Campos, B.; Lisboa, M.S.; Almeida, T.B.; Costa, A.O.S.; Lins, V.F.C. Analysis of a Brazilian thermal plant operation applying energetic and exergetic balances. *Braz. J. Chem. Eng.* **2018**, *35*, 1395–1403. [CrossRef]

53. Nitzsche, R.; Budzinski, M.; Gröngröft, A. Techno-economic assessment of a wood-based biorefinery concept for the production of polymer-grade ethylene, organosolv lignin and fuel. *Bioresour. Technol.* **2016**, *200*, 928–939. [CrossRef] [PubMed]

54. Green, D.W.; Southard, M.Z. *Perry's Chemical Engineers' Handbook*, 9th ed.; McGraw Hill: New York, NY, USA, 2018.

55. König, D.H.; Baucks, N.; Dietrich, R.-U.; Wörner, A. Simulation and evaluation of a process concept for the generation of synthetic fuel from CO_2 and H_2. *Energy* **2015**, *91*, 833–841. [CrossRef]

56. Albrecht, F.G.; König, D.H.; Baucks, N.; Dietrich, R.-U. A standardized methodology for the techno-economic evaluation of alternative fuels—A case study. *Fuel* **2017**, *194*, 511–526. [CrossRef]

57. Peters, M.S.; Timmerhaus, K.D.; West, R.E. *Plant Design and Economics for Chemical Engineers*, 5th ed.; McGraw-Hill Education Ltd.: Boston, MA, USA, 2002; p. 1008.

58. Hennig, M.; Haase, M. Techno-economic analysis of hydrogen enhanced methanol to gasoline process from biomass-derived synthesis gas. *Fuel Processing Technol.* **2021**, *216*, 106776. [CrossRef]

59. Travelex. Worldwide Money. Available online: https://www.travelex.com/currency/currency-pairs/usd-to-eur (accessed on 4 April 2022).

60. Towler, G.; Sinnott, R. *Chemical Engineering Design—Principles, Practice and Economics of Plant and Process Design*, 2nd ed.; Elsevier: Amsterdam, The Netherlands, 2013.
61. Aden, A.; Ruth, M.; Ibsen, K.; Jechura, J.; Neeves, K.; Sheehan, J.; Wallace, B.; Montague, L.; Slayton, A.; Lukas, J. *Lignocellulosic Biomass to Ethanol Process Design and Economics Utilizing Co-Current Dilute Acid Prehydrolysis and Enzymatic Hydrolysis for Corn Stover*; National Renewable Energy Laboratory: Golden, CO, USA, 2002; 154p.
62. Cheremisinoff, N.P. *Clean Electricity Through Advanced Coal Technologies*; Elsevier: Amsterdam, The Netherlands, 2012. [CrossRef]
63. Bustillo-Lecompte, C.F.; Mehrvar, M.; Quiñones-Bolaños, E. Cost-effectiveness analysis of TOC removal from slaughterhouse wastewater using combined anaerobic–aerobic and UV/H$_2$O$_2$ processes. *J. Environ. Manag.* **2014**, *134*, 145–152. [CrossRef]
64. Tan, E.C.; Talmadge, M.; Dutta, A.; Hensley, J.; Snowden-Swan, L.J.; Humbird, D.; Schaidle, J.; Biddy, M. Conceptual process design and economics for the production of high-octane gasoline blendstock via indirect liquefaction of biomass through methanol/dimethyl ether intermediates. *Biofuels Bioprod. Biorefining* **2016**, *10*, 17–35. [CrossRef]
65. Ashraf, M.T.; Schmidt, J.E. Process simulation and economic assessment of hydrothermal pretreatment and enzymatic hydrolysis of multi-feedstock lignocellulose—Separate vs. combined processing. *Bioresour. Technol.* **2018**, *249*, 835–843. [CrossRef] [PubMed]
66. Granjo, J.F.O.; Oliveira, N.M.C. Process simulation and techno-economic analysis of the production of sodium methoxide. *Ind. Eng. Chem. Res.* **2016**, *55*, 156–167. [CrossRef]
67. Prašnikar, A.; Pavlišič, A.; Ruiz-Zepeda, F.; Kovač, J.; Likozar, B. Mechanisms of copper-based catalyst deactivation during CO$_2$ reduction to methanol. *Ind. Eng. Chem. Res.* **2019**, *58*, 13021–13029. [CrossRef]
68. Argus Media. European Methanol Spot and Contract Prices. Available online: https://www.argusmedia.com/en/blog/2022/february/4/european-methanol-spot-and-contract-prices (accessed on 4 April 2022).
69. Methanol Institute—Methanol Prices, Supply and Demand. Available online: https://www.methanol.org/methanol-price-supply-demand/ (accessed on 4 April 2022).

Article

Model-Based Analysis for Ethylene Carbonate Hydrogenation Operation in Industrial-Type Tubular Reactors

Hai Huang [1], Chenxi Cao [2,*], Yue Wang [3], Youwei Yang [3], Jianning Lv [4] and Jing Xu [1,5,*]

[1] State Key Laboratory of Chemical Engineering, School of Chemical Engineering, East China University of Science and Technology, Shanghai 200237, China; haielvis@163.com

[2] Key Laboratory of Smart Manufacturing in Energy Chemical Process, Ministry of Education, East China University of Science and Technology, Shanghai 200237, China

[3] Key Laboratory for Green Chemical Technology of Ministry of Education, Collaborative Innovation Center of Chemical Science and Engineering, School of Chemical Engineering and Technology, Tianjin University, Tianjin 300072, China; yuewang@tju.edu.cn (Y.W.); yangyouwei@tju.edu.cn (Y.Y.)

[4] Wison Engineering Ltd., 633 Zhongke Rd., Shanghai 201210, China; lvjianning@wison.com

[5] Guangxi Key Laboratory of Petrochemical Resource Processing and Process Intensification Technology, School of Chemistry and Chemical Engineering, Guangxi University, Nanning 530004, China

* Correspondence: caocx@ecust.edu.cn (C.C.); xujing@ecust.edu.cn (J.X.)

Abstract: Hydrogenation of ethylene carbonate (EC) to co-produce methanol (MeOH) and ethylene glycol (EG) offers an atomically economic route for CO_2 utilization. Herein, aided with bench and pilot plant data, we established engineering a kinetics model and multiscale reactor models for heterogeneous EC hydrogenation using representative industrial-type reactors. Model-based analysis indicates that single-stage adiabatic reactors, despite a moderate temperature rise of 12 K, suffer from a narrow operational window delimited by EC condensation at lower temperatures and intense secondary EG hydrogenation at higher temperatures. Boiling water cooled multi-tubular reactors feature near-isothermal operation and exhibit better operability, especially under high pressure and low space velocity. Conduction oil-cooled reactors show U-type axial temperature profiles, rendering even wider operational windows regarding coolant temperatures than the water-cooled reactor. The revelation of operational characteristics of EC hydrogenation under industrial conditions will guide further improvement in reactor design and process optimization.

Keywords: ethylene carbonate hydrogenation; methanol; ethylene glycol; multiscale reactor model; reactor analysis; operation window

Citation: Huang, H.; Cao, C.; Wang, Y.; Yang, Y.; Lv, J.; Xu, J. Model-Based Analysis for Ethylene Carbonate Hydrogenation Operation in Industrial-Type Tubular Reactors. *Processes* **2022**, *10*, 688. https://doi.org/10.3390/pr10040688

Academic Editor: Blaž Likozar

Received: 21 March 2022
Accepted: 28 March 2022
Published: 31 March 2022

Publisher's Note: MDPI stays neutral with regard to jurisdictional claims in published maps and institutional affiliations.

1. Introduction

Efficient chemical conversion and utilization of CO_2 is envisioned as an important vector in future carbon-neutral and carbon-negative human society [1,2]. Researchers are in hot pursuit of disposing the energy- and industry-related CO_2 emissions by renewable electricity or discarded H_2 to produce feedstock or commodity chemicals [3–5]. In particular, thermo-/electro- chemical CO_2 reduction to oxygenates, for instance, aldehydes, alcohols, and carboxyl acids, exerts great prospects in high-atom-economy CO_2-based chemical manufacture by reducing the by-production of H_2O [6–10]. Direct hydrogenation of CO_2 into methanol (MeOH) has attracted intense research attention in recent years in that MeOH serves as a commodity chemical as well as a crucial platform to various downstream products [11–15]. However, this process is inherently limited by the kinetic inertness and thermodynamic stability of the CO_2 molecule, resulting in extremely low per-pass CO_2 conversion (~10~20%), and thereby a poor atom economy [16–18].

A feasible solution to by-pass the thermodynamic limitation is to "bridge" the low-energy CO_2 reactant and the high-energy MeOH product with a CO_2-derivable, medium-energy intermediate, such as ethylene carbonate (EC), dimethyl carbonate, methyl formate, and so on [19–23]. Such indirect green methanol production processes are advantageous in milder reaction conditions as well as higher per-pass CO_2 conversion (over 95%) and overall methanol yield. Amongst these processes, EC hydrogenation as illustrated in Figure 1 is of great interest:

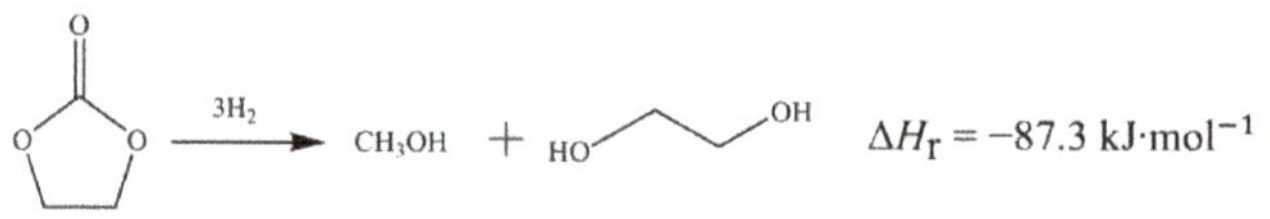

$$\text{EC} \xrightarrow{3H_2} CH_3OH + HO\text{-}CH_2CH_2\text{-}OH \qquad \Delta H_r = -87.3 \text{ kJ·mol}^{-1}$$

Figure 1. EC hydrogenation reaction.

This process co-produces methanol and ethylene glycol (EG), another value-added bulk chemical, with theoretically 100% atom economy. In addition, EC production using CO_2 and ethylene oxide has already been demonstrated [24,25], making the subsequent EC hydrogenation process a promising enabler for industrially relevant green alcohols production.

Development of catalyst systems for EC hydrogenation has gained considerable progress. Since Han et al. [26] first proposed a Ru II PNP catalyst for homogeneous EC hydrogenation, the catalyst system has evolved from homogeneous nobel metal complex [27] to heterogenous supported copper (Cu) based catalysts [28–33], in view of difficult separation and recovery of the former and, in contrast, convenient deployment of the latter. As the early-version Cu-based catalysts are inferior to the homogeneous catalysts in performance, continuous efforts are being paid to develop more active and selective candidates suitable for industrial application [34–36]. Song et al. [37] prepared MoO_x-promoted Cu/SiO_2 catalyst, which achieved 89% MeOH yield and 99% EG yield within 150 h time-on-stream at a H_2/EC ratio of 20 and a WLHSVEC (weight liquid hourly space velocity of EC) of 0.64 $g_{EC} \cdot g_{cat}^{-1} \cdot h^{-1}$. Carbon modified Cu catalyst proposed by Chen et al. [38] displayed good activity and stability after using for 264 h, owing to improved dispersion of Cu particles, showing EC conversion of 100%, and EG and MeOH selectivities up to 99.9% and 85.8%, respectively. Appropriate synergy between Cu^0 and Cu^+ species are believed to play an important role in the activation of H_2, providing adsorption sites for the carbonyl groups of EC, and stabilizing the surface acyl and methoxy species, leading to improved EC conversion and alcohol yields [39,40].

Despite the progress in catalyst investigation, to date, the heterogeneous EC hiydrogenation process has not been studied at a pilot scale. Tailoring the transport characteristics of industrial-scale reactors in accordance with the intrinsic reaction kinetics is crucial for commercial operation of catalytic processes [41,42]. In this respect, industrial methanol synthesis serves as a reference given a similar reaction enthalpy change and the employment of similar Cu-based catalysts. Multi-stage adiabatic fixed bed reactors and boiling water-cooled multi-tubular reactors are the most commonly applied types of reactors in methanol synthesis [42]. The design and operation optimization of these reactors has been well aided by model-based analysis accounting for intrinsic reaction kinetics, internal/external mass transfer at the scale of catalyst particles, and heat and mass transfer at the reactor-scale [43]. Recently, Samimi et al. [44] compared water cooled, gas cooled, and doubled cooled reactors for direct CO_2 hydrogenation to MeOH by one-dimensional reactor models focusing on the phase stability. The gas cooled reactor exhibits the lowest possibility of methanol condensation and the highest methanol yield. Cui et al. [45] assessed the potentials of adiabatic, water cooled and gas cooled reactors for direct CO_2 hydrogenation to MeOH using coupled two-dimensional (2D) computational fluid dynamics (CFD) models for the reactor and single catalyst particles. The water-cooled reactors demonstrate outstanding temperature control at the expense of a higher capital cost. It is apparent that, to promote the process towards commercialization, understanding of the operational behaviors of EC hydrogenation in industrial-type reactors is urgently needed.

Herein, we investigated the operational characteristics of industrial-type adiabatic, water-cooled and oil-cooled tubular reactors for a 3×10^4 t/a EC hydrogenation process by model-based comparative analysis. Two-dimensional (2D) pseudo-homogeneous CFD models were established with engineering kinetic models developed for heterogeneous EC hydrogenation over an industrial Cu-based catalyst, and validated by bench-scale and pilot-scale reaction data. The operation windows of key reactor operating variables, including reactant/coolant inlet temperature, total pressure and space velocity, were delineated for the representative industrial-type reactors, considering both the reactor performance and the EC phase change boundaries. The presented results will pave the way to future industrial design and optimization of reactors and process for the green alcohols production.

2. Methods

2.1. Method Description

Figure 2 illustrates the industrial reactors for ethylene carbonate hydrogenation considered and the modeling scheme used in this study. Two types of industrial reactors are modeled, namely, adiabatic and multi-tubular heat-exchange reactors; for the heat-exchange reactors, two different coolants are considered: boiling water and heat conduction oil (Figure 2a). The adiabatic reactor is 2 m in diameter and 8 m in length, whereas the heat-exchange reactor comprises 1682 tubes of 0.05 m diameter and 8 m length, corresponding to a total capacity of 3×10^4 t/a methanol and ethylene glycol. Key geometric parameters and operating conditions are listed in Table 1. The 2D asymmetrical steady-state multi-scale reactor models are built under the following assumptions:

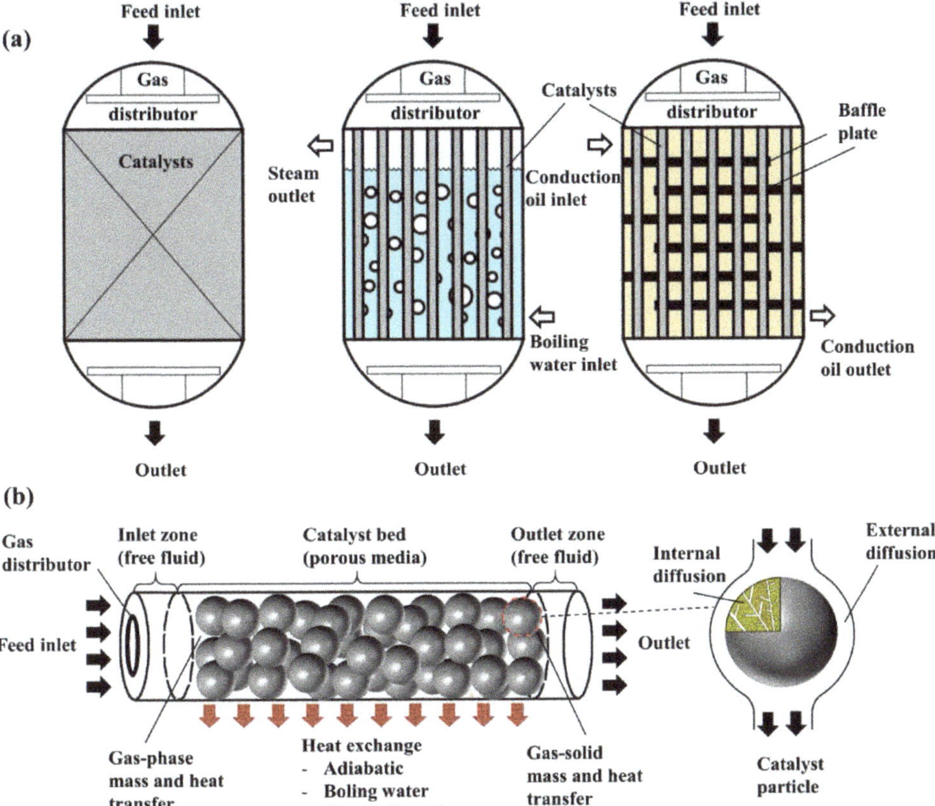

Figure 2. Schematic diagram of (**a**) industrial adiabatic and heat-exchange reactors for ethylene carbonate hydrogenation and (**b**) the multi-scale reactor model.

Table 1. Model parameters.

Geometric Parameters	Adiabatic FBR	Heat-Exchange FBR
Bed/Tube diameter, D (m)	2	0.05
Developing zone length	2 m	
Reaction zone length	8 m	
Catalyst type	Cu/SiO_2 [30]	
Catalyst geometry	3/5/7 mm (sphere); $3 \times 3/5 \times 5/7 \times 7$ mm (cylinder)	
Catalyst loading mass (kg)	13730 kg for (5 mm sphere)	
Average bed voidage	0.4 (5 mm sphere)	0.43 (5 mm sphere)
Gas distributor geometry	1 m diameter, ring	/
Operating conditions (Nominal value/range)		
Temperature, T_{in} (K)	463/433–483	
Coolant temperature, T_c (K)	/	463/423–463
Pressure, P_{op} (MPa)	3/1–5	
Space velocity, SV $(g_{EC} \cdot g_{cat}^{-1} \cdot h^{-1})$	0.3/0.1–0.5	
Molar ratio of H_2 to EC	200/80–200	

Details of the multi-scale reactor models, including governing equations, heat and mass transfer correlations, chemical kinetics and model implementation are introduced in the following text.

1. Reactants are fed into the adiabatic reactor through inlet manifolds, which are simplified into a ring.
2. The inlet gas distributor of the heat-exchange multi-tubular reactor ensures uniform distribution of the reactants to all reaction tubes, so that only a single reaction tube is modeled as representative.
3. The wall temperatures of all reaction tubes of the boiling water-cooled reactor are equal to the boiling temperature of pressurized water.
4. The coolant temperature and external heat transfer coefficient are the same for all reaction tubes of the oil-cooled reactor.

2.2. Governing Equations

The fluid computational mass conservation equation in the bed is

$$\frac{\partial(\varepsilon_b \rho_f)}{\partial t} + \nabla \cdot (\varepsilon_b \rho_f v) = 0 \tag{1}$$

The momentum Equation (2), the energy Equation (7) and the species transport equation Equation (10) for flow in the catalyst bed are listed below.

Momentum equation:

$$\frac{\partial(\varepsilon_b \rho_f v)}{\partial t} + \nabla \cdot (\varepsilon_b \rho_f v v) = -\varepsilon_b \nabla p + \nabla \cdot (\varepsilon_b \tau) + S_\phi \tag{2}$$

where p is the static pressure and τ is the stress tensor. S_ϕ is the momentum source term for fluid flow in porous media,

$$S_\phi = B_f - \left(\frac{\varepsilon_b^2 \mu}{K} v + \frac{\varepsilon_b^3 C_2}{2} \rho_f |v| v \right) \tag{3}$$

representing the viscous and inertial drag forces imposed on the fluid flow by the pore walls within the porous media, in which C_2 is the inertial loss coefficient.

$$C_2 = \frac{3.5}{d_p^v} \frac{(1 - \varepsilon_b)}{\varepsilon_b^3} \tag{4}$$

The k-ω SST turbulence model is used with the turbulence kinetic energy k and the specific dissipation rate ω obtained from the following equations:

$$\frac{\partial(\rho_f k)}{\partial t} + \frac{\partial(\rho_f k u_i)}{\partial x_i} = \frac{\partial}{\partial x_j}\left(\Gamma_k \frac{\partial k}{\partial x_j}\right) + G_k - Y_k + S_k + G_{kb} \tag{5}$$

$$\frac{\partial(\rho_f \omega)}{\partial t} + \frac{\partial(\rho_f \omega u_i)}{\partial x_i} = \frac{\partial}{\partial x_j}\left(\Gamma_\omega \frac{\partial \omega}{\partial x_j}\right) + G_\omega - Y_\omega + S_\omega + G_{\omega b} \tag{6}$$

Energy equation:

$$\frac{\partial}{\partial t}\left(\varepsilon_b \rho_f E_f + (1 - \varepsilon_b)\rho_s E_s\right) + \nabla\cdot(v(\rho_f E_f + p)) = S_f^h + \nabla\cdot\left[k_e \nabla T - \left(\sum_i h_i J_i\right) + (\tau\cdot v)\right] \tag{7}$$

where E_f is total fluid energy and E_s is total solid medium energy. The energy source S_f^h represents the chemical reaction heat.

$$S_f^h = -\sum_j \frac{h_j^0}{M_{w,j}} R_j \tag{8}$$

where h_j^0 is the enthalpy of formation of species and R_j is the volumetric rate of creation of species j. The effective heat conduction of bed k_e is computed in Equation (9).

$$k_e = \begin{cases} k_{e,\,ax} = \dfrac{u_0 \rho_f c_{p,f} d_p^v}{Pe_{h,ax}} \\[2ex] k_{e,r} = \dfrac{k_r^0}{k_f} + k_f\cdot\dfrac{Pe_h^0}{Pe_{h,r}^\infty} \end{cases} \tag{9}$$

Species equation:

$$\frac{\partial}{\partial t}(\rho_f Y_i) + \nabla\cdot(\rho_f v Y_i) = -\nabla\cdot J_i + S_i \tag{10}$$

where Y_i is mass fraction of each species, and J_i is the diffusive flux of species i arising from gradients of species concentration and temperature.

$$J_i = -\rho_f D_e \sum_{j=1}^{N-1} \nabla Y_j - D_{T,i} \frac{\nabla T}{T} \tag{11}$$

S_i, the net source of species i due to chemical reactions, S_i is computed as the sum of the reaction rates:

$$S_i = M_{w,i} \sum_{r=1}^{N_R} \hat{R}_{i,r} \tag{12}$$

where $M_{w,i}$ is the molecular weight of species i and $\hat{R}_{i,r}$ is the molar rate of creation/destruction of species i in reaction r. In Equation (11), the calculation of thermal diffusion coefficients $D_{T,i}$ adopts the following empirically based expression [46].

$$D_{T,i} = -2.59 \times 10^{-7} T^{0.659} \left[\frac{M_{w,i}^{0.511} X_i}{\sum_{i=1}^{N} M_{w,i}^{0.511} X_i} - Y_i\right]\cdot\left[\frac{\sum_{i=1}^{N} M_{w,i}^{0.511} X_i}{\sum_{i=1}^{N} M_{w,i}^{0.489} X_i}\right] \tag{13}$$

where X_i is mole fraction of species i; the effective mass transfer coefficient of bed D_e is computed as:

$$D_e = \begin{cases} D_{e,ax} = \dfrac{d_p^v u_0}{Pe_{m,ax}} \\[2ex] D_{e,r} = \dfrac{d_p^v u_0}{Pe_{m,r}} \end{cases} \tag{14}$$

2.3. Bed Voidage and Pressure Drop

When the ratio of tube diameter to the catalyst's volume-equivalent diameter (denoted by R in the following paper) is less than 10, the wall effect cannot be ignored [47]. In this work, R values for the adiabatic fixed bed (2 m in diameter) and the heat-exchange reactor (single tube 0.05 m in diameter) filled with 5 mm spherical catalyst are 400 and 10, respectively. Therefore, without the wall effect, the bed voidage of the adiabatic reactor is set as constant in the radial direction (Supplementary Materials Table S5). Meanwhile, the fluctuations of bed voidage in the radial direction is considered for the heat-exchange reactors (Table S5) [48–50].

The bed pressure drop is computed by the Ergun equation in Equation (15) [51].

$$\frac{|\Delta P|}{L} = \frac{150}{d_p^{s\,2}} \frac{(1-\varepsilon_b)^2 \mu}{\varepsilon_b{}^3} u_0 + \frac{1.75\rho_f}{d_p^s} \frac{(1-\varepsilon_b)}{\varepsilon_b{}^3} u_0{}^2 \tag{15}$$

2.4. Heat and Mass Transfer in the Catalyst Bed

The axial effective heat conduction is computed as [52]:

$$k_{e,ax} = \frac{u_0 \rho_f c_{p,f} d_p^v}{Pe_{h,ax}} \tag{16}$$

in which the Peclet number for axial heat conduction $Pe_{h,ax}$ equals 2.

The radial effective heat conduction is computed by:

$$k_{e,r} = \frac{k_r^0}{k_f} + k_f \cdot \frac{Pe_h^0}{Pe_{h,r}^\infty} \tag{17}$$

in which the molecular Peclet number Pe_h^0 is [53]:

$$Pe_h^0 = RePr = \frac{u_0 \rho_f c_{p,f} d_p^v}{k_f} \tag{18}$$

and $Pe_{h,r}^\infty$ is [54]:

$$Pe_{h,r}^\infty = 8[2 - (1 - \frac{2}{R})^2] \tag{19}$$

The following expression for the effective thermal stagnant conductivity $\frac{k_r^0}{k_f}$ is obtained by Equation (20) [55]:

$$\frac{k_r^0}{k_f} = (1 - \sqrt{1 - \varepsilon_b}) + \frac{2\sqrt{1-\varepsilon_b}}{1 - B\kappa^{-1}}$$
$$\times \left[\frac{B(1-\kappa^{-1})}{(1-B\kappa^{-1})^2} \ln\left(\frac{\kappa}{B}\right) - \frac{B-1}{1-B\kappa^{-1}} - \frac{B+1}{2} \right] \tag{20}$$

where $B = C_f \left(\frac{1-\varepsilon_b}{\varepsilon_b} \right)^{1.11}$; $C_f = 1.25$ (sphere), $C_f = 2.5$ (cylinder).

Under turbulent conditions ($Re > 100$), axial mixing of mass can be approximated as mixing in a cascade of L/d_p ideal mixers [56]. The Peclet number for axial mass dispersion approximately equals 2.

$$Pe_{m,ax} = \frac{d_p^v u_0}{D_{e,ax}} = 2 \tag{21}$$

The effective radial mass transfer coefficient of bed $D_{e,r}$ is computed as [57]:

$$Pe_{m,r} = \frac{d_p^v u_0}{D_{e,r}} = C\left(1 + \frac{19.4}{R^2}\right) \tag{22}$$

where $C = 8.65 \frac{d_p^v}{d_p^a}$ [58].

2.5. Heat and Mass Transfer of Catalyst Particles

Due to mild exothermicity of the ethylene carbonate hydrogenation reaction, the temperature difference between the catalyst surface and the gas phase is estimated to be less than 1 K in Equation (23). Consequently, the temperature of catalyst surface and the gas phase is taken as identical. The gas–solid heat transfer coefficient α_f is calculated as [59]:

$$\Delta T_{ex} = \frac{Q_{reaction}}{\alpha_f \cdot S_{cat}} \tag{23}$$

$$Nu = 2 + 1.1 Re^{0.6} \cdot Pr^{\frac{1}{3}} \tag{24}$$

$$Nu = \frac{\alpha_f d_p^v}{k_f} \tag{25}$$

For heat transfer inside the catalyst particle, the temperature gradient within the catalyst particle is related to the heat of reaction $Q_{reaction}$ and the effective thermal conductivity of the catalyst particle $\lambda_{eff,cat}$ (Equation (26)) and can be calculated by Equation (27) [60,61].

$$\Delta T_{in} = \frac{D_{eff,EC} \cdot (C_{EC,s} - C_{EC,center}) \cdot (-\Delta_r H)}{\lambda_{eff,cat}} \tag{26}$$

$$\lambda_{eff,cat} = k_f \left(\frac{k_s}{k_f}\right)^{1-\varepsilon_{cat}} \tag{27}$$

Under the reaction conditions considered in this article, ΔT_{in} is estimated to be less than 1 K. Therefore, the catalyst particle is considered isothermal.

The mass transfer between the catalyst surface and the gas phase is described by a mass transport coefficient k_g:

$$k_g \cdot a \cdot (C_{EC,gas} - C_{EC,s}) = R_{EC,s} \tag{28}$$

where $R_{EC,s}$ is the effective consumption rate of EC at the particle surface. k_g is related to the catalyst shape, the Reynolds number and the Schmidt number, as shown in Equations (29)–(31) [62].

$$Sh = 2 + 0.6 Re^{0.5} Sc^{\frac{1}{3}} \text{ (sphere)} \tag{29}$$

$$Sh = 0.61 Re^{0.5} Sc^{\frac{1}{3}} \text{ (cylinder)} \tag{30}$$

$$Sh = \frac{k_g d_p^v}{D_{EC,m}} \tag{31}$$

The effect of mass transfer inside the catalyst particles, namely, internal mass transfer, is accounted for with the generalized Thiele modulus approach, which is applicable to a broad range of rate equations [63]. The generalized Thiele modulus with respect to EC consumption rate ($\phi_{gen,EC}$) is expressed as

$$\phi_{gen,EC} = \frac{V_{cat}}{S_{cat}} \sqrt{\frac{k_v}{D_{eff,EC}} \cdot \frac{n+1}{2} \cdot C_{EC,s}^{n-1}} \tag{32}$$

The effectiveness factor for internal mass transfer is then computed as:

$$\eta_{EC} = \frac{\tanh(\phi_{gen,EC})}{\phi_{gen,EC}} \tag{33}$$

Note that Equations (28)–(31) must be solved simultaneously in an iterative manner because the two sides in Equation (28) are related by the EC concentration at catalyst particle surface, which is unknown in advance.

2.6. Coolant Heat Transfer

For the heat exchange by pressurized boiling water, a large heat transfer coefficient (4000–6000 $W \cdot m^{-2} \cdot K^{-1}$) and a constant wall temperature (T_w) are considered:

$$T_w = T_b \tag{34}$$

where T_b is the boiling point of pressurized water.

In contrast, the heat conduction oil has a low specific heat capacity, and thus is heated up easily. The oil temperature is therefore considered as a function of the transferred heat. The oil-to-tube heat transfer coefficient is set as a function of the space in between adjacent baffle plates t and the coolant mass flow rate $\dot{m}_{oil}$.

$$\alpha_{oil} = \overline{\alpha}_{oil}\left(t, \dot{m}_{oil}\right) \tag{35}$$

The detailed expressions are given in the Supporting Information. The heat transferred between the oil and the tube wall is

$$Q_{oil} = \overline{\alpha}_{oil} \cdot (T_{oil} - T_w) \cdot S_{wall} \tag{36}$$

As the oil temperature T_{oil} varies with the axial position, the steady-state coolant temperature profile is determined by iteration. The single reaction tube is divided into 10 segments (each 0.8 m long) in the flow direction with the oil temperature in each segment calculated as

$$T_{oil,i+1} = \begin{cases} T_{oil,in}, & i = -1 \\ T_{oil,i} + \dfrac{Q_{oil,i}}{3600 c_{p,oil} \cdot \dot{m}_{oil}}, & 0 \leq i \leq 9 \end{cases} \tag{37}$$

when the iteration converges, the oil temperature as a continuous function of the the axial position is calculated by linear interpolation of the above discrete temperatures.

$$T_{oil}(x) = T_{oil,i-1} + \frac{T_{oil,i} - T_{oil,i-1}}{x_i - x_{i-1}} \cdot (x - x_{i-1}), \quad (x_{i-1} < x < x_i) \tag{38}$$

2.7. Chemical Reactions

The global reaction of ethylene carbonate hydrogenation to ethylene glycol and methanol is described by three separate reactions as shown in (R1)–(R3), respectively. This scheme allows for investigation of the effect of operating variables on the reactant conversion and product selectivity. The intrinsic kinetics of these reactions are modeled by power-law equations. The kinetic parameters were fitted to bench-scale experimental data covering a wide range of reaction conditions of T = 175–220 °C, P_{op} = 3.0 MPa, H_2/EC = 120–200 and SV = 0.5–2.2 h^{-1}. Details regarding the kinetic equations and parameters are given in the Supporting Information (Section S1).

R1: $\quad + 3H_2 \rightarrow OH(CH_2)_2OH + CH_3OH$

R2: $\quad + H_2 \rightarrow OH(CH_2)_2OH + CO$

R3: $\quad OH(CH_2)_2OH + H_2 \rightarrow CH_3CH_2OH + H_2O$

Since the two EC hydrogenation reactions ((R1) and (R3)) are dependent on the gas phase EC concentration only, their effective reaction rates over the catalyst particle surface are:

$$r_{\text{obs}} = \eta_{\text{EC}} \cdot r_{\text{chem}}(C_{\text{EC,s}}, T_{\text{s}}) \tag{39}$$

The EG hydrogenation reaction (R3) is zeroth order; therefore, its effective reaction rate is equal to the intrinsic reaction rate:

$$r_{\text{obs}} = r_{\text{chem}}(C_{\text{EG,s}}, T_{\text{s}}) \tag{40}$$

The performance metrics of the reaction, EC conversion X_{EC}, EG selectivity S_{EG}, MeOH selectivity S_{MeOH} and total alcohols selectivity S_{Alcohol}, are defined by the following definitions:

$$X_{\text{EC}} = \frac{\dot{m}_{\text{EC,in}} - \dot{m}_{\text{EC,out}}}{\dot{m}_{\text{EC,in}}} \tag{41}$$

$$S_{\text{EG}} = \frac{\dot{m}_{\text{EG,out}} / M_{\text{EG}}}{(\dot{m}_{\text{EC,in}} - \dot{m}_{\text{EC,out}}) / M_{\text{EC}}} \tag{42}$$

$$S_{\text{MeOH}} = \frac{\dot{m}_{\text{MeOH,out}} / M_{\text{MeOH}}}{(\dot{m}_{\text{EC,in}} - \dot{m}_{\text{EC,out}}) / M_{\text{EC}}} \tag{43}$$

$$S_{\text{Alcohol}} = \frac{S_{\text{EG}} + S_{\text{MeOH}}}{2} \tag{44}$$

2.8. Model Implementation

The multi-scale reactor models were implemented in the FLUENT software. The computational domains were discretized using rectangular meshes refined in the inlet regions and adjacent to the walls. Spatial discretization was conducted by the second-order upwind differencing scheme. The pressure–velocity coupling was performed with the SIMPLEC algorithm. User defined function (UDF) of FLUENT was used to couple the heat and mass transport models and chemical kinetics to the reactor-level models. The inlet boundary was set as uniform velocity. The outlet boundary was set as constant pressure (operating pressure) and zero normal gradients of velocity, temperature and species mass fractions. A no-slip, adiabatic or thermally coupled boundary was imposed on the reactor walls depending on the reactor type. Axisymmetric boundaries were imposed on the centerline of the reactor. The convergence criteria were set as all residuals below 10^{-3}. Simulations were performed on PC with Intel i9-9900X processors (Intel Corporation, Santa Clara, CA, USA) and 64 GB RAM (G.skill International Enterprise Co., Ltd., Taipei, China). The typical computation time under each operating condition is between 0.5 and 2 h.

3. Results and Discussion

3.1. Model Validation

Table 2 shows results of the grid independence study using EC conversion as the most sensitive key variable. When the grid size exceeds 10,000, the EC conversions remain unchanged for both adiabatic and heat-exchange reactor models with sphere/cylinder catalysts. Therefore, the following work was carried out with a grid size of 10,000.

To validate the established multi-scale reactor and chemical kinetics model, we compared the model predictions with pilot plant data. The EC hydrogenation pilot reactor was operated at a production capacity of 1000 t_{alcohol}/a and had steadily run for over 72 h to allow collection of all reaction products. The structure and working conditions of the pilot reactor are shown in Table 3. Simulations were conducted with the same parameter settings as the pilot reactor, including the reactor geometry, operating conditions and catalyst properties. Figure 3 shows the mass flow rates of the converted EC and the produced EG, MeOH and by-products measured in pilot plant run and those predicted by our model. The model predictions are in quantitative agreement with the pilot plant data with relative errors for both the converted EC and the produced EG < 5%. The amount of produced MeOH is underestimated by ~20%, which is accompanied by an overestimated amount of by-products. The reason is suspected to be incomplete measurement of the volatile MeOH in bench-scale experiments used for deriving the kinetics parameters. Nonetheless, the multi-scale reactor model is reliable in predicting the generic reaction performance of EC hydrogenation in industrial reactors and will be used to explore the operational behaviors of different types of tubular reactors.

Table 2. Dependence of EC conversion on grid sizes in simulation.

	Adiabatic Reactor Model			
	Mesh Size	1250	10,000	20,000
EC conversions (%)	20% length (sphere)	37.6	39.0	39.1
	outlet (sphere)	94.1	94.0	94.0
	20% length (cylinder)	29.7	30.2	30.1
	outlet (cylinder)	86.0	85.1	84.8
	Heat-Exchange Reactor Model			
	Mesh Size	2500	10,000	40,000
EC conversions (%)	20% length (sphere)	36.8	38.0	38.3
	outlet (sphere)	93.2	93.2	93.2
	20% length (cylinder)	33.7	34.8	35.1
	outlet (cylinder)	89.4	89.4	89.4

Table 3. Structure and working conditions of the pilot reactor for EC hydrogenation.

Pilot Reactor Configuration	
Tube Inner Diameter (m)	0.034
Tube length (m)	4
Tube number	95
Operating Conditions	
EC mass flow (kg·h^{-1})	107.3
Feed EC to H_2 molar ratio	170
Feed temperature (K)	458.2
Coolant type	Conduction oil
Coolant mass flow rate (kg·h^{-1})	20,000
Coolant temperature (K)	453.2
Catalyst Information	
Catalyst composition	Cu/SiO_2
Size and shape	3×5 mm cylinder

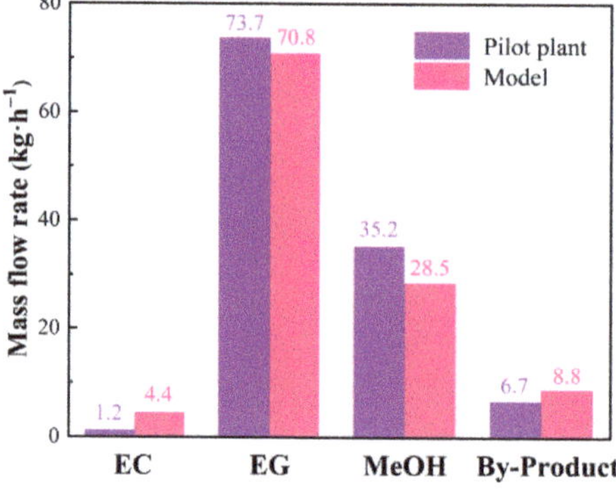

Figure 3. Comparison between pilot plant and simulation data for EC hydrogenation in the conduction oil-cooled multi-tubular reactor.

3.2. Reactor Profiles

3.2.1. Adiabatic

As the simplest form of fixed bed reactors, adiabatic reactors are often designed with multi-stage feeding in order to control the temperature rise of the catalytic bed by inter-stage cooling. For the EC hydrogenation reaction, however, the adiabatic temperature rise is estimated as only 10~20 K under typical operating conditions because of the high H_2/EC ratio, despite a considerable reaction enthalpy change (-87.3 kJ/mol). Therefore, a one-stage adiabatic reactor is considered herein in view of a facile temperature control that could be envisioned.

For the adiabatic reactor loaded with sphere catalysts (5 mm), Figure 4 shows the contours of bed temperature and partial pressures of EC and EG under three different inlet temperatures. The adiabatic reactor exhibits a plug-flow behavior with minimal radial gradients of velocity, temperature and species concentrations. In the axial (flow) direction, a mild adiabatic temperature rise of c.a. 12 K is observed, which is scarcely affected by the variation of inlet temperatures from 443 to 483 K (Figure 4a−c). With the increase of the inlet temperature, the average bed temperature increases, thus accelerating EC transformation as reflected by the decrease of average P_{EC} inside the catalyst bed. Consistently, in the front part of the catalyst bed, P_{EG}, is found to increase with the inlet temperature. P_{EG} near the reactor outlet, however, is lower for higher inlet temperatures as a result of the intensified secondary EG hydrogenation reaction (R3). This side reaction has a high activation energy of 114 kJ/mol (Table S1) and is sensitive to temperature rise than the main reaction (R1). The consumption of EG in the rear part of the catalyst bed is remarkable when the local bed temperature rises to above 473 K (Figure S3), corresponding to inlet temperatures above 463 K. Therefore, maintaining a moderate bed temperature is necessary to maximize the selectivity to desired alcohol products.

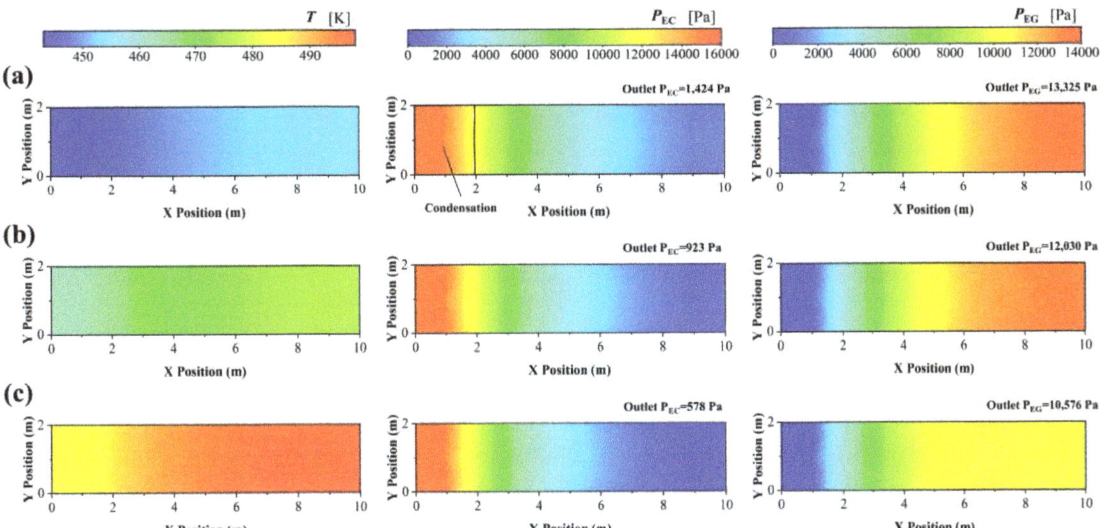

Figure 4. Contours of T, P_{EC} and P_{EG} in the adiabatic reactor with different reactant inlet temperatures of (**a**) 443 K, (**b**) 463 K and (**c**) 483 K under the conditions of 3 MPa, 0.3 $g_{EC} \cdot gcat^{-1} \cdot h^{-1}$ and H_2/EC = 200.

Another necessary practical consideration for reactor operation is that EC is a high boiling-point organic liquid with a dew-point temperature of c.a. 450 K under the inlet condition (Figure S2), which overlaps with the range of operating temperatures. If EC liquefies within the catalyst bed, liquid film would cover the catalyst particle, not only blocking the active sites but also causing coking in the longer term. At the inlet temperature of 443 K, EC condensation is predicted to occur in the first c.a. 20% of the catalyst bed from the inlet (marked by a vertical line in Figure 4a) as judged by the local partial pressure of EC. From the condensation boundary downstream to the reactor outlet, EC remains in the gas phase as a consequence of both the increasing temperature profile of the adiabatic reactor and the decreasing EC partial pressure with the reaction proceeding.

3.2.2. Boiling Water Cooling

Boiling water-cooled fixed-bed reactors often adopt a shell-and-tube structure in which multiple parallel reaction tubes are immersed in flowing boiling water in a shell. To enhance heat transfer from the tube to the coolant, a small tube diameter is chosen. Figure 5 shows the contours of bed temperature and partial pressures of EC and EG within a single reaction tube of the boiling-water cooled reactor under three different coolant temperatures. Within the first 10% of the catalyst bed, the bed temperature approaches the coolant temperature with the temperature gradients in the radial direction reducing from 20 K to zero (Figures 5a–c and S7a,c). No exotherm is observed within the entire catalyst bed in the axial direction for the three coolant temperatures. Because of the relatively low heat release from the reaction and the high convective heat transfer coefficient of boiling water, the coolant temperature rather than the reactant inlet temperature is the determining factor for the bed temperature (Figure S6a). When the coolant temperature increases, the average P_{EC} decreases with the accelerated EC hydrogenation reaction. Different from the adiabatic reactor, an upward trend is observed in P_{EG} with an increasing coolant temperature from 423 K to 463 K because EC conversion to EG is promoted whilst the EG hydrogenation side-reaction remains negligible in the temperature range. For the inlet temperature of 463 K, it is remarkable that the boiling water-cooled reactor gives rise to an EG yield ~10% higher than the adiabatic reactor, with only 12 K difference in the bed temperature.

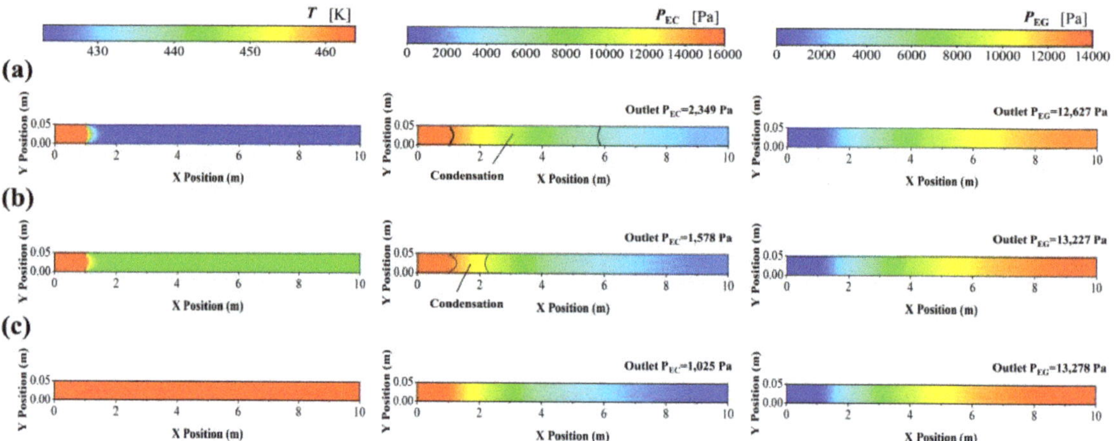

Figure 5. Contours of T, P_{EC} and P_{EG} in boiling water-cooled reactor with different coolant temperatures of (**a**) 423 K, (**b**) 443 K and (**c**) 463 K under the conditions of 3 MPa, 0.3 $g_{EC} \cdot gcat^{-1} \cdot h^{-1}$, $H_2/EC = 200$ and reactant inlet temperature of 463 K.

The lower bed temperatures relative to the adiabatic reactor increase the risk of EC condensation. Figure 5a,b illustrate that the condensation zones are more extended in the near-wall region than around the reactor centerline. Due to an oscillatorily increasing bed porosity towards the reactor wall (Table S5), the EC flowrate increase towards the wall (Figures S4 and S5), contributing to higher local partial pressures of EC. Such a behavior is prominent for the multi-tubular heat-exchange reactors with high tube-to-particle diameter ratios and thus considerable channeling flows near the wall.

3.2.3. Conduction Oil Cooling

In the pilot plant reactor, conduction oil is selected as the coolant in consideration of easiness of control and its satisfactory performance of heat removal. The heat transfer coefficients of conduction oils are significantly lower than the boiling water and depend on geometrical and operational parameters such as the oil flow rates, the distance between adjacent baffle plates, the type of baffle, etc. The specific relationships are given in the Supporting Information (Section S3.2). In general, conduction oil is not as effective as boiling water in controlling the reactor temperature, especially under low oil flow rates. Taking 25% open baffle, 1 m distance between adjacent baffle plates and the 200,000 kg·h^{-1} mass flow rate as the reference case, the resulting heat transfer coefficient is calculated to be 619 W·m^{-2}·K^{-1}. These values will be used in the following study.

Figure 6 shows that the bed temperature drops quickly close to the inlet of the oil-cooled reactor as a consequence of the low inlet temperature of the conduction oil. Then, with the oil being heated up continuously by the exothermic reaction, the bed temperature gradually rises towards the reactor outlet (Figure 6a–c and Figure S7a). The radial temperature gradient in the catalyst bed reduces from 9 K at the inlet to zero after 10% of the catalyst bed. Distinct from that of the boiling water-cooled reactor, the bed temperature of the oil-cooled reactor is also affected by the inlet temperature of reactants: both the increase of reactant inlet temperature and the coolant inlet temperature decrease the EG selectivity (Figure S6b). Different also from both the adiabatic and the boiling water-cooled reactors is that P_{EG} increases slightly with the bed temperature for oil inlet temperatures below 463 K, but decreases slightly with the bed temperature when the oil inlet temperature exceeds 463 K. The outlet flow rates of EG are the highest among the studied reactor types for reactant inlet temperatures between 423 and 463 K thanks to the moderate temperature profile, which balances between promoting EC conversion and preventing secondary EG hydrogenation. Meanwhile, the predicted condensation region of EC is significantly smaller than that of the boiling-water cooled reactor, and the condensation only appears under the lowest coolant temperature of 423 K.

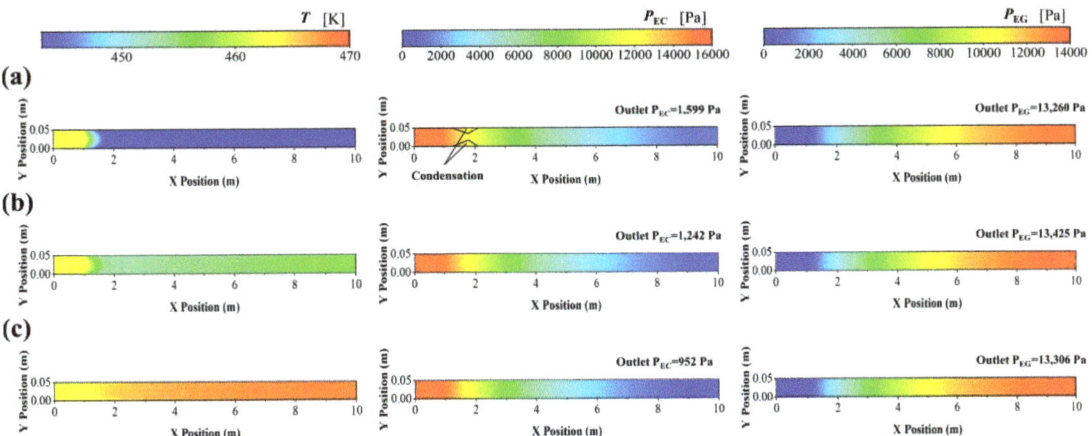

Figure 6. Contours of T, P_{EC} and P_{EG} in oil-cooled reactor with different coolant temperatures of (**a**) 423 K, (**b**) 443 K and (**c**) 463 K under the conditions of 3 MPa, 0.3 g$_{EC}$·gcat^{-1}·h^{-1}, H$_2$/EC = 200 and reactant inlet temperature of 463 K.

3.3. Effects of Key Operating Variables

To elaborate the reactor's operation behaviors, the effects of major operation variables on the reactor performance are further depicted. Five operation variables are studied, including the inlet temperature T_{in}, the coolant temperature T_c, the operating pressure P_{op}, the space velocity SV, and the inlet H_2/EC. The chosen performance metrics are EC conversion X_{EC}, EG selectivity S_{EG}, MeOH selectivity S_{MeOH} and total alcohols selectivity $S_{Alcohol}$.

3.3.1. Temperature

For all reactor types, Figure 7 shows that X_{EC} and S_{MeOH} increase while S_{EG} and $S_{Alcohol}$ decrease with the increase of inlet temperature. At higher temperatures, the faster reaction results in a greater amount of EC being converted (X_{EC} rises). Given that the activation energies of Reaction 1, 2 and 3 (R1)–(R3) are 30.1, 28.1 and 113.8 kJ·mol^{-1}, respectively (Table S1), higher temperatures promote hydrogenation of EG to by-products (S_{EG} decreases) but inhibits EC conversion to CO instead of MeOH (S_{MeOH} increases). Comparison among different reactor types (Figure 7a–d) show slight differences in the EC conversion. X_{EC} of the adiabatic reactor is higher than that of the other two types of reactor owing to its high average temperature of the catalyst bed. X_{EC} of the boiling water-cooled reactor is the lowest among the three types of reactor, but the difference from that of the oil-cooled reactor diminishes as the coolant temperature is raised above 453 K. Since the reactant inlet temperature also affects the performance of the oil-cooled reactor, Figure 7d additionally illustrates the effect of reactant inlet temperature on X_{EC}, S_{EG}, S_{MeOH} and $S_{Alcohol}$. The variations in the reactant inlet temperature and the oil temperature exhibit similar influences on the reactor performance.

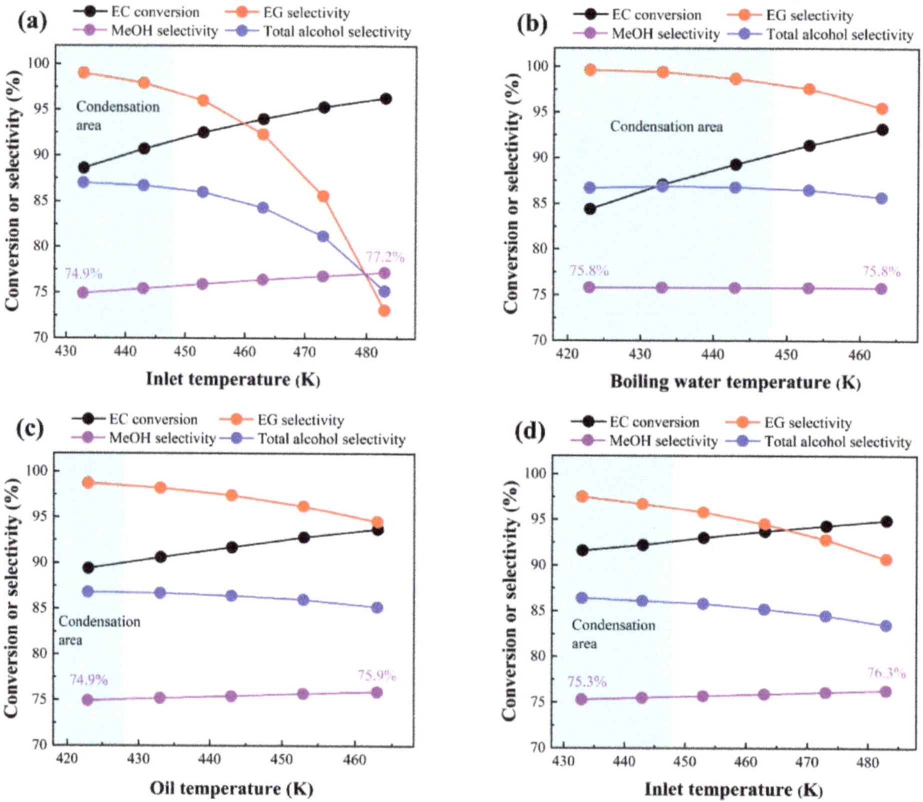

Figure 7. Reactor performance under different reactant/coolant inlet temperatures: (**a**) adiabatic reactor with T_{in} = 433–483 K, (**b**) boiling water-cooled reactor with T_{in} = 463 K and T_c = 423–463 K, (**c**) conduction oil-cooled reactor with T_{in} = 463 K and T_c = 423–463 K, (**d**) conduction oil-cooled reactor with T_c = 463 K and T_{in} = 433–483 K. Operating conditions: 3 MPa, 0.3 g$_{EC}$·gcat^{-1}·h^{-1}, H_2/EC = 200.

EC condensation occurs in the boiling water-cooled reactor when the boiling water temperature is 448 K (Figure 7b). This critical temperature is consistent with the reactant inlet temperature leading to condensation in the adiabatic and oil-cooled reactors (Figure 7a,d) because their inlet regions are the most prone to condensation. For the oil-cooled reactor, moderate heat removal by the conduction oil combined with its relatively low heat capacity yields a bed temperature higher than the initial oil temperature. Therefore, condensation with a fixed reactant inlet temperature of 463 K does not occur until the oil temperature is lowered to 428 K (Figure 7c).

3.3.2. H_2/EC

In this work, a higher H_2/EC means lower EC partial pressures under almost the same H_2 partial pressure (the variation in the H_2 partial pressure is negligible for $H_2/EC > 80$). As shown in Figure 8a–c, higher H_2/EC inhibits EC conversion because the EC consumption reactions (R1 and R2) are positive-order with respect to EC. For the adiabatic reactor, the MeOH selectivity increases slightly with H_2/EC (Figure 8a). The reason is that S_{MeOH} is related to parallel reactions R1 and R2, which are 0.65-order and 1-order with respect to EC, respectively. The lower the EC partial pressure, the lower the relative rate of Reaction 1 to Reaction 2, and therefore the higher the MeOH selectivity. The EG selectivity increases more significantly with H_2/EC, because enhanced axial convective heat transfer by the gas mixture under high H_2/EC alleviates the adiabatic rise of bed temperature, which favors EG selectivity. For the boiling water-cooled reactor, the variation of S_{MeOH} with H_2/EC in Figure 8b resembles that for the adiabatic reactor. However, S_{EG} and $S_{Alcohol}$ of the water-cooled reactor only increase by 1% and 3%, respectively, for H_2/EC from 80 to 200, because the catalyst bed is close to isothermal. In Figure 8c, almost identical behavior is observed for the oil-cooled reactor as the boiling water-cooled reactor, except that X_{EC} of the latter is slightly higher for $H_2/EC > 160$, and that S_{EG} and $S_{Alcohol}$ of the latter is slightly lower for $H_2/EC < 120$. EC condensation happens when inlet H_2/EC is smaller than 120 regardless of the type of the reactor, provided that EC most likely liquefies at the reactor inlet when its partial pressure is the highest.

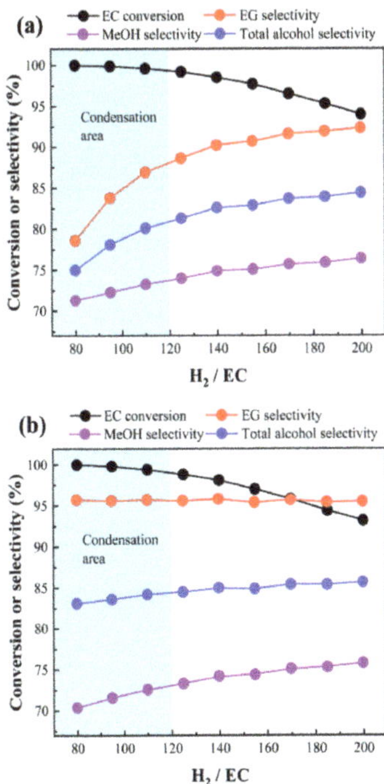

Figure 8. *Cont.*

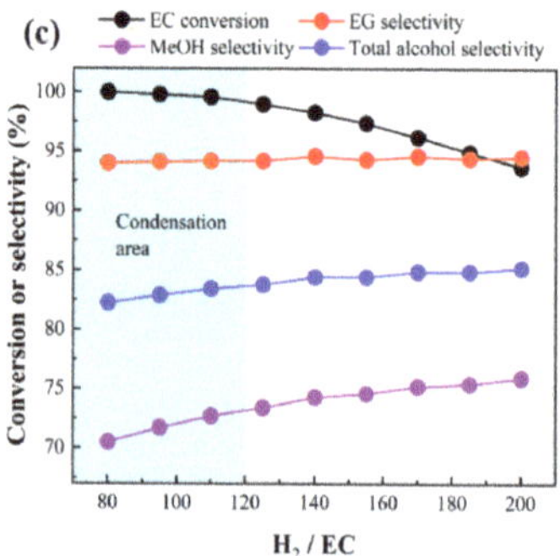

Figure 8. Reactor performance under different H_2/EC: (**a**) adiabatic reactor, (**b**) boiling water-cooled reactor, (**c**) conduction oil-cooled reactor. Operating conditions: reactant and coolant inlet temperatures 463 K, 3 MPa, 0.3 $g_{EC}{\cdot}g_{cat}^{-1}{\cdot}h^{-1}$, 80–200 H_2/EC.

3.3.3. Pressure and Space Velocity

Pressure and space velocity have a great impact on the catalytic reaction but a minor influence on the transport characteristics. The influence of pressure and space velocity on the reactor performance is exemplified by the case of the boiling water-cooled reactor in view of the consistent behaviors among the different types of reactors. Figure 9a shows that the increase of operating pressures accelerates EC hydrogenation as the partial pressure of EC is increased (R1) and (R3). Meanwhile, since R1 and R2 are boosted at higher pressure while remaining unaffected, an increase in EG selectivity is observed. The MeOH selectivity is determined by the relative rates between R1 and R2. With a higher total pressure and EC partial pressure, the rate of R1 increases less than that of R2 (EC orders 0.65 and 1), resulting in lower MeOH selectivity. Although a higher operating pressure favors both EC conversion and the selectivity of EG—the more value-added product—it may also lead to EC condensation when exceeding 4.8 MPa. The optimal operating pressure is around 3 MPa when all performance metrics achieves a reasonable trade-off.

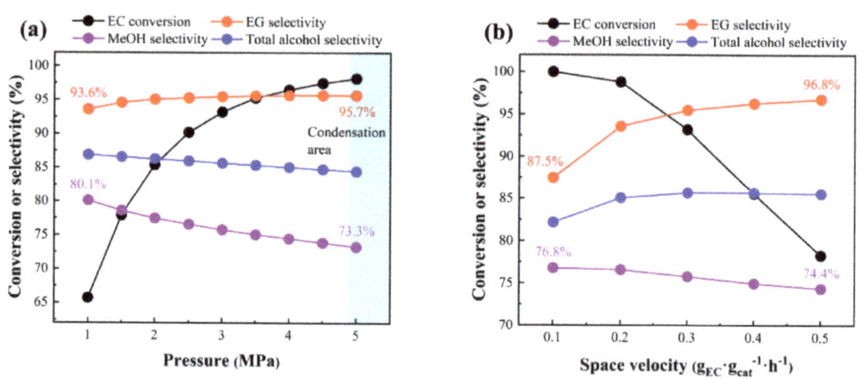

Figure 9. Reactor performance of the boiling water-cooled reactor under the conditions of (**a**) 1–5 MPa, 0.3 $g_{EC}{\cdot}g_{cat}^{-1}{\cdot}h^{-1}$, (**b**) 3 MPa, 0.1–0.5 $g_{EC}{\cdot}g_{cat}^{-1}{\cdot}h^{-1}$ (T_{in} = 463 K, T_c = 463 K and H_2/EC = 200).

Figure 9b depicts the influence of space velocity on the reactor performance. Lower space velocity is beneficial for both X_{EC} and S_{MeOH}. To ensure a conversion over 90%, the space velocity should be maintained below 0.3 $g_{EC}{\cdot}g_{cat}^{-1}{\cdot}h^{-1}$. The slightly higher MeOH selectivity at low SV and high X_{EC} originates from the fact that the main reaction R1 has a smaller reaction order of EC than the parallel side-reaction R2 (i.e., R1 decays slower at high EC conversion than R2). In contrast to MeOH, the EG selectivity is boosted at high SV when the secondary EG hydrogenation reaction proceeds inadequately. Since high space velocity results in too large bed pressure drop (Figure S5b), a space velocity of around 0.3 $g_{EC} \cdot gcat^{-1} \cdot h^{-1}$ is deemed feasible with balanced operational cost and reaction performance.

The specific effects of key operating variables T, P_{op}, SV and H_2/EC on the reactor performance are summarized in the Table S7.

3.4. Operation Windows

Informed by the preceding illustrations, the different types of reactors response differently to variations of operating parameters. The major variables that affect the reactor performance and could be frequently changed in industrial practice include the inlet temperature of reactants and coolant (T), the operating pressure (P_{op}), and space velocity (SV). Therefore, the T, P_{op} and SV values required to reach certain X_{EC}, S_{EG} and $S_{Alcohol}$ are further calculated for the different types of reactors to shed light on their respective operating windows.

Figure 10 shows the demanded temperature and operating pressure for X_{EC}, S_{EG} and $S_{Alcohol}$ to reach 90/95/98%, respectively. In brief, the contour lines for $X_{EC} = 90\%$ and 98% indicate that higher temperatures are required should the EC conversion be kept constant at decreasing operating pressures. The contour lines for $S_{EG} = 90\%$ and 95% indicate that the operating pressure has negligible effects on S_{EG}, which only decreases with the increase of temperature. The 85% $S_{Alcohol}$ contour line shows a downward trend with temperature for $P_{op} > 3.5$ MPa. This is because higher pressures decrease S_{MeOH} with S_{EG} unaffected; thus, lower temperatures are demanded to compensate for the decrease in $S_{Alcohol}$. Additionally, EC condensation restrains the allowable scope of the operation variables; the condensation regimes for the different types of reactors are marked in cyan in Figure 10. As P_{op} increases, the dewpoint temperature of EC rises, thus enlarging the condensation regime. In Figure 10b,c, EC condensation is unavoidable regardless of the coolant temperature for $P_{op} > 4.9$ MPa because the reactants condensate at the reactor inlet with the given inlet temperature of 463 K.

For all types of reactors under investigation, the contour lines divide the reactor's operating window (T and P_{op}) into different regimes with distinct reactor performance. For instance, the red, triangular zone in Figure 10a denotes the T_{in} and P_{op} window for the adiabatic reactor to achieve $X_{EC} > 90\%$ and $S_{EG} > 95\%$ under the SV of 0.3 $g_{EC} \cdot gcat^{-1} \cdot h^{-1}$ and inlet H_2/EC of 200. Figure 10a also indicates that the adiabatic reactor is not operable if $X_{EC} > 98\%$ and $S_{EG} > 95\%$ are desired, as the corresponding regimes would overlap inside the EC condensation regime with pressures over 5 MPa and inlet temperature below 450 K. It can be concluded that the adiabatic reactor is difficult to operate under pressures >4 MPa owing to a very narrow range of feasible inlet temperatures for adequate product yields. In contrast to the adiabatic reactor, Figure 10b−d demonstrate that the boiling water and conduction oil-cooled reactors exhibit wider operating windows of the inlet/coolant temperatures under higher pressures. The reason is twofold: First, heat removal uplifts the contour lines of S_{EG} in the case of boiling water cooling and in the case of oil cooling with varying inlet temperatures (Figure 10b,d). Second, the ascending temperature profile along the oil-cooled reactor drastically lowers the EC condensation line on the oil inlet temperature-reactor pressure diagram (Figure 10c). The reactors with heat exchange would allow $X_{EC} > 98\%$ and $S_{EG} > 95\%$ with the demanded T_c and P_{op} being 463–465 K and 4.8–4.9 MPa for the boiling water-cooled reactor (green zone, Figure 10b), and 458–461 K and 4.8–4.9 MPa for $T_{in} = 463$ K, or 466–474 K and 4.8–5.0 MPa for $T_c = 453$ K, respectively, for the conduction oil-cooled reactor (Figure 10c,d). In all, the conduction oil-cooled reactor demonstrates the best operability in terms of allowable reactant/coolant inlet temperatures, especially under lower pressures.

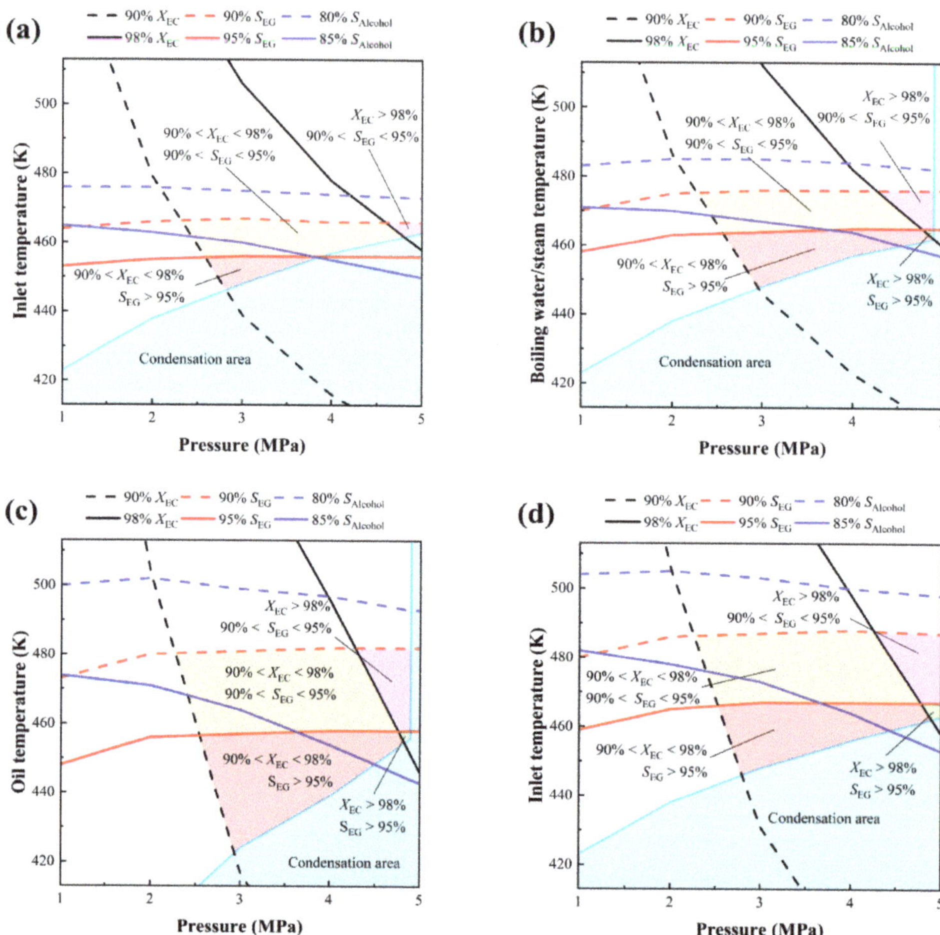

Figure 10. Operation windows of reactors for demanded T_{in}/T_c -P_{op} parameters on outlet EC conversion, EG selectivity and total alcohol selectivity under $SV = 0.3$ $g_{EC} \cdot g_{cat}^{-1} \cdot h^{-1}$ and $H_2/EC = 200$. Reactor type: (**a**) adiabatic, (**b**) boiling water-cooled, (**c**) oil-cooled, $T_{in} = 463$ K, (**d**) oil-cooled, $T_c = 453$ K.

As the production capacity of reactors might be varied during industrial operation, we further investigated the reactors' operating windows with respect to various T_{in}/T_c and SV in with P_{op} and H_2/EC set to 3 MPa and 200, respectively. Figure 11 shows that increasing temperature with the space velocity is required in order to maintain certain EC conversions, and in contrast, the maximum temperature limits to keep the EG selectivity and total alcohols selectivity above thresholds are uplifted under a higher SV. An exception is found for the oil-cooled reactor with varying inlet temperatures, where the maximum inlet temperature allowable for 85% total alcohols selectivity decreases again with the increase of space velocity at above c.a. 0.4 $g_{EC} \cdot g_{cat}^{-1} \cdot h^{-1}$. The EC condensation regime remains invariant with the space velocity for the adiabatic and boiling water-cooled reactor, as well as for the oil-cooled reactor with varying reactant inlet temperatures. In these cases, whether EC condensates in the catalyst bed is determined by the inlet conditions. For the oil-cooled reactor with varying oil inlet temperatures, the EC condensation regime shifts to lower temperatures with increasing space velocity, under which conditions the initial sink of bed temperature due to cooling is less prominent.

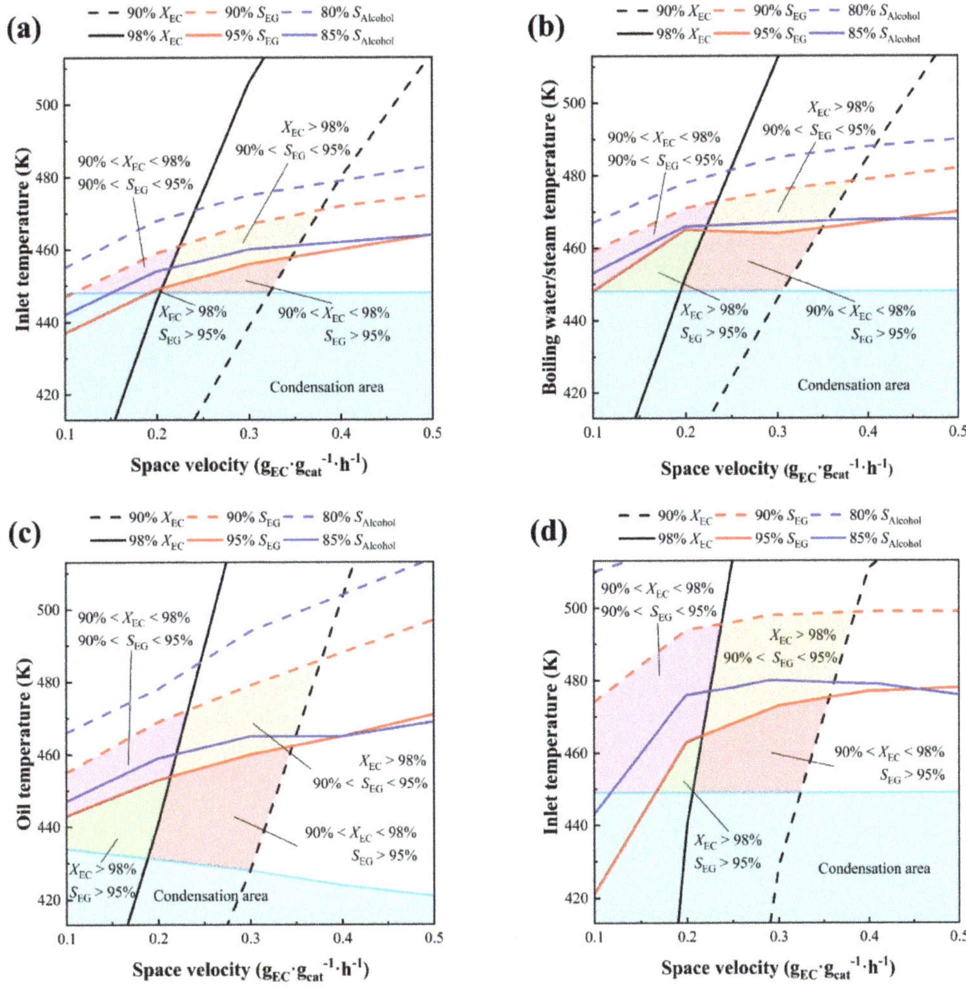

Figure 11. Operation windows of reactors for demanded T_{in}/T_c -SV parameters on outlet EC conversion, EG selectivity and total alcohol selectivity under 3 MPa and 200 H_2/EC. Reactor type: (**a**) adiabatic, (**b**) boiling water-cooled, (**c**) oil-cooled, T_{in} = 463 K, (**d**) oil-cooled, T_c = 453 K.

A very small operation window for $X_{EC} > 98\%$ and $S_{EG} > 95\%$ (green zone; T_{in} of 448–449 K, SV of 0.19–0.20 $g_{EC} \cdot g_{cat}^{-1} \cdot h^{-1}$) is shown for the adiabatic reactor in Figure 11a. However, common fluctuations in industry makes it difficult to keep T_{in} and SV static in the range. For the boiling water-cooled reactor, the corresponding ranges of T_c and SV are c.a. 448–463 K and 0.10–0.22 $g_{EC} \cdot g_{cat}^{-1} \cdot h^{-1}$, respectively, while for the conduction oil-cooled reactor with varying oil inlet temperatures, the ranges of T_c and SV are c.a. 432–454 K and 0.10–0.21 $g_{EC} \cdot g_{cat}^{-1} \cdot h^{-1}$. As in Figure 11b–d, the larger T_c-SV window of the boiling water-cooled reactor benefits from an isothermal temperature profile that avoids deteriorating S_{EG} at higher water temperatures, while that of the conduction oil-cooled reactor from a wide non-condensable regime. To sum up, the conduction oil-cooled reactor has the greatest allowable scope of space velocity for a wide range of reactant/coolant inlet temperatures.

4. Conclusions

In this work, multiscale reactor models for heterogenous EC hydrogenation to co-produce MeOH and EG in industrial-type adiabatic, water-cooled and oil-cooled tubular fixed-bed reactors are established and validated with bench-scale and pilot plant data. The main and side reactions occurring during the heterogeneous EC hydrogenation over Cu-based catalysts are described by a power-law engineering kinetics model involving three independent reactions.

EC hydrogenation under typical operating conditions in the adiabatic reactor renders a mild temperature rise of c.a. 12 K. However, bed temperatures above 473 K would significantly reduce the selectivity to the primary product EG due to its secondary hydrogenation, which is more temperature-sensitive than the main EC hydrogenation reaction. If the inlet temperature is lower than 453 K, however, EC condensation might happen and deactivate the catalyst, especially under low H_2/EC and high operating pressures. The boiling water-cooled reactor behaves close to isothermally regardless of the reactant inlet temperature, with temperature gradients only existing in the first 10% of the catalyst bed, which is beneficial for the improving the yield of EG. The conduction oil-cooled reactor shows a minimum bed temperature near the bed entrance as the cold oil gets constantly heated up towards the bed outlet. Such a U-shape temperature profile allows a relatively wide scope of both the reactant inlet temperature and the oil inlet temperature with adequate reactor performance.

Model-based operational analysis of the three different types of reactors further suggests that the application of the adiabatic reactor in EC hydrogenation is restrained by a very narrow operating window of the inlet temperature, especially under higher pressures and space velocities, if practical EC conversion and alcohols selectivity are to be acquired within the non-condensable regime of EC. The boiling water-cooled reactor exhibits no restraint on the reactant inlet temperature and a relatively wide window of the coolant temperature under different pressures and space velocities. The conduction oil-cooled reactor has a sufficiently wide window of the reactant inlet temperature and a larger operating window of the coolant temperature than the water-cooled reactor. This enables reactor operation under higher pressures and space velocities, thus providing a greater production flexibility. Understanding of these operational characteristics of representative industrial-type reactors for EC hydrogenation not only reveals the key in reactor design, but will also pave the way to further process optimization.

Supplementary Materials: The following supporting information can be downloaded at: https://www.mdpi.com/article/10.3390/pr10040688/s1, Figure S1: Model and experiment values deviation of EC hydrogenation reaction; Figure S2: EC condensation zone under the condition of T = 410–530 K, P_{op} = 1–5 MPa and H_2/EC = 40–200; Figure S3: Contours of P_{EG} in different heat-exchange-type reactors: (a) (d) (g) adiabatic, (b) (e) (h) boiling water-cooled, (c) (f) (i) oil-cooled, under the conditions of 3 MPa, 0.3 $g_{EC} \cdot g_{cat}^{-1} \cdot h^{-1}$, 200 H_2/EC and inlet/coolant temperatures of (a) (d) (g) inlet temperatures 443, 463, 483 K, (b)/(c) (e)/(f) (h)/(i) inlet temperature 463 K, coolant temperatures 423, 443, 463 K; Figure S4: Contours of u_0 in different heat exchange type reactor: (a) adiabatic (b) boiling water-cooled (c) oil-cooled under the conditions of 3 MPa, 0.3 $g_{EC} \cdot g_{cat}^{-1} \cdot h^{-1}$, 200 H_2/EC, 463 K inlet temperature and 463 K coolant temperature; Figure S5: The influence of SV on (a) superficial velocity and (b) bed pressure drop, in boiling water-cooled reactor under the conditions of 3 MPa, 0.3 $g_{EC} \cdot g_{cat}^{-1} \cdot h^{-1}$ and 200 H_2/EC; Figure S6: The influence of reactant/coolant inlet temperatures on S_{EG} under the conditions of 3 MPa, 0.3 $g_{EC} \cdot g_{cat}^{-1} \cdot h^{-1}$ and 200 H_2/EC: (a) boiling water-cooled, (b) conduction oil-cooled reactor; Figure S7: Under the conditions of 3 MPa, 0.3 $g_{EC} \cdot g_{cat}^{-1} \cdot h^{-1}$ and 200 H_2/EC, (a) bed and coolant temperatures in the axial direction; bed temperatures in the radial direction: (b) adiabatic, T_{in} = 443 K, (c) boiling water-cooled, T_{in} = 463 K and T_c = 443 K (d) conduction oil-cooled reactors, T_{in} = 463 K and T_c = 423 K; Table S1: Intrinsic kinetic parameters; Table S2: Specific heat capacity; Table S3: Thermal conductivity; Table S4: Viscosity; Table S5: Bed voidage; Table S6: Influence of catalyst sizes and shapes on reactor performance; Table S7: Effect of key operating variables on X_{EC}, S_{EG}, S_{MeOH}. Refs. [62,64,65] are cited in supplementary materials.

Author Contributions: Conceptualization, C.C.; Funding acquisition, Y.W., J.L. and J.X.; Investigation, H.H., C.C. and Y.Y.; Methodology, C.C. and Y.W.; Project administration, Y.W., J.L. and J.X.; Supervision, C.C. and J.X.; Writing—original draft preparation, H.H.; Writing—review and editing, C.C., Y.W., J.L. and J.X. All authors have read and agreed to the published version of the manuscript.

Funding: This research was funded by the National Natural Science Fund for Distinguished Young Scholars (61725301), International (Regional) Cooperation and Exchange Project (61720106008), National Natural Science Foundation of China (21878080, 61973124), and the Dean/Opening Project of Guangxi Key Laboratory of Petrochemical Resource Processing and Process Intensification Technology.

Data Availability Statement: Data available on request due to restrictions e.g., privacy or ethical.

Conflicts of Interest: The authors declare no conflict of interest.

Notations

a	Surface area of particles per unit volume (m^{-1})
$c_{p,\mathrm{f}}$	Specific heat capacity of fluid ($\mathrm{J \cdot kg^{-1} \cdot K^{-1}}$)
$c_{p,\mathrm{oil}}$	Specific heat capacity of conduction oil ($\mathrm{J \cdot kg^{-1} \cdot K^{-1}}$)
C_2	Inertial loss coefficient (m^{-1})
$C_{\mathrm{EC,center}}$	Concentration of EC in the center of catalyst ($\mathrm{kmol \cdot m^{-3}}$)
$C_{\mathrm{EC,gas}}$	Concentration of EC in the gas phase ($\mathrm{kmol \cdot m^{-3}}$)
$C_{\mathrm{EC,s}}$	Concentration of EC on the surface of catalyst ($\mathrm{kmol \cdot m^{-3}}$)
d_{p}	Diameter of sphere (m)
$d_{\mathrm{p}}^{\mathrm{a}}$	Diameter of sphere with equal specific surface area (m)
$d_{\mathrm{p}}^{\mathrm{s}}$	Diameter of sphere with equal surface area (m)
$d_{\mathrm{p}}^{\mathrm{v}}$	Diameter of sphere with equal volume (m)
D	Bed or tube diameter (m)
$D_{\mathrm{e,ax}}$	Effective axis diffusion coefficient ($\mathrm{m^2 \cdot s^{-1}}$)
$D_{\mathrm{e,r}}$	Effective radial diffusion coefficient ($\mathrm{m^2 \cdot s^{-1}}$)
$D_{\mathrm{eff,\ EC}}$	Effective diffusion coefficient of EC in the catalyst particle ($\mathrm{m^2 \cdot s^{-1}}$)
$D_{\mathrm{EC,m}}$	Molecular diffusivity of EC ($\mathrm{m^2 \cdot s^{-1}}$)
$D_{T,i}$	Thermal diffusion coefficient ($\mathrm{kg \cdot m^{-1} \cdot s^{-1}}$)
E_{f}	Total fluid energy ($\mathrm{m^2 \cdot s^{-2}}$)
E_{s}	Total solid energy ($\mathrm{m^2 \cdot s^{-2}}$)
h_j^0	Enthalpy of formation of species j ($\mathrm{J \cdot kmol^{-1}}$)
$\mathrm{H_2/EC}$	Molar ratio of EC to $\mathrm{H_2}$
J_i	Diffusion flux of species i vector ($\mathrm{kg \cdot m^{-2} \cdot s^{-1}}$)
k	Turbulence kinetic energy
$k_{\mathrm{e,ax}}$	Effective axial thermal conductivity ($\mathrm{W \cdot m^{-1} \cdot K^{-1}}$)
$k_{\mathrm{e,r}}$	Effective radial thermal conductivity ($\mathrm{W \cdot m^{-1} \cdot K^{-1}}$)
k_{f}	Thermal conductivity of fluid ($\mathrm{W \cdot m^{-1} \cdot K^{-1}}$)
k_{g}	Gas-solid mass transfer coefficient ($\mathrm{m \cdot s^{-1}}$)
k_{s}	Thermal conductivity of catalyst particle ($\mathrm{W \cdot m^{-1} \cdot K^{-1}}$)
k_{v}	Pre-exponential factor
L	Bed length (m)
$\dot{m}_{\mathrm{EC,in}}$	EC inlet mass flow ($\mathrm{kg \cdot h^{-1}}$)
$\dot{m}_{\mathrm{EC,out}}$	EC outlet mass flow ($\mathrm{kg \cdot h^{-1}}$)
$\dot{m}_{\mathrm{EG,out}}$	EG outlet mass flow ($\mathrm{kg \cdot h^{-1}}$)
$\dot{m}_{\mathrm{MeOH,out}}$	MeOH outlet mass flow ($\mathrm{kg \cdot h^{-1}}$)
$\dot{m}_{\mathrm{oil}}$	Conduction oil mass flow ($\mathrm{kg \cdot h^{-1}}$)
M_{w}	Molecular weight ($\mathrm{kg \cdot kmol^{-1}}$)
n	Reaction order
Nu	Nusselt number
p	Static pressure (Pa)
P_{EC}	EC partial pressure (Pa)
P_{EG}	EG partial pressure (Pa)

P_{op}	Operating pressure (MPa)
Pe_h^0	Fluid Peclet number for heat transfer
$Pe_{h,ax}$	Peclet number for axial heat conduction
$Pe_{h,r}^\infty$	Peclet radial heat transfer for fully developed turbulence flow
$Pe_{m,ax}$	Peclet number for axial mass dispersion
$Pe_{m,r}$	Peclet number for radial mass dispersion
Pr	Prandtl number
Q_{oil}	Heat flux to oil (W)
$Q_{reaction}$	Reaction heat (W)
r_{chem}	Intrinsic reaction rate (kmol·m^{-3}·s^{-1})
r_{obs}	Effective reaction rate (kmol·m^{-3}·s^{-1})
R	Ratio of tube diameter to catalyst's volume-equivalent diameter
$R_{EC,s}$	Effective consumption rate of EC at the particle surface (kmol·m^{-3}·s^{-1})
R_j	Volumetric rate of creation of species j (kmol·m^{-3}·s^{-1})
Re	Reynolds number based on particle diameter
$S_{Alcohol}$	Total alcohol selectivity
S_{cat}	Surface area of catalyst particle (m^2)
S_{EG}	EG selectivity
S_f^h	Energy source term (W·m^{-3})
S_i	Mass source term (kg·m^{-3}·s^{-1})
S_{MeOH}	MeOH selectivity
S_{wall}	Surface area of wall (m^2)
S_ϕ	Momentum source term (kg·m^{-2}·s^{-2})
Sc	Schmidt number
Sh	Sherwood number
SV	Space velocity (g$_{EC}$·g$_{cat}^{-1}$·h^{-1})
t	Time (s) or adjacent baffle plate space (m)
T	Temperature (K)
T_b	Boiling point of pressurized water (K)
T_c	Coolant temperature (K)
T_{in}	Reactant inlet temperature (K)
T_{oil}	Conduction oil temperature (K)
T_s	Catalyst surface temperature (K)
T_w	Wall temperature (K)
u_0	Superficial velocity (m·s^{-1})
v	Fluid flow velocity vector (m·s^{-1})
v_{cat}	Volume of catalyst particle (m^3)
x	X position along the bed axial direction (m)
X_{EC}	EC conversion
X_i	Mass fraction of species i
Y_i	Mole fraction of species i
Greek letters	
α_f	Gas-solid heat transfer coefficient (W·m^{-2}·K^{-1})
$\Delta_r H$	Enthalpy of reaction (J·kmol^{-1})
ΔP	Bed pressure drop (Pa)
ΔT_{ex}	Temperature difference between catalyst surface and gas phase (K)
ΔT_{in}	Temperature difference between catalyst surface and center (K)
ε_b	Bed voidage
ε_{cat}	Internal porosity of catalyst particle
η_{EC}	Effectiveness factor for EC internal mass transfer
$\lambda_{eff,cat}$	Effective thermal conductivity of catalyst particle (W·m^{-1}·K^{-1})
μ	Fluid viscosity (Pa·s)
ρ_f	Fluid density (kg·m^{-3})
τ	Stress tensor (Pa)
$\phi_{gen,EC}$	Generalized Thiele modulus of EC
ω	Specific dissipation rate

References

1. Hepburn, C.; Adlen, E.; Beddington, J.; Carter, E.A.; Fuss, S.; Mac Dowell, N.; Minx, J.C.; Smith, P.; Williams, C.K. The technological and economic prospects for CO_2 utilization and removal. *Nature* **2019**, *575*, 87–97. [CrossRef]
2. Zhang, Z.; Pan, S.; Li, H.; Cai, J.; Olabi, A.G.; Anthony, E.J.; Manovic, V. Recent advances in carbon dioxide utilization. *Renew. Sustain. Energy Rev.* **2020**, *125*, 109799. [CrossRef]
3. Ronda-Lloret, M.; Rothenberg, G.; Shiju, N.R. A Critical Look at Direct Catalytic Hydrogenation of Carbon Dioxide to Olefins. *ChemSusChem* **2019**, *12*, 3896–3914. [CrossRef] [PubMed]
4. Guo, L.; Sun, J.; Ge, Q.; Tsubaki, N. Recent advances in direct catalytic hydrogenation of carbon dioxide to valuable C_{2+} hydrocarbons. *J. Mater. Chem. A* **2018**, *6*, 23244–23262. [CrossRef]
5. Yang, H.; Zhang, C.; Gao, P.; Wang, H.; Li, X.; Zhong, L.; Wei, W.; Sun, Y. A review of the catalytic hydrogenation of carbon dioxide into value-added hydrocarbons. *Catal. Sci. Technol.* **2017**, *7*, 4580–4598. [CrossRef]
6. Ye, K.; Zhou, Z.; Shao, J.; Lin, L.; Gao, D.; Ta, N.; Si, R.; Wang, G.; Bao, X. In Situ Reconstruction of a Hierarchical Sn-Cu/SnO$_x$ Core/Shell Catalyst for High-Performance CO_2 Electroreduction. *Angew. Chem. Int. Ed.* **2020**, *59*, 4814–4821. [CrossRef]
7. Wu, C.; Lin, L.; Liu, J.; Zhang, J.; Zhang, F.; Zhou, T.; Rui, N.; Yao, S.; Deng, Y.; Yang, F.; et al. Inverse ZrO_2/Cu as a highly efficient methanol synthesis catalyst from CO_2 hydrogenation. *Nat. Commun.* **2020**, *11*, 5767. [CrossRef]
8. Pedersen, J.K.; Batchelor, T.A.A.; Bagger, A.; Rossmeisl, J. High-Entropy Alloys as Catalysts for the CO_2 and CO Reduction Reactions. *ACS Catal.* **2020**, *10*, 2169–2176. [CrossRef]
9. Jiang, Z.; Wang, T.; Pei, J.; Shang, H.; Zhou, D.; Li, H.; Dong, J.; Wang, Y.; Cao, R.; Zhuang, Z.; et al. Discovery of main group single Sb–N_4 active sites for CO_2 electroreduction to formate with high efficiency. *Energy Environ. Sci.* **2020**, *13*, 2856–2863. [CrossRef]
10. Hu, J.; Yu, L.; Deng, J.; Wang, Y.; Cheng, K.; Ma, C.; Zhang, Q.; Wen, W.; Yu, S.; Pan, Y.; et al. Sulfur vacancy-rich MoS2 as a catalyst for the hydrogenation of CO_2 to methanol. *Nat. Catal.* **2021**, *4*, 242–250. [CrossRef]
11. Wang, J.; Li, G.; Li, Z.; Tang, C.; Feng, Z.; An, H.; Liu, H.; Liu, T.; Li, C. A highly selective and stable ZnO-ZrO_2 solid solution catalyst for CO_2 hydrogenation to methanol. *Sci. Adv.* **2017**, *3*, e1701290. [CrossRef] [PubMed]
12. Sen, R.; Goeppert, A.; Kar, S.; Prakash, G.K.S. Hydroxide Based Integrated CO_2 Capture from Air and Conversion to Methanol. *J. Am. Chem. Soc.* **2020**, *142*, 4544–4549. [CrossRef] [PubMed]
13. Hartadi, Y.; Widmann, D.; Behm, R.J. CO_2 Hydrogenation to Methanol on Supported Au Catalysts under Moderate Reaction Conditions: Support and Particle Size Effects. *ChemSusChem* **2015**, *8*, 456–465. [CrossRef] [PubMed]
14. Frei, M.S.; Mondelli, C.; Cesarini, A.; Krumeich, F.; Hauert, R.; Stewart, J.A.; Curulla Ferré, D.; Pérez-Ramírez, J. Role of Zirconia in Indium Oxide-Catalyzed CO_2 Hydrogenation to Methanol. *ACS Catal.* **2020**, *10*, 1133–1145. [CrossRef]
15. Dang, S.; Qin, B.; Yang, Y.; Wang, H.; Cai, J.; Han, Y.; Li, S.; Gao, P.; Sun, Y. Rationally designed indium oxide catalysts for CO_2 hydrogenation to methanol with high activity and selectivity. *Sci. Adv.* **2020**, *6*, eaaz2060. [CrossRef] [PubMed]
16. Wang, W.; Wang, S.; Ma, X.; Gong, J. Recent advances in catalytic hydrogenation of carbon dioxide. *Chem. Soc. Rev.* **2011**, *40*, 3703–3727. [CrossRef]
17. Wang, S.; Zhao, L.; Wang, W.; Zhao, Y.; Zhang, G.; Ma, X.; Gong, J. Morphology control of ceria nanocrystals for catalytic conversion of CO_2 with methanol. *Nanoscale* **2013**, *5*, 5582–5588. [CrossRef]
18. Song, Q.; Zhou, Z.; He, L. Efficient, selective and sustainable catalysis of carbon dioxide. *Green Chem.* **2017**, *19*, 3707–3728. [CrossRef]
19. Tamura, M.; Kitanaka, T.; Nakagawa, Y.; Tomishige, K. Cu Sub-Nanoparticles on Cu/CeO_2 as an Effective Catalyst for Methanol Synthesis from Organic Carbonate by Hydrogenation. *ACS Catal.* **2016**, *6*, 376–380. [CrossRef]
20. Li, Y.; Junge, K.; Beller, M. Improving the Efficiency of the Hydrogenation of Carbonates and Carbon Dioxide to Methanol. *ChemCatChem* **2013**, *5*, 1072–1074. [CrossRef]
21. Du, X.; Sun, X.; Jin, C.; Jiang, Z.; Su, D.; Wang, J. Efficient Hydrogenation of Alkyl Formate to Methanol over Nanocomposite Copper/Alumina Catalysts. *ChemCatChem* **2014**, *6*, 3075–3079. [CrossRef]
22. Balaraman, E.; Gunanathan, C.; Zhang, J.; Shimon, L.J.W.; Milstein, D. Efficient hydrogenation of organic carbonates, carbamates and formates indicates alternative routes to methanol based on CO_2 and CO. *Nat. Chem.* **2011**, *3*, 609–614. [CrossRef] [PubMed]
23. Du, X.; Jiang, Z.; Su, D.; Wang, J. Research Progress on the Indirect Hydrogenation of Carbon Dioxide to Methanol. *ChemSusChem* **2016**, *9*, 322–332. [CrossRef] [PubMed]
24. Yu, B.-Y.; Chen, M.-K.; Chien, I.L. Assessment on CO_2 Utilization through Rigorous Simulation: Converting CO_2 to Dimethyl Carbonate. *Ind. Eng. Chem. Res.* **2018**, *57*, 639–652. [CrossRef]
25. Gu, X.; Zhang, X.; Zhang, X.; Deng, C. Simulation and assessment of manufacturing ethylene carbonate from ethylene oxide in multiple process routes. *Chin. J. Chem. Eng.* **2021**, *31*, 135–144. [CrossRef]
26. Han, Z.; Rong, L.; Wu, J.; Zhang, L.; Wang, Z.; Ding, K. Catalytic Hydrogenation of Cyclic Carbonates: A Practical Approach from CO_2 and Epoxides to Methanol and Diols. *Angew. Chem. Int. Ed.* **2012**, *51*, 13041–13045. [CrossRef]
27. Wu, X.; Ji, L.; Ji, Y.; Elageed, E.H.M.; Gao, G. Hydrogenation of ethylene carbonate catalyzed by lutidine-bridged N-heterocyclic carbene ligands and ruthenium precursors. *Catal. Commun.* **2016**, *85*, 57–60. [CrossRef]
28. Lian, C.; Ren, F.; Liu, Y.; Zhao, G.; Ji, Y.; Rong, H.; Jia, W.; Ma, L.; Lu, H.; Wang, D.; et al. Heterogeneous selective hydrogenation of ethylene carbonate to methanol and ethylene glycol over a copper chromite nanocatalyst. *Chem. Commun.* **2015**, *51*, 1252–1254. [CrossRef]

29. Zhang, M.; Yang, Y.; Li, A.; Yao, D.; Gao, Y.; Fayisa, B.A.; Wang, M.-Y.; Huang, S.; Lv, J.; Wang, Y.; et al. Nanoflower-like Cu/SiO$_2$ Catalyst for Hydrogenation of Ethylene Carbonate to Methanol and Ethylene Glycol: Enriching H$_2$ Adsorption. *ChemCatChem* **2020**, *12*, 3670–3678. [CrossRef]

30. Yang, Y.; Yao, D.; Zhang, M.; Li, A.; Gao, Y.; Fayisa, B.A.; Wang, M.; Huang, S.; Wang, Y.; Ma, X. Efficient hydrogenation of CO$_2$-derived ethylene carbonate to methanol and ethylene glycol over Mo-doped Cu/SiO$_2$ catalyst. *Catal. Today* **2021**, *371*, 113–119. [CrossRef]

31. Ding, Y.; Tian, J.; Chen, W.; Guan, Y.; Xu, H.; Li, X.; Wu, H.; Wu, P. One-pot synthesized core/shell structured zeolite@copper catalysts for selective hydrogenation of ethylene carbonate to methanol and ethylene glycol. *Green Chem.* **2019**, *21*, 5414–5426. [CrossRef]

32. Zhou, M.; Shi, Y.; Ma, K.; Tang, S.; Liu, C.; Yue, H.; Liang, B. Nanoarray Cu/SiO$_2$ Catalysts Embedded in Monolithic Channels for the Stable and Efficient Hydrogenation of CO$_2$-Derived Ethylene Carbonate. *Ind. Eng. Chem. Res* **2018**, *57*, 1924–1934. [CrossRef]

33. Chen, X.; Cui, Y.; Wen, C.; Wang, B.; Dai, W. Continuous synthesis of methanol: Heterogeneous hydrogenation of ethylene carbonate over Cu/HMS catalysts in a fixed bed reactor system. *Chem. Commun.* **2015**, *51*, 13776–13778. [CrossRef] [PubMed]

34. Li, F.; Wang, L.; Han, X.; Cao, Y.; He, P.; Li, H. Selective hydrogenation of ethylene carbonate to methanol and ethylene glycol over Cu/SiO$_2$ catalysts prepared by ammonia evaporation method. *Int. J. Hydrogen Energy* **2017**, *42*, 2144–2156. [CrossRef]

35. Chen, W.; Song, T.; Tian, J.; Wu, P.; Li, X. An efficient Cu-based catalyst for the hydrogenation of ethylene carbonate to ethylene glycol and methanol. *Catal. Sci. Technol.* **2019**, *9*, 6749–6759. [CrossRef]

36. Tian, J.; Chen, W.; Wu, P.; Zhu, Z.; Li, X. Cu–Mg–Zr/SiO$_2$ catalyst for the selective hydrogenation of ethylene carbonate to methanol and ethylene glycol. *Catal. Sci. Technol.* **2018**, *8*, 2624–2635. [CrossRef]

37. Song, T.; Qi, Y.; Jia, A.; Ta, N.; Lu, J.; Wu, P.; Li, X. Continuous hydrogenation of CO$_2$-derived ethylene carbonate to methanol and ethylene glycol at Cu-MoO$_x$ interface with a low H$_2$/ester ratio. *J. Catal.* **2021**, *399*, 98–110. [CrossRef]

38. Chen, X.; Wang, L.; Zhang, C.; Tu, W.; Cao, Y.; He, P.; Li, J.; Li, H. The effective and stable Cu–C@SiO$_2$ catalyst for the syntheses of methanol and ethylene glycol via selective hydrogenation of ethylene carbonate. *Int. J. Hydrogen Energy* **2021**, *46*, 17209–17220. [CrossRef]

39. Deng, F.; Li, N.; Tang, S.; Liu, C.; Yue, H.; Liang, B. Evolution of active sites and catalytic consequences of mesoporous MCM-41 supported copper catalysts for the hydrogenation of ethylene carbonate. *Chem. Eng. J.* **2018**, *334*, 1943–1953. [CrossRef]

40. Poels, E.K.; Brands, D.S. Modification of Cu/ZnO/SiO$_2$ catalysts by high temperature reduction. *Appl. Catal. A: Gen.* **2000**, *191*, 83–96. [CrossRef]

41. Meyer, J.J.; Tan, P.; Apfelbacher, A.; Daschner, R.; Hornung, A. Modeling of a Methanol Synthesis Reactor for Storage of Renewable Energy and Conversion of CO$_2$—Comparison of Two Kinetic Models. *Chem. Eng. Technol.* **2016**, *39*, 233–245. [CrossRef]

42. Bozzano, G.; Manenti, F. Efficient methanol synthesis: Perspectives, technologies and optimization strategies. *Prog. Energy Combust. Sci.* **2016**, *56*, 71–105. [CrossRef]

43. Haid, J.; Koss, U. Lurgi's Mega-Methanol technology opens the door for a new era in down-stream applications. In *Studies in Surface Science and Catalysis*; Iglesia, E., Spivey, J.J., Fleisch, T.H., Eds.; Elsevier Science Pub. Co. Inc.: Amsterdam, The Netherlands, 2001; Volume 136, pp. 399–404.

44. Samimi, F.; Feilizadeh, M.; Ranjbaran, M.; Arjmand, M.; Rahimpour, M.R. Phase stability analysis on green methanol synthesis process from CO$_2$ hydrogenation in water cooled, gas cooled and double cooled tubular reactors. *Fuel Processing Technol.* **2018**, *181*, 375–387. [CrossRef]

45. Cui, X.; Kær, S.K. A comparative study on three reactor types for methanol synthesis from syngas and CO$_2$. *Chem. Eng. J.* **2020**, *393*, 124632. [CrossRef]

46. Kuo, K.K. *Principles of Combustion*; Elsevier Science Pub. Co. Inc.: Amsterdam, The Netherlands, 1986.

47. Verman, L.C.; Banerjee, S. Effect of Container Walls on Packing Density of Particles. *Nature* **1946**, *157*, 584. [CrossRef]

48. Theuerkauf, J.; Witt, P.; Schwesig, D. Analysis of particle porosity distribution in fixed beds using the discrete element method. *Powder Technol.* **2006**, *165*, 92–99. [CrossRef]

49. Roshani, S. *Elucidation of Local and Global Structural Properties of Packed Bed Configurations*; The University of Leeds: Leeds, UK, 1990.

50. de Klerk, A. Voidage variation in packed beds at small column to particle diameter ratio. *AIChE J.* **2003**, *49*, 2022–2029. [CrossRef]

51. Ergun, S. Fluid Flow Through Packed Columns. *Chem. Eng. Prog.* **1952**, *48*, 89–94.

52. Koning, B. *Heat and Mass Transport in Tubular Packed Bed Reactors at Reacting and Non-Reacting Conditions*; University of Twente: Enschede, The Netherlands, 2002.

53. Agnew, J.B.; Potter, O. Heat transfer properties of packed tubes of small diameter. *Trans. Inst. Chem. Eng.* **1970**, *48*, T15.

54. Bauer, R.; Schluender, E.U. Effective Radial Thermal Conductivity of Packings in Gas Flow-1. Convective Transport Coefficient. *Int. Chem. Eng.* **1978**, *18*, 181–188.

55. Bauer, R.; Schluender, E.U. Effective Radial Thermal Conductivity of Packings in Gas Flow-2. Thermal Conductivity of the Packing Fraction without Gas Flow. *Int. Chem. Eng.* **1978**, *18*, 189–204.

56. Westerterp, K.R.; Swaaij, W.P.M.V.; Beenackers, A.A.C.M. *Chemical Reactor Design and Operation*, 2nd ed.; John Wiley & Sons, Ltd.: Hoboken, NJ, USA, 1987; p. 800.

57. Fahien, R.W.; Smith, J.M. Mass transfer in packed beds. *AIChE J.* **1955**, *1*, 28–37. [CrossRef]

58. Specchia, V.; Baldi, G.; Sicardi, S. Heat transfer in packed bed reactors with one phase flow. *Chem. Eng. Commun.* **1980**, *4*, 361–380. [CrossRef]

59. Wakao, N.; Funazkri, T. Effect of fluid dispersion coefficients on particle-to-fluid mass transfer coefficients in packed beds: Correlation of sherwood numbers. *Chem. Eng. Sci.* **1978**, *33*, 1375–1384. [CrossRef]
60. Bai, P.T.; Manokaran, V.; Saiprasad, P.S.; Srinath, S. Studies on Heat and Mass Transfer Limitations in Oxidative Dehydrogenation of Ethane Over Cr_2O_3/Al_2O_3 Catalyst. *Procedia Eng.* **2015**, *127*, 1338–1345. [CrossRef]
61. Soomro, M.; Hughes, R. The thermal conductivity of porous catalyst pellets. *Can. J. Chem. Eng.* **1979**, *57*, 24–28. [CrossRef]
62. McCabe, W.L.; Smith, J.C.; Harriott, P. Heat Tranfer to Fluids without Phase Change. In *Unit Operations in Chemical Engineering*; Clark, B.J., Castellano, E., Eds.; Mcgraw Hill, Inc.: New York, NY, USA, 1993; pp. 647–685.
63. Ross, J.R.H. Chapter 8-Mass and Heat Transfer Limitations and Other Aspects of the Use of Large-Scale Catalytic Reactors. In *Contemporary Catalysis*; Ross, J.R.H., Ed.; Elsevier Science Pub. Co. Inc.: Amsterdam, The Netherlands, 2019; pp. 187–213.
64. Plessis, J.P.; Diedericks, G. Fluid transport in porous media: Pore-scale modelling of interstitial transport phenomena. *Comput. Mech. Publ.* **1997**, *11*, 61–104.
65. Bird, R.B.; Stewart, W.E.; Lightfoot, E.N. *Transport Phenomena*; John Wiley & Sons, Ltd.: New York, NY, USA, 1960.

Article

Modeling and Experimental Studies on Carbon Dioxide Absorption with Sodium Hydroxide Solution in a Rotating Zigzag Bed

Zhibang Liu [1], Arash Esmaeili [1], Hanxiao Zhang [1], Dan Wang [2,3], Yuan Lu [4] and Lei Shao [1,*]

[1] Research Center of the Ministry of Education for High Gravity Engineering and Technology, Beijing University of Chemical Technology, Beijing 100029, China; 2019400002@mail.buct.edu.cn (Z.L.); ar.esmaeli@mail.buct.edu.cn (A.E.); 2018200070@mail.buct.edu.cn (H.Z.)

[2] State Key Laboratory of Petroleum Pollution Control, Beijing 102206, China; wang-dan@cnpc.com.cn

[3] CNPC Research Institute of Safety and Environmental Technology, Beijing 102206, China

[4] CenerTech Oilfield Chemical Co., Ltd., Tianjin 300450, China; luyuan2@cnooc.com.cn

[*] Correspondence: shaol@mail.buct.edu.cn; Tel.: +86-10-6442-1706

Abstract: The enhancement of mass transfer is very important in CO_2 absorption, and a rotating zigzag bed (RZB) is a promising device to intensify the gas–liquid mass transfer efficiency. In this study, the mass transfer characteristics in an RZB in relation to the overall gas-phase volumetric mass-transfer coefficient (K_Ga) were investigated with a CO_2–NaOH system. A mathematical model was established to illustrate the mechanism of the gas–liquid mass transfer with irreversible pseudo-first-order reaction in the RZB. The effects of various operating conditions on K_Ga were examined. Experimental results show that a rise in the liquid flow rate, inlet gas flow rate, rotational speed, absorbent temperature, and absorbent concentration was conducive to the mass transfer between gas and liquid in the RZB. It was found that the rotational speed had the largest impact on K_Ga in the RZB. The K_Ga predicted by the model agreed well with that by the experiments, with deviations generally within 10%. Therefore, this model can be employed to depict the mass transfer process between gas and liquid in an RZB and provide guidance for the application of RZBs in CO_2 absorption.

Keywords: rotating zigzag bed; modeling; gas–liquid mass transfer; carbon dioxide; absorption

Citation: Liu, Z.; Esmaeili, A.; Zhang, H.; Wang, D.; Lu, Y.; Shao, L. Modeling and Experimental Studies on Carbon Dioxide Absorption with Sodium Hydroxide Solution in a Rotating Zigzag Bed. *Processes* **2022**, *10*, 614. https://doi.org/10.3390/pr10030614

Academic Editors: Elio Santacesaria, Riccardo Tesser and Vincenzo Russo

Received: 21 February 2022
Accepted: 17 March 2022
Published: 21 March 2022

Publisher's Note: MDPI stays neutral with regard to jurisdictional claims in published maps and institutional affiliations.

1. Introduction

The rotating zigzag bed (RZB) is an emerging high-gravity device with an innovative zigzag passage structure to enhance fluid turbulence and mass transfer [1,2]. Liquid disperses and coalesces repeatedly by colliding with the rotating and static baffles continuously in the zigzag passage, leading to a high surface renewal frequency of the liquid [3]. Correspondingly, a significant increase in the gas–liquid contact time and liquid holdup is achieved in the RZB, which is in favor of mass transfer [4]. Compared to the conventional rotating packed bed (RPB), the RZB also exhibits the advantages of no demand for the liquid distributor and static seal, easy realization of the intermediate feed, and configuration of the multi-stage rotor in one casing [5,6]. Therefore, RZBs have been applied to stripping [7], distillation [5], extraction [8], and gas–liquid reaction [9,10].

Li et al. [10] studied the mass transfer characteristics of CO_2 absorption into NaOH solution in an RZB and found that the efficiency of liquid-side mass transfer in the RZB was significantly superior to that in the conventional RPB. Other research has demonstrated that the effective interfacial area and the efficiency of gas-side mass transfer were on par with those in RPBs with stainless wire mess packing. In consideration of the rotor structure with zigzag channel, pressure drop and power requirement in RZB are higher than those in RPB [1,11]. However, in terms of the study of Liu et al. [12] on the features of CO_2 absorption and mass transfer with diethylenetriamine-based absorbents in an RZB, the

RZB exhibited a better performance on CO_2 absorption and mass transfer with decreased device size and absorbent usage in comparison with an RPB, which could make up for the above disadvantages of RZB and reduce the operating cost. Moreover, a smaller volume of RZB with higher gas–liquid mass transfer efficiency makes it more suitable for application in confined spaces, such as vessels and offshore platforms, relative to conventional columns and RPBs [12], suggesting that the RZB is a promising apparatus for intensifying the gas–liquid contact processes.

The overall volumetric mass transfer coefficient is a crucial index to assess the mass transfer performance of a gas–liquid contact device. Extensive studies on the modelling of the overall volumetric mass transfer coefficient between gas and liquid in RPBs or other high-gravity devices have been conducted [13–17]. Additionally, the modelling of local mass transfer in RPBs has also been proposed [18,19]. However, there are only a few reports on the performance of mass transfer in the RZB, such as experimental investigations on the effective interfacial area and local mass transfer coefficient [10] and qualitative analysis of the mass transfer characteristics [2]. To the best of our knowledge, there is no report on the modelling of overall volumetric mass transfer coefficient in the RZB. Therefore, modelling the gas–liquid mass transfer behavior described by the overall volumetric mass transfer coefficient in the RZB can provide a theoretical basis for process intensification in the RZB.

Herein, based on fluid flow in different regions of the rotor in an RZB, a mechanism model has been proposed for the first time to describe the reactive mass transfer process in the CO_2–NaOH system and predict the mass transfer performance in the RZB. The effects of various operating conditions on the overall gas-phase volumetric mass-transfer coefficient ($K_G a$) were studied. The model was validated by the agreement of the predicted values with the experimental results.

2. Materials and Methods

2.1. Materials

Granulated sodium hydroxide (purity 98.0%) was provided by Shanghai Macklin Biochemical Co., Ltd., Shanghai, China. A carbon dioxide cylinder (purity 99.0%) was supplied by Beijing Shunanqite Gas Co., Ltd., Beijing, China. Deionized water was used through the experiments.

2.2. Experimental Apparatuses and Procedure

The RZB used in this work is schematically shown in Figure 1 [12], and the specifications of the RZB are given in Table 1 [12].

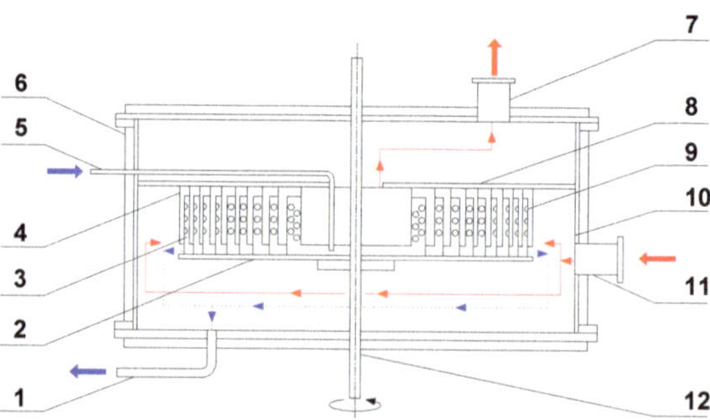

Figure 1. Structure of the rotating zigzag bed (RZB): (**1**) liquid outlet; (**2**) rotating disk; (**3**) rotating baffles; (**4**) static baffles; (**5**) liquid inlet; (**6**) insulation materials; (**7**) gas outlet; (**8**) static disk; (**9**) perforations; (**10**) shell; (**11**) gas inlet; (**12**) shaft. (Blue and red lines indicate liquid and gas streams, respectively).

Table 1. Specifications of the rotating zigzag bed (RZB).

Item	Value
Inner diameter of the rotor (cm)	5.7
Outer diameter of the rotor (cm)	18.3
Inner diameter of the shell (cm)	23.1
Diameter of the static baffles (cm)	5.7, 8.1, 9.9, 11.4, 12.7, 14.0, 15.2, 16.3, 17.3, 18.3
Diameter of the rotating baffles (cm)	7.2, 9.0, 10.7, 12.2, 13.5, 14.7, 15.8, 16.8, 17.8
Volume of the rotor (cm^3)	665
Axial depth of the rotor (cm)	2.8
Axial depth of the static baffles (cm)	2.5
Axial depth of the rotating baffles (cm)	2.4
Diameter of perforation in the rotating baffles (mm)	2.1

The rotor is the main component of the RZB. Static and rotating baffles are mounted alternately in the radial direction on the static and rotating disks, respectively, in the rotor. The upper section of the rotating baffles is perforated. The annular space between the static and rotating baffles forms the zigzag passage for fluid streams.

The shell of the RZB was covered with insulation materials to keep the reaction temperature in the RZB at a certain level, and an infrared thermometer (F-380, Shenzhen Flank Electronic Co., Ltd.) was employed to measure the temperature of the insulation materials.

The experimental setup is illustrated in Figure 2 [12]. The RZB was preheated by using a NaOH solution until a designated temperature was reached. A mixed gas stream containing air and CO_2 was then directed into the RZB via the gas inlet. When CO_2 concentration at the gas inlet reached 4%, the NaOH solution with a preset temperature was pumped into the RZB through the liquid inlet. The gas stream flowed inwards from the periphery of the rotor and made contact in a countercurrent with the liquid stream flowing outwards in the rotor to realize the absorption of CO_2 by the NaOH solution. Finally, the liquid and gas flows left the RZB via the liquid and gas outlets, respectively.

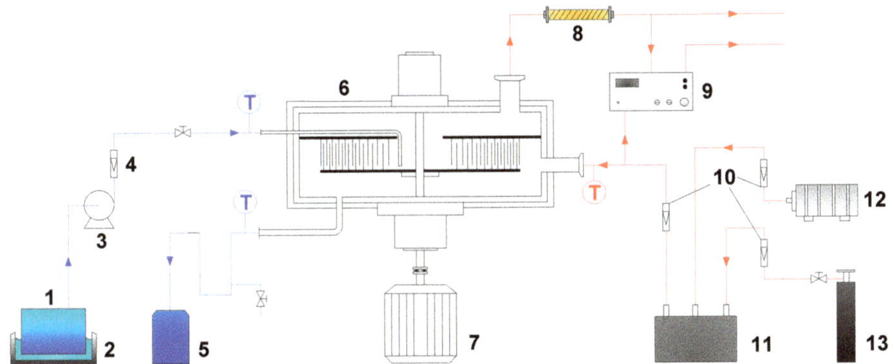

Figure 2. Experimental setup: (**1**) absorbent container; (**2**) thermostatic bath; (**3**) liquid pump; (**4**) liquid rotameter; (**5**) effluent tank; (**6**) RZB; (**7**) motor; (**8**) dryer; (**9**) CO_2 analyzer; (**10**) gas rotameter; (**11**) gas mixing tank; (**12**) air compressor; (**13**) CO_2 cylinder.

The experiments were performed at ambient pressure, and the experimental data were collected when a stable state of CO_2 absorption process was achieved. The concentration of CO_2 in gas streams entering and leaving the RZB was measured by an infrared CO_2 analyzer (GXH-3010F, Beijing Huayun Analytical Instrument Institution Co., Ltd., Beijing, China). All the experiments were repeated to validate the reproducibility of the results.

3. Theory

3.1. Reactions of CO_2 in Sodium Hydroxide Solution

When CO_2 is absorbed into the NaOH solution, the reactions between CO_2 with NaOH take place. These reactions can be expressed as follows [20]:

$$CO_2 + NaOH \leftrightarrow NaHCO_3 \tag{1}$$

$$NaHCO_3 + NaOH \leftrightarrow Na_2CO_3 + H_2O \tag{2}$$

The rate of the above reactions can be calculated by the following second-order reaction rate equation [21]:

$$r_{CO_2} = k_2 C_{NaOH} C_{CO_2} \tag{3}$$

The reaction can be regarded as a pseudo-first-order one when in Equation (4) is satisfied [20,22].

$$\sqrt{1 + \frac{D_{CO_2} k_2 C_{NaOH}}{k_L^2}} - 1 \ll \frac{C_{NaOH}}{2C_0} \tag{4}$$

where D_{CO2} is the diffusion coefficient of CO_2 in NaOH solution.

In this study, in Equation (4) held because the left-side of in Equation (4) was over 100 times smaller than the right-side. Therefore, the rate constant of the pseudo-first-order reaction can be written as:

$$k_{app} = k_2 C_{NaOH} \tag{5}$$

3.2. Gas and Liquid Flow in the RZB Rotor

Gas and liquid flows in the RZB rotor are presented in Figure 3. Gas flows inwards from the periphery to the center of the RZB rotor, and the gas velocity can be divided into tangential component (v_θ), radial component (v_r), and axial component (v_z) in this process. The ratio of the tangential velocity to the total velocity is 81.6% to 99.3% [23], suggesting that gas mainly moves in a spiral between the rotating and static baffles.

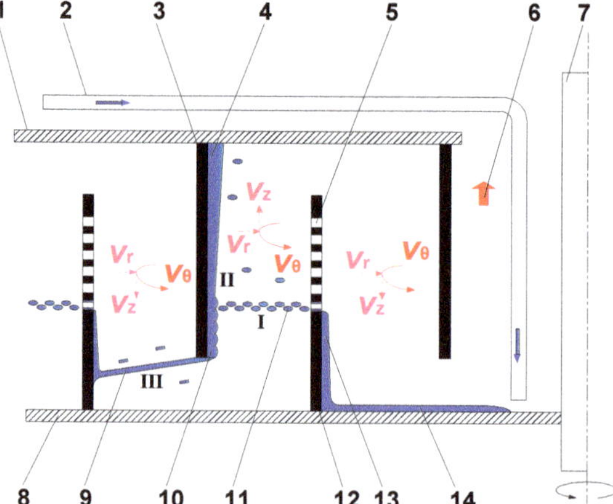

Figure 3. Side view of gas and liquid flow in RZB rotor: (**1**) static disk; (**2**) liquid inlet; (**3**) static baffle; (**4**) stagnant filmy liquid; (**5**) perforations; (**6**) gas stream; (**7**) shaft; (**8**) rotating disk; (**9**) flying filmy liquid; (**10**) turbulent filmy liquid; (**11**) flying fine droplets; (**12**) rotating baffle; (**13**) filmy liquid climbing up on the internal surface of rotating baffle; (**14**) filmy liquid on the rotating disk. (Red and blue symbols represent gas and liquid flows, respectively.).

The liquid flow moves outwards through the zigzag passage in the rotor. At the center of the rotating disk, liquid flows tangentially outwards under the action of centrifugal force until it reaches the internal surface of the rotating baffle and then climbs up to the perforated area. The liquid is dispersed into fine droplets when passing through the perforations in the upper section of the rotating baffle and flies along a tangential direction to the static baffle. The fine droplets are captured by the static baffle to form a filmy liquid, which flows spirally downwards on the internal surface of the static baffle due to gravity and tangential movement of the liquid. The filmy liquid departs tangentially from the static baffle as a sheet and reaches the next rotating baffle without falling onto the rotating disk because the tangential liquid velocity is much larger than the axial liquid velocity caused by gravity. The liquid is then captured by the next rotating baffle and repeats the above process in the zigzag passage until leaving the RZB rotor [10].

In terms of the above gas and liquid flow characteristics in the RZB rotor, the gas–liquid contact area can be divided into three different zones, as shown in Figure 3, as zones I, II, and III, and all of them make contributions to the mass transfer [3]. In zone I, fine droplets contact with gas in a spiral movement in the space formed by the rotating and static baffles, contributing to the gas–liquid mass transfer. In zone II, mass transfer is scarcely attributed to contact between gas and the stagnant filmy liquid formed by droplets on the upper section of the static baffle, but mainly results from the surface renewal caused by droplets continually impinging on the turbulent filmy liquid on the lower section of the static baffle [2]. In zone III, the liquid exists as an intact sheet when there is no gas flow, but some fine droplets may result from rupture of the sheet when gas flows through the passage. The contact of the flying liquid and gas in zone III also contributes to mass transfer. Hence, the rotor of the RZB can be regarded as a series of wetted-wall columns with the centrifugal field [3].

It has been reported that mass transfer in zone II contributes most to the overall mass transfer because the continued horizontal impingement by liquid jets on the turbulent filmy liquid on the static baffles brings about a great surface renewal rate. Mass transfer in zones I and III is at least one order of magnitude less than that in zone II because the renewal time of the liquid surface in zones I and III is longer than that in zone II according to the penetration theory [2].

3.3. Model Development

According to the flow characteristics of gas and liquid in the RZB rotor, model assumptions for CO_2 absorption into the NaOH solution in the RZB were made as follows:

(1)　Velocity of droplets leaving the rotating baffle along the tangential direction in zone I was equal to the circumferential velocity of the rotating baffle [11].

(2)　Only the tangential velocity of gas phase in the RZB rotor was considered, and the radial and axial velocities of gas were ignored because the tangential velocity was the main component of the gas velocity [23].

(3)　It should be noted that the tangential gas velocity is smaller than the circumferential velocity of rotating baffles when the gas flow rate is low [23]. Because the gas flow rate was less than 1200 L/h in this study, the tangential gas velocity was much lower than the tangential liquid velocity in the space between the static baffle and rotating baffle. Thus, it was assumed that the effect of gas flow on the liquid form could be ignored in zone III, where the flying liquid remained in the form of an intact sheet [11]. Meanwhile, the liquid form in zones I and II was not affected by gas flow either.

(4)　Liquid existed as droplets in zone I and filmy liquid in zone II [10]. Hence, the overall gas–liquid effective interfacial area was the sum of the surface area of droplets in zone I, turbulent filmy liquid on the static baffle in zone II, and the flying intact liquid sheet in zone III [2,11].

(5)　Liquid left the rotating baffle only through the lowest circle of perforations in the rotating baffles because the perforations were closely spaced [24]. Droplets formed by every perforation had the same amount, lifetime, velocity, and diameter.

(6) Liquid film is a thin surface layer in the filmy liquid. The liquid film in the turbulent filmy liquid in zone II was renewed by impingement of fine droplets from zone I, and the liquid film in the filmy liquid in every zone of the RZB had the same lifetime [2].

(7) The reaction between CO_2 and NaOH was considered to be a pseudo-first-order reaction. The concentration of OH^- in the liquid film was assumed to be constant during the CO_2 absorption process, and thus k_{app} was regarded as a constant in the liquid element.

For the chemical absorption of CO_2 into NaOH solution, CO_2 is absorbed from the gas phase to the liquid phase via a gas–liquid interface. During this process, the gas–liquid mass transfer depends both on the liquid-phase and gas-phase mass transfer. Therefore, the gas–liquid effective interfacial area (a), liquid-side mass-transfer coefficient (k_L), and gas-side mass-transfer coefficient (k_G) should be considered for calculating $K_G a$.

The space in the rotor of the RZB was divided into certain annular regions for model development (Figure 4). Nine rotating baffles ($1, 2, \ldots, i_a, i_a + 1, \ldots, 9$) and ten static baffles ($0', 1', \ldots, i_b, i_b + 1, \ldots, 9'$) were radially alternately arranged. Region i was the annular region between the rotating baffle i_a of radius $r_{a,i}$ and static baffle i_b of radius $r_{b,i}$, and region i' was the annular region between the static baffle i_b of radius $r_{b,i}$ and rotating baffle $i_a + 1$ of radius $r_{a,i+1}$. Mass transfer zones I and II were located in region i, while zone III was in region i'.

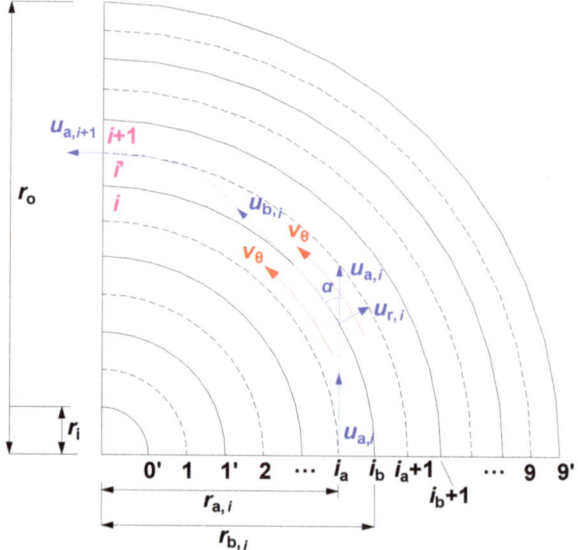

Figure 4. Top view of liquid and gas flows in the rotor of RZB.

In zone I of region i, droplets left the rotating baffle i_a of radius $r_{a,i}$ with a tangential velocity $u_{a,i}$ along the tangential direction to the static baffle i_b. The lifetime of droplets in zone I $t_{I,i}$ is written as:

$$t_{I,i} = \frac{\sqrt{r_{b,i}^2 - r_{a,i}^2}}{u_{a,i}} \tag{6}$$

$$u_{a,i} = \omega r_{a,i} \tag{7}$$

where ω represents the angular velocity of the RZB rotor.

The liquid holdup of droplets $\varepsilon_{L,i}$ in this area can be calculated by the following equation:

$$\varepsilon_{L,i} = \frac{Q_L t_{L,i}}{\pi\left(r_{b,i}^2 - r_{a,i}^2\right)z} = \frac{Q_L}{\pi \omega z r_{a,i}\sqrt{r_{b,i}^2 - r_{a,i}^2}} \tag{8}$$

where Q_L is the liquid volumetric flow rate in the RZB rotor and z is the axial depth of the rotor.

The correlation of the effective mass transfer area of droplets $A_{I,i}$ in zone I of region i is as follows [16]:

$$A_{I,i} = a_{I,i}\pi\left(r_{b,i}^2 - r_{a,i}^2\right)z = \frac{6\varepsilon_{L,i}}{d_i}\pi\left(r_{b,i}^2 - r_{a,i}^2\right)z \tag{9}$$

where d_i is the average diameter of droplets in zone I of region i.

In zone II of region i, the surface area of the turbulent filmy liquid is equal to the effective mass transfer area in this zone, which can be obtained by the following equation:

$$A_{II,i} = 2\pi r_{b,i} A_2 h_{II} \tag{10}$$

where A_2 is a turbulent coefficient used to correct the variation of surface area of the filmy liquid caused by the impingement of droplets (with a value of 2.19) [25] and h_{II} represents the axial length of the turbulent filmy liquid.

In zone III of region i', the liquid remained in the form of an intact sheet between the static and rotating baffles according to assumption (4). The filmy liquid tangentially left the static baffle of radius $r_{b,i}$ with a tangential velocity of $u_{b,i}$, and thus the lifetime of the flying filmy liquid in zone III of region i' $t_{III,i'}$ is expressed as follows:

$$t_{III,i'} = \frac{\sqrt{r_{a,i+1}^2 - r_{b,i}^2}}{u_{b,i}} = \frac{\sqrt{r_{a,i+1}^2 - r_{b,i}^2}}{u_{a,i}\frac{r_{a,i}}{r_{b,i}}} \tag{11}$$

Thus, the surface area of the flying filmy liquid equals the gas–liquid effective mass transfer area in this zone:

$$A_{III,i'} = (2\pi r_{b,i} + 2\pi r_{a,i+1})\sqrt{(r_{a,i+1} - r_{b,i})^2 + \left(u_{g,i}t_{III,i'} + \frac{1}{2}g t_{III,i'}^2\right)^2} \tag{12}$$

where $u_{g,i}$ is the axial component of the turbulent filmy liquid velocity in zone II.

Therefore, the gas–liquid effective interfacial area in the RZB rotor can be obtained as follows:

$$a = a_I + a_{II} + a_{III} = \frac{\sum_{i=1}^{9} A_{I,i} + \sum_{i=1}^{9} A_{II,i} + \sum_{i'=1}^{8} A_{III,i'}}{\pi(r_o^2 - r_i^2)z} \tag{13}$$

The mass transfer in the liquid phase consists of that in the droplets and filmy liquid. Because the droplets were treated as rigid balls without inside circulation during the flight due to a small distance in zone I of region i [26], the following mass partial differential equation can be used to present the CO_2 diffusion process from gas bulk to droplets based on the pseudo-first-order irreversible chemical reaction:

$$\frac{\partial C_{CO_2}}{\partial t_{L,i}} = D_{CO_2}\frac{\partial^2 C_{CO_2}}{\partial R^2} + \frac{2D_{CO_2}}{R}\frac{\partial C_{CO_2}}{\partial R} - k_{app}\left(C_{CO_2} - C_{CO_2}^*\right) \tag{14}$$

$$\text{I.C. } t_{L,i} = 0,\ R \geq 0:\ C_{CO_2} = C_{CO_2}^*;$$

$$\text{B.C. } R = 0,\ t_{L,i} \geq 0:\ C_{CO_2} = C_{CO_2}^*;$$

$$R = \frac{d_i}{2},\ t_{L,i} \geq 0:\ C_{CO_2} = C_0; \tag{15}$$

where C_0 $(=P_{CO2,0}/H)$ and C_{CO2}^* are the molar concentration of CO_2 at the gas–liquid interface and the equilibrium molar concentration of CO_2 in the liquid bulk, respectively. Letting $C_A = C_{CO2} - C_{CO2}^*$, the following expressions can be obtained:

$$\frac{\partial C_A}{\partial t_{I,i}} = D_{CO_2}\frac{\partial^2 C_A}{\partial R^2} + \frac{2D_{CO_2}}{R}\frac{\partial C_A}{\partial R} - k_{app}C_A \tag{16}$$

$$I.C.\ t_{I,i} = 0,\ R \geq 0:\ C_A = 0;$$

$$B.C.\ R = 0,\ t_{I,i} \geq 0:\ C_A = 0;$$

$$R = \frac{d_i}{2},\ t_{I,i} \geq 0:\ C_A = C_0 - C_{CO_2}^*; \tag{17}$$

After Laplace transform, the following ordinary differential equation is derived:

$$\frac{d^2u}{dR^2} + \frac{2}{R}\frac{du}{dR} - \frac{k_{app}+s}{D_{CO_2}}u = 0 \tag{18}$$

Letting $\beta = u \times R$, Equation (22) can be obtained by Equations (19)–(21).

$$\frac{d\beta}{dR} = \frac{du}{dR}R + u \tag{19}$$

$$\frac{d^2\beta}{dR^2} = \frac{d^2u}{dR^2}R + 2\frac{du}{dR} \tag{20}$$

$$\frac{1}{R}\frac{d^2\beta}{dR^2} = \frac{d^2u}{dR^2} + \frac{2}{R}\frac{du}{dR} \tag{21}$$

$$\frac{d^2\beta}{dR^2} = \frac{k_{app}+s}{D_{CO_2}}\beta \tag{22}$$

The general solution of β can be calculated by Equation (23):

$$\beta = u \times R = C_1 e^{-R\sqrt{\frac{k_{app}+s}{D_{CO_2}}}} + C_2 e^{R\sqrt{\frac{k_{app}+s}{D_{CO_2}}}} \tag{23}$$

Then, the following equation is obtained by Laplace inverse transform:

$$\begin{aligned}
C_A = {}&\frac{C_1}{2R}exp\left(R\sqrt{\frac{k_{app}}{D_{CO_2}}}\right)erfc\left(-\frac{R}{2\sqrt{D_{CO_2}t_{I,i}}} - \sqrt{k_{app}t_{I,i}}\right)\\
&+\frac{C_1}{2R}exp\left(-R\sqrt{\frac{k_{app}}{D_{CO_2}}}\right)erfc\left(-\frac{R}{2\sqrt{D_{CO_2}t_{I,i}}} + \sqrt{k_{app}t_{I,i}}\right)\\
&+\frac{C_2}{2R}exp\left(-R\sqrt{\frac{k_{app}}{D_{CO_2}}}\right)erfc\left(\frac{R}{2\sqrt{D_{CO_2}t_{I,i}}} - \sqrt{k_{app}t_{I,i}}\right)\\
&+\frac{C_2}{2R}exp\left(R\sqrt{\frac{k_{app}}{D_{CO_2}}}\right)erfc\left(\frac{R}{2\sqrt{D_{CO_2}t_{I,i}}} + \sqrt{k_{app}t_{I,i}}\right)
\end{aligned} \tag{24}$$

where $erfc(x)$ is the excess error function.

To simplify Equation (24), B is defined as follows:

$$B = exp\left(\frac{d_i}{2}\sqrt{\frac{k_{app}}{D_{CO_2}}}\right)erf\left(\frac{\frac{d_i}{2}}{2\sqrt{D_{CO_2}t_{I,i}}} + \sqrt{k_{app}t_{I,i}}\right) + exp\left(-\frac{d_i}{2}\sqrt{\frac{k_{app}}{D_{CO_2}}}\right)erf\left(\frac{\frac{d_i}{2}}{2\sqrt{D_{CO_2}t_{I,i}}} - \sqrt{k_{app}t_{I,i}}\right) \tag{25}$$

where $erf(x)$ is the error function.

Letting $C_1 = -C_2$ and substituting Equation (17) into Equation (24), C_1 and C_2 can be expressed as follows:

$$C_1 = \frac{C_0 - C_{CO_2}^*}{2B}d_i \tag{26}$$

$$C_2 = -C_1 = -\frac{C_0 - C^*_{CO_2}}{2B} d_i \tag{27}$$

Therefore, Equation (28) can be obtained by combining Equations (24), (26) and (27):

$$\begin{aligned}
C_A = \frac{\left(C_0 - C^*_{CO_2}\right) d_i}{2RB} \ exp\left(R\sqrt{\frac{k_{app}}{D_{CO_2}}}\right) erf\left(\frac{R}{2\sqrt{D_{CO_2} t_{I,i}}} + \sqrt{k_{app} t_{I,i}}\right) \\
+ \frac{\left(C_0 - C^*_{CO_2}\right) d_i}{2RB} \ exp\left(-R\sqrt{\frac{k_{app}}{D_{CO_2}}}\right) erf\left(\frac{R}{2\sqrt{D_{CO_2} t_{I,i}}}\right) \\
- \sqrt{k_{app} t_{I,i}}\Big)
\end{aligned} \tag{28}$$

In terms of Fick's first law, the rate equation describing mass transfer at the gas–liquid interface can be written as:

$$k_{L-I,i}\left(C_{CO_2} - C^*_{CO_2}\right) = D_{CO_2} \frac{\partial C_A}{\partial R}\Big|_{R=\frac{d_i}{2}} \tag{29}$$

It was assumed that CO_2 was completely consumed by chemical reaction of CO_2 and NaOH in the liquid bulk. Thus, $C_{CO_2}{}^*$ can be ignored, and $k_{L-I,i}$ in zone I is given by Equation (30):

$$k_{L-I,i} = \sqrt{k_{app} D_{CO_2}} - \frac{2D_{CO_2}}{d_i} \tag{30}$$

In regard to liquid film in the turbulent filmy liquid in zone II of region i, CO_2 diffusion into the liquid film with a pseudo-first-order reaction can be expressed as:

$$\frac{\partial C_{CO_2}}{\partial t_{II,i}} = D_{CO_2} \frac{\partial^2 C_{CO_2}}{\partial x^2} - k_{app}\left(C_{CO_2} - C^*_{CO_2}\right) \tag{31}$$

$$I.C. \ t_{II,i} = 0, \ x \geq 0: \ C_{CO_2} = C^*_{CO_2};$$

$$B.C. \ x = 0, t_{II,i} \geq 0: \ C_{CO_2} = C_0;$$

$$x = \delta(\to \infty), t_{II,i} \geq 0: \ C_{CO_2} = C^*_{CO_2}; \tag{32}$$

where δ is the average thickness of the liquid film in zone II and $t_{II,i}$ is the lifetime of the liquid film, which is the average time consumed for renewing the liquid film once.

The following equation can be obtained from Equations (31) and (32):

$$\frac{\partial C_A}{\partial t_{II,i}} = D_{CO_2} \frac{\partial^2 C_A}{\partial x^2} - k_{app} C_A \tag{33}$$

$$I.C. \ t_{II,i} = 0, \ x \geq 0: \ C_A = 0;$$

$$B.C. \ x = 0, t_{II,i} \geq 0: \ C_A = C_0 - C^*_{CO_2};$$

$$x = \delta(\to \infty), t_{II,i} \geq 0: \ C_A = 0; \tag{34}$$

By Laplace transform, the following equation is derived:

$$\frac{d^2 u}{dx^2} - \frac{k_{app} + s}{D_{CO_2}} u = 0 \tag{35}$$

$$B.C. \ x = 0, \ s \geq 0: \ u = \frac{C_0 - C^*_{CO_2}}{s};$$

$$x = \delta(\to \infty), \ s \geq 0: \ u = 0; \tag{36}$$

Substituting the boundary conditions in Equation (36) into Equation (35), the general solution of u can be obtained by Equation (37):

$$u = \frac{C_0 - C_{CO_2}^*}{s} exp\left(-x\sqrt{\frac{k_{app} + s}{D_{CO_2}}}\right) \tag{37}$$

By the Laplace inverse transform, the relationship of the CO_2 concentration distribution, lifetime, and liquid film thickness can be described by the following equation:

$$\begin{aligned}
C_A = \frac{\left(C_0 - C_{CO_2}^*\right)}{2} \; & exp\left(x\sqrt{\frac{k_{app}}{D_{CO_2}}}\right) erfc\left(\frac{x}{2\sqrt{D_{CO_2}t_{II,i}}} + \sqrt{k_{app}t_{II,i}}\right) \\
& + \frac{\left(C_0 - C_{CO_2}^*\right)}{2} exp\left(-x\sqrt{\frac{k_{app}}{D_{CO_2}}}\right) erfc\left(\frac{x}{2\sqrt{D_{CO_2}t_{II,i}}} \right. \\
& \left. - \sqrt{k_{app}t_{II,i}}\right)
\end{aligned} \tag{38}$$

Hence, the mass transfer rate at the gas–liquid interface in zone II can be obtained based on Fick's first law.

$$R_{CO_2-II,i} = D_{CO_2}\frac{\partial C_A}{\partial x}\Big|_{x=0} = \left(C_0 - C_{CO_2}^*\right)\left(\sqrt{k_{app}D_{CO_2}}\, erf\sqrt{k_{app}t_{II,i}} + \sqrt{\frac{D_{CO_2}}{\pi t_{II,i}}}exp\left(-k_{app}t_{II,i}\right)\right) \tag{39}$$

According to assumptions (5) and (6), the renewal of the liquid film in zone II was caused by the impingement of droplets, which came from the lowest circle of the perforated area in the rotating baffle. It was assumed that the droplets formed by every perforation had the same amount, lifetime, velocity, and diameter. Therefore, the renewal frequency, S_i, defined as the number of droplets leaving the rotating baffles per unit time, in zone II of region i is expressed as:

$$S_i = \frac{36Q_L}{\pi n_{a,i} d_i^3} \tag{40}$$

where $n_{a,i}$ is the number of perforations in the rotating baffle i_a.

Then the lifetime of the liquid film in the turbulent filmy liquid can be calculated by Equation (41) [18]:

$$t_{II,i} = \frac{1}{S_i} \tag{41}$$

It was assumed that $t_{II,i}$ caused by every droplet was the same. Thus, Equation (42) can be used to express the Higbie distribution function of the lifetime of the liquid film [16]:

$$\psi(t_{II,i}) = \frac{1}{t_{II,i}} \tag{42}$$

Hence, the relationship between the liquid-side mass-transfer coefficient in zone II ($k_{L-II,i}$) and the mass transfer rate of liquid film at the gas–liquid interface can be written as:

$$\int_0^{t_{II,i}} R_{CO_2-II,i}\psi(t_{II,i})dt = k_{L-II,i}\left(C_0 - C_{CO_2}^*\right) \tag{43}$$

Based on the same assumption for droplets, $C_{CO_2}^*$ can be ignored. Therefore, $k_{L-II,i}$ related to k_{app}, D_{CO2}, and $t_{II,i}$ can be deduced as follows:

$$k_{L-II,i} = \frac{\sqrt{k_{app}D_{CO_2}}}{t_{II,i}}\left[t_{II,i}erf\left(\sqrt{k_{app}t_{II,i}}\right) + \sqrt{\frac{t_{II,i}}{\pi k_{app}}}exp\left(-k_{app}t_{II,i}\right) + \frac{1}{2k_{app}}erf\left(\sqrt{k_{app}t_{II,i}}\right)\right] \tag{44}$$

As for the liquid film in the flying filmy liquid in zone III of region i', the model development process of $k_{L-III,i'}$ was the same as that of $k_{L-II,I}$, and thus the expression of

$k_{L-III,i'}$ was similar to that of $k_{L-II,i}$, except for the lifetime of the liquid film in the flying filmy liquid, which was calculated by Equation (11).

Hence, $k_L a$ in the RZB rotor is written as follows:

$$k_L a = k_{L-I} a_I + k_{L-II} a_{II} + k_{L-III} a_{III} = \frac{\sum_{i=1}^{9} k_{L-I,i} A_{I,i} + \sum_{i=1}^{9} k_{L-II,i} A_{II,i} + \sum_{i'=1}^{8} k_{L-III,i'} A_{III,i'}}{\pi(r_o^2 - r_i^2)z} \tag{45}$$

To date, there is no available expression to calculate k_G in the RZB rotor. Thus, a surface renewal model reported by Guo et al. [27] was employed to calculate k_G in the RZB rotor as follows:

$$k_G = \sqrt{D_G k_s v_\theta^2} \tag{46}$$

where D_G is the diffusion coefficient of CO_2 in gas phase, k_s is the proportionality coefficient, and v_θ is the average tangential gas velocity in the RZB rotor.

The $K_G a$ of the RZB can be calculated by [15]:

$$\frac{1}{K_G a} = \frac{1}{k_G a} + \frac{H}{k_L a} \tag{47}$$

The experimental $K_G a$ can be determined by Equation (48), which was obtained in the previous study on CO_2 absorption [12]:

$$K_G a = \frac{G'}{\pi P z (r_o^2 - r_i^2)} \left[\ln \frac{y_{CO_2-in}(1 - y_{CO_2-out})}{y_{CO_2-out}(1 - y_{CO_2-in})} + \left(\frac{y_{CO_2-in}}{1 - y_{CO_2-in}} - \frac{y_{CO_2-out}}{1 - y_{CO_2-out}} \right) \right] \tag{48}$$

where G' is the inlet gas flow rate of inert gas (without reaction or dissolution), P is the total pressure, and y_{CO_2-in} and y_{CO_2-out} denote the molar fraction of CO_2 in the inlet and outlet gas streams, respectively.

By substituting the experimental $K_G a$ from Equation (48) and the calculated $k_L a$ and a from Equations (45) and (13), respectively, into Equation (47), k_G can be obtained. Consequently, k_s can be obtained, which is necessary for the establishment of the mass transfer model. Thus, the calculated k_G, k_L, a, and $K_G a$ for the RZB were obtained from this mass transfer model.

The density and viscosity values were obtained from reference [28], while the surface tension value was from reference [29]. The equilibrium, kinetics, and transport parameters used for modeling are tabulated in Table 2.

Table 2. Parameters for model development.

Parameter	Expression
H [30]	$\log \frac{H}{H_W} = \sum hI$
H_W [30]	$H_W = 101.3 * (23.9 + 0.757(T/^\circ C - 18))$
D_{CO2} [31]	$D_{CO_2} \mu_L = D_W \mu_W$
D_W [31]	$\log D_W = -8.1764 + \frac{712.5}{T} - \frac{2.591 \times 10^5}{T^2}$
D_G [32]	$D_G = 2.189 \times 10^{-5} - 0.393 \times 10^{-5} \left[OH^- \right]$
k_2 [31]	$\log \frac{k_2}{k_{OH^-}^\infty} = 0.221I - 0.016I^2$
$k_{OH^-}^\infty$ [31]	$\log k_{OH^-}^\infty = 11.895 - \frac{2382}{T}$
d_i [27]	$d_i = 0.7284 \left(\frac{\sigma}{\omega^2 r_{a,i} \rho_L} \right)^{0.5}$
v_θ [23]	$\frac{v_\theta}{v_{in}} = 101.9987 Re_G^{-0.7004} \left(\frac{r_m}{r_a} \right)^{-1.6356} \left(\frac{\omega^2 r_a}{g} \right)^{-0.4384}$
$u_{g,i}$ [25]	$u_{g,i} = \left(\frac{\rho_L g Q_L^2}{3\pi^2 \mu_L (2r_{b,i})^2} \right)^{\frac{1}{3}}$

4. Results and Discussion

4.1. Model Validation

The RZB had an average value for k_s of 6.17×10^{-8} $kmol^2$ s/kPa^2 m^8 under the experimental conditions. By using k_L, k_G, and a obtained from the above model and the average value of k_s, the calculated K_Ga of the RZB was obtained according to Equation (47). The comparison of k_L, k_G, and a between this work and reference [10] can be found in Table S1 in the Supplementary Materials file.

Figure 5 is the diagonal diagram of the predicted and experimental values of K_Ga. It is shown that this model provided good predictions on K_Ga of CO_2 absorption in the RZB, with deviations generally less than 10% in comparison with the experimental results. Additionally, in the Figures shown in Sections 4.2–4.6, the curves for the predicted values were consistent with those for the experimental results. In addition, the values of individual and overall mass transfer parameters have been presented in Table S2 in the Supplementary Materials file.

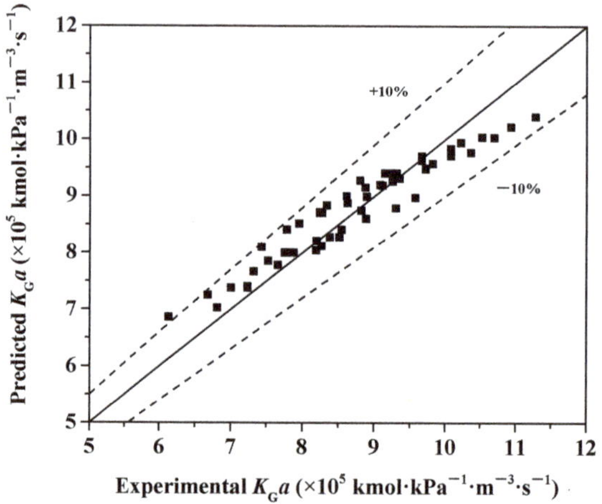

Figure 5. Diagonal diagram of experimental and predicted K_Ga.

4.2. Effect of Liquid Flow Rate

Figure 6 presents the effect of the liquid flow rate on K_Ga in the RZB. It can be seen that K_Ga in NaOH solution obviously increased, with a rise in the liquid flow rate from 25 to 40 L/h.

With a rising liquid flow rate, more fine droplets from the rotating baffles are produced. The falling velocity of the filmy liquid on the static baffles also increases in terms of the expression of $u_{g,i}$ in Table 2, causing an increasing area of the flying liquid sheet between the static and rotating baffles according to Equation (12). These factors enhance the liquid holdup in the RZB, thereby causing a rising gas–liquid contact area based on Equation (8). Meanwhile, with the increase of the liquid flow rate, the circumferential distribution of droplets in zone I is improved [10], and more droplets continually impinge on the filmy liquid on the lower section of the static baffles, leading to an enhancement of liquid turbulence and a higher surface renewal rate of liquid film in the turbulent filmy liquid in zone II, as suggested by Equation (40), which are conducive to the liquid-side mass transfer. Therefore, a higher liquid flow rate brought about a larger K_Ga in this study because a and k_L contributed notably to K_Ga for CO_2 absorption into NaOH solution.

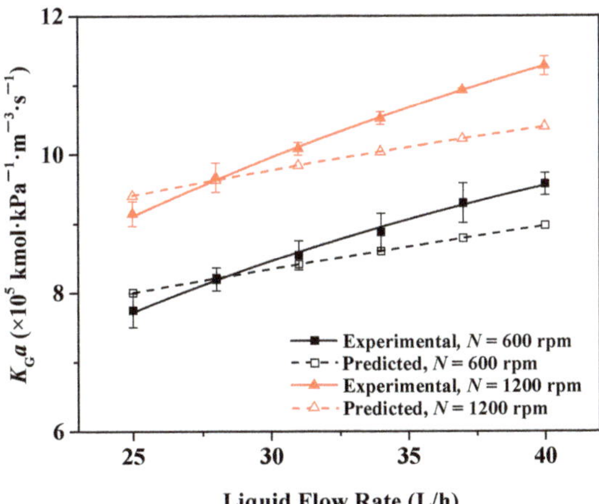

Figure 6. Effect of liquid flow rate on K_Ga in the RZB (G = 1000 L/h, T = 298.15 K, T_{gas} = 298.15 K, $y_{CO2\text{-in}}$ = 4%, C_{NaOH} = 0.15 kmol/m^3).

4.3. Effect of Inlet Gas Flow Rate

The influence of the inlet gas flow rate on K_Ga in the RZB is given in Figure 7, which shows that K_Ga in the NaOH solution rose from 6.13×10^{-5} to 8.26×10^{-5} kmol/kPa m^3 s and from 6.67×10^{-5} to 9.65×10^{-5} kmol/kPa m^3 s, with the inlet gas flow rate increasing from 200 to 1200 L/h at the rotational speed of 600 rpm and 1200 rpm, respectively.

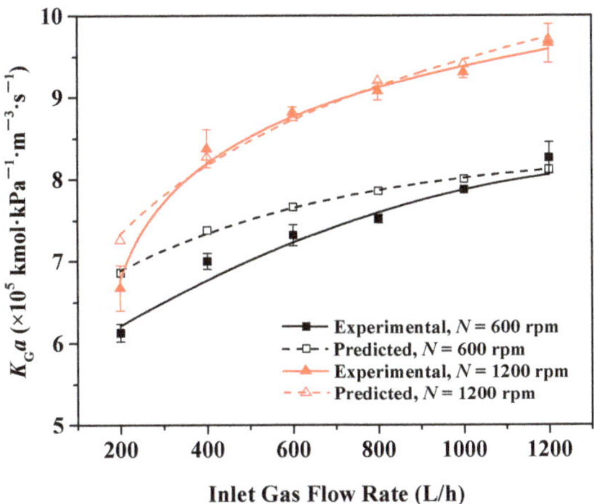

Figure 7. Effect of inlet gas flow rate on K_Ga in the RZB (L = 25 L/h, T = 298.15 K, T_{gas} = 298.15 K, $y_{CO2\text{-in}}$ = 4%, C_{NaOH} = 0.15 kmol/m^3).

Because K_Ga in the NaOH solution is affected by k_Ga, a rising inlet gas flow rate in the RZB increases the tangential gas velocity in the annular space between the rotating and static baffles, which leads to the increase of gas turbulence and a decrease of gas film thickness [16]. Therefore, the mass transfer resistance in the gas side reduces and consequently k_Ga increases, thereby enhancing K_Ga in the RZB. Hence, an increasing inlet gas flow rate brought about a rise in K_Ga in the experimental range.

4.4. Effect of Rotational Speed

The variation of K_Ga with the rotational speed is shown in Figure 8. K_Ga augmented with a rise in the rotational speed from 400 to 1200 rpm.

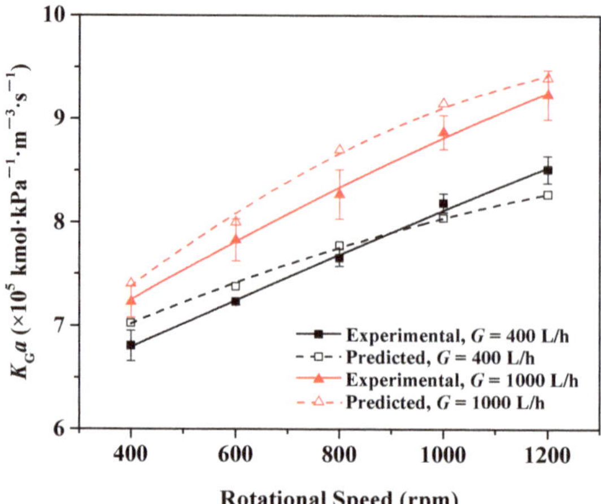

Figure 8. Effect of rotational speed on $K_G a$ in the RZB (L = 25 L/h, T = 298.15 K, T_{gas} = 298.15 K, $y_{CO2\text{-}in}$ = 4%, C_{NaOH} = 0.15 kmol/m^3).

When the rotational speed rises, a larger centrifugal force is created by the rotating baffles, leading to an enhanced liquid turbulence and higher tangential velocity of droplets departing from the rotating baffles according to Equation (7), which reduces the size of droplets in the space between the rotating and static baffles based on the expression of droplet diameter in Table 2. Therefore, a in zone I increase.

At the same time, rising rotational speed causes an increase in the number of droplets impinging on the turbulent filmy liquid as a result of the reduction in droplet size and a larger circumferential velocity of the turbulent filmy liquid in zone II, which is equal to the tangential velocity of the flying liquid sheet leaving the static baffles in zone III [11], resulting in a rise in the surface renewal frequency of the liquid film in the turbulent filmy liquid on the static baffles in terms of Equation (40) and a decrease in the lifetime of the flying liquid sheet in zone III according to Equation (11). The former factor brought about an increase in k_{L-II} in zone II, and the latter factor was conducive to $k_{L\text{-III}}$ in zone III according to Equation (44).

The above analysis indicates that a higher rotational speed led to higher a and k_L, which markedly an enhanced $K_G a$ of the CO$_2$-NaOH system in the RZB.

4.5. Effect of Absorbent Temperature

The effect of the absorbent temperature on $K_G a$ is presented in Figure 9. The figure indicates that a higher temperature of NaOH solution is favorable for $K_G a$, which increased from 7.95×10^{-5} to 8.80×10^{-5} kmol/kPa m^3 s and from 9.35×10^{-5} to 1.02×10^{-4} kmol/kPa m^3 s with an increase in temperature from 293.15 to 313.15 K at the liquid flow rate of 25 L/h and 35 L/h, respectively.

In accordance with the Arrhenius equation for the reaction rate constant in Table 2, a higher temperature of NaOH solution increases the second-order rate constant k_2, thereby leading to a larger pseudo-first-order reaction rate constant k_{app}, which is favorable for the liquid-side mass transfer performance. Moreover, the resistance in liquid-side mass transfer reduces and the diffusion of CO$_2$ in the liquid phase is improved with increasing temperature [12]. These factors promoted mass transfer and caused a higher k_L, thereby leading to an increasing $K_G a$.

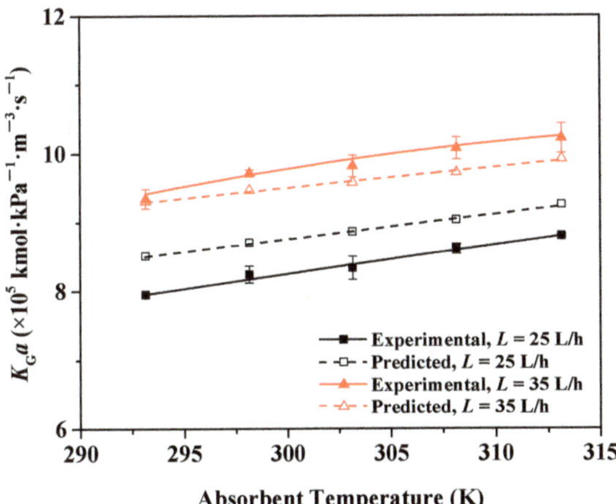

Figure 9. Effect of absorbent temperature on K_Ga in the RZB (G = 1000 L/h, N = 800 rpm, T_{gas} = 298.15 K, $y_{CO2\text{-in}}$ = 4%, C_{NaOH} = 0.15 kmol/m^3).

4.6. Effect of Absorbent Concentration

Figure 10 illustrates the effect of NaOH absorbent concentration on K_Ga in the RZB. It is noted that K_Ga increased from 7.43×10^{-5} to 9.25×10^{-5} kmol/kPa m^3 s with an increase in the absorbent concentration from 0.1 to 0.2 mol/L at the liquid flow rate of 25 L/h, while K_Ga increased from 8.62×10^{-5} to 1.07×10^{-4} kmol/kPa m^3 s as the absorbent concentration rose from 0.1 to 0.2 kmol/m^3 at the liquid flow rate of 35 L/h.

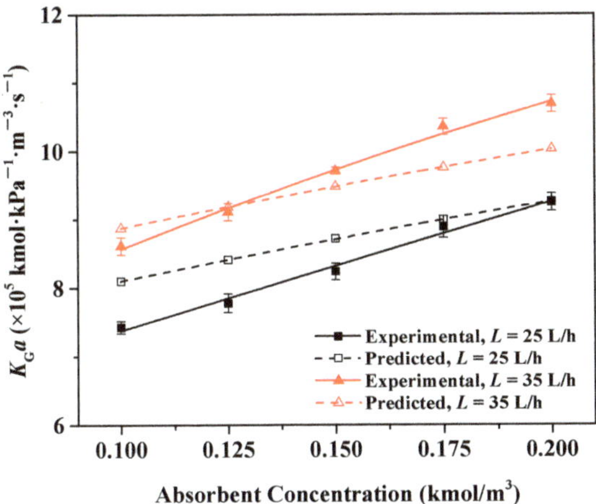

Figure 10. Effect of absorbent concentration on K_Ga in the RZB (G = 1000 L/h, N = 800 rpm, T = 298.15 K, T_{gas} = 298.15 K, $y_{CO2\text{-in}}$ = 4%).

When the concentration of NaOH solution increases, the second-order reaction rate constant between CO_2 and the absorbent rises according to the expression of k_2 shown in Table 2. Meanwhile, a higher k_2 and C_{NaOH} can concurrently enhance the pseudo-first-order reaction rate constant k_{app}, causing a higher k_L in every mass transfer zone of the RZB. Hence, K_Ga increased with an increasing NaOH concentration.

4.7. Comparison between Spray Column and RZB

Table 3 reveals comparative results of mass transfer efficiency between a spray column and the RZB. It was found that the experimental K_Ga in the spray column was less than that in the RZB on the basis of a similar gas–liquid volume ratio. Additionally, the RZB possessed a greater mass transfer efficiency accompanied by a much lower NaOH solution concentration compared to the spray column, suggesting that an obvious enhancement of mass transfer in CO_2 absorption can be realized by the RZB.

Table 3. Comparison between spray column and RZB.

	Javed et al. [33]	**This Work**
Reactor	Spray column	RZB
Mass transfer zone volume (cm^3)	9813	665
Liquid flow rate (L/h)	120–300	25–40
Gas–liquid volume ratio	32–42	25–40
Rotational speed (rpm)	/	400–1200
Absorbent	1.25 $kmol/m^3$ NaOH	0.10–0.20 $kmol/m^3$ NaOH
Inlet CO_2 concentration (%)	2.5	4
Absorbent temperature (K)	ca. 298 K	298 K
K_Ga ($10^5 \times kmol/kPa\ m^3\ s$)	2.65–4.56	6.81–11.28

Considering an enhanced dispersion and coalescence of the liquid phase in the annular regions between the static and rotating baffles, a shorter lifetime of the liquid element in the RZB is gained compared to the spray column, and the intensification of mass transfer is thus expected in the RZB, conducing to the CO_2 absorption process.

5. Conclusions

In this study, by obtaining the analytical expressions of the gas–liquid effective interfacial area, liquid-side and gas-side mass-transfer coefficients in an RZB, a mathematic model of CO_2 absorption into NaOH solution with irreversible pseudo-first-order reaction in the RZB was established to quantitatively describe the gas–liquid mass transfer process and predict K_Ga.

The K_Ga calculated by the model was consistent with the experimental data under different operating conditions. The calculated K_Ga exhibited deviations generally less than 10% in comparison with the experimental data, which demonstrated the excellent predictability of this model for CO_2 absorption in an RZB. Meanwhile, the influences of various operating conditions on K_Ga in the RZB were predicted reasonably by this model. Experimental results indicate that higher liquid flow rate, inlet gas flow rate, rotational speed, absorbent temperature, and absorbent concentration favored gas–liquid mass transfer in the RZB. It was found that the rotational speed had the largest impact on K_Ga in the RZB. This study provides the theoretical basis for potential application of RZBs in CO_2 absorption.

Supplementary Materials: The following supporting information can be downloaded at: https://www.mdpi.com/article/10.3390/pr10030614/s1, Table S1: Comparison of mass transfer performance between RZB in this work and that in Reference 10; Table S2: Values of individual and overall mass transfer factors in RZB.

Author Contributions: Conceptualization, L.S.; methodology, Z.L.; validation, A.E.; formal analysis, H.Z.; investigation, D.W.; resources, Y.L.; data curation, Z.L.; writing—original draft preparation, Z.L.; writing—review and editing, Z.L.; supervision, L.S.; project administration, L.S.; funding acquisition, L.S. All authors have read and agreed to the published version of the manuscript.

Funding: This research was funded by the National Natural Science Foundation of China, grant number 22178021.

Institutional Review Board Statement: Not applicable.

Informed Consent Statement: Not applicable.

Data Availability Statement: Not applicable.

Acknowledgments: The authors gratefully acknowledge financial support from the National Natural Science Foundation of China (No. 22178021).

Nomenclature

a	gas–liquid effective interfacial area, m^2/m^3
a_I, a_{II}, a_{III}	gas–liquid effective interfacial area in zone I, II, and III, respectively, m^2/m^3
$a_{I,i}$	gas–liquid effective interfacial area in zone I of region i, m^2/m^3
$A_{I,i}$, $A_{II,i}$	effective mass transfer area in zone I and II of region i, respectively, m^2
$A_{III,i'}$	effective mass transfer area in zone III of region i', m^2
A_2	turbulent coefficient
C_A	difference between actual and equilibrium CO_2 concentration in liquid phase, $kmol/m^3$, $C_A = C_{CO2} - C_{CO2}^*$
C_{CO2}	concentration of CO_2 in liquid phase, $kmol/m^3$
C_{CO2}^*	equilibrium concentration of CO_2 in liquid bulk, $kmol/m^3$
C_0	concentration of CO_2 at gas–liquid interface, $kmol/m^3$
C_{NaOH}	concentration of NaOH solution, $kmol/m^3$
d_i	average droplet diameter in zone I of region i, m
D_{CO2}	diffusion coefficient of CO_2 in NaOH solution, m^2/s
D_G	diffusion coefficient of CO_2 in gas phase, m^2/s
D_W	diffusion coefficient of CO_2 in water, m^2/s
g	acceleration of gravity, m/s^2
G	inlet gas flow rate, L/h
G'	inlet flow rate of inert gas (without reaction and dissolution), kmol/s
h	constant related to h_+, h_-, h_g, $m^3/kmol$
h_+, h_-, h_g	constant of cation, anion, and gas, respectively, $m^3/kmol$
h_{II}	axial length of turbulent filmy liquid, m
H	Henry's constant of NaOH solution, $kPa\ m^3/kmol$
H_W	Henry's constant of water, $kPa\ m^3/kmol$
I	ionic strength, $kmol/m^3$
k_2	second-order reaction rate constant, $m^3/kmol\ s$
k_{app}	pseudo-first-order rate constant, $1/s$
k_G	gas-side mass-transfer coefficient, $kmol/kPa\ m^2\ s$
$K_G a$	overall gas-phase volumetric mass-transfer coefficient, $kmol/kPa\ m^3\ s$
k_L	liquid-side mass-transfer coefficient, m/s
k_{L-I}, k_{L-II}, k_{L-III}	liquid-side mass-transfer coefficient in zone I, II, and III, respectively, m/s
$k_{L-I,i}$, $k_{L-II,i}$	liquid-side mass-transfer coefficient in zone I and II of region i, respectively, m/s
$k_{L-III,i'}$	liquid-side mass-transfer coefficient in zone III of region i', m/s
$k_L a$	liquid-side volumetric mass-transfer coefficient, $1/s$
$k_{OH^-}^\infty$	reaction rate constant in infinitely dilute NaOH solution, $m^3/kmol\ s$
k_s	proportionality coefficient, $kmol^2\ s/kPa^2\ m^8$
L	liquid flow rate, L/h
$n_{a,i}$	number of perforations in rotating baffle i_a
P	gas phase pressure, kPa
$P_{CO2,0}$	partial pressure of CO_2 in gas phase at gas–liquid interface, kPa
Q_G	gas volumetric flow rate, m^3/s
Q_L	liquid volumetric flow rate, m^3/s
$r_{a,i}$	radius of rotating baffle i_a, m
$r_{b,i}$	radius of static baffle i_b, m
r_h	hydraulic radius of annular region between rotating and static baffles, m
r_i	inner radius of rotor, m
r_o	outer radius of rotor, m
r_m	radius of logarithmic mean, m
r_{CO2}	reaction rate of CO_2 with NaOH solution, $kmol/m^3\ s$
R	radial coordinate of a droplet, m

$R_{CO2\text{-}II,i}$	mass transfer rate of liquid film at the gas–liquid interface in zone II of region i, mol/m^2 s
Re_G	gas Reynolds number, $Re_G = \frac{4r_h v_G \rho_G}{\mu_G}$
s	complex variable
S_i	renewal frequency of liquid film in zone II of region i, $1/s$
$t_{I,i}$	lifetime of droplets in zone I of region i, s
$t_{II,i}$	lifetime of liquid film in turbulent filmy liquid in zone II of region i, s
$t_{III,i}$	lifetime of liquid film in flying filmy liquid in zone III of region i', s
T	absorbent temperature, K
T_{gas}	gas temperature at gas inlet of RZB, K
$u_{a,i}$	tangential velocity of droplets leaving rotating baffle i_a, m/s
$u_{b,i}$	tangential velocity of flying filmy liquid leaving static baffle i_b, m/s
$u_{r,i}$	radial component of $u_{a,i}$, m/s
$u_{g,i}$	axial component of turbulent filmy liquid velocity on static baffle i_b, m/s
v_G	gas velocity in annular region between rotating and static baffles, m/s
v_θ	average tangential component of gas velocity, m/s
v_r	radial component of gas velocity, m/s
v_z	axial component of gas velocity, m/s
v_{in}	gas velocity at gas inlet of RZB, m/s
x	liquid film thickness of filmy liquid from gas–liquid interface, m
$y_{CO2\text{-}in}$	molar fraction of CO_2 in gas inlet of RZB, %
$y_{CO2\text{-}out}$	molar fraction of CO_2 in gas outlet of RZB, %
z	axial depth of rotor, m
α	angle between $u_{a,i}$ and tangent to static baffle i_b, °
ω	angular velocity, rad/s
$\varepsilon_{I,i}$	liquid holdup of droplets in zone Iof, region i
δ	average thickness of liquid film in zone II, m
ρ_G	density of gas phase, kg/m^3
ρ_L	density of NaOH solution, kg/m^3
μ_G	viscosity of gas phase, pa s
μ_L	viscosity of NaOH solution, pa s
μ_W	viscosity of water, pa s
σ	surface tension of NaOH solution, N/m

References

1. Li, Y.; Liu, P.; Wang, G.; Ji, J. Enhanced mass transfer and reduced pressure drop in a compound rotating zigzag bed. *Sep. Purif. Technol.* **2020**, *250*, 117188. [CrossRef]
2. Wang, G.; Zhou, Z.; Li, Y.; Ji, J. Qualitative relationships between structure and performance of rotating zigzag bed in distillation. *Chem. Eng. Process.-Process Intensif.* **2019**, *135*, 141–147. [CrossRef]
3. Wang, G.; Xu, Z.; Ji, J. Progress on higee distillation—Introduction to a new device and its industrial applications. *Chem. Eng. Res. Des.* **2011**, *89*, 1434–1442. [CrossRef]
4. Wang, G.; Xu, Z.; Yu, Y.; Ji, J. Performance of a rotating zigzag bed—A new Higee. *Chem. Eng. Process.-Process Intensif.* **2008**, *47*, 2131–2139. [CrossRef]
5. Wang, G.; Xu, O.; Xu, Z.; Ji, J. New Higee-rotating zigzag bed and its mass transfer performance. *Ind. Eng. Chem. Res.* **2008**, *47*, 8840–8846. [CrossRef]
6. Li, Y.; Li, X.; Wang, Y.; Chen, Y.; Ji, J.; Yu, Y.; Xu, Z. Distillation in a counterflow concentric-ring rotating bed. *Ind. Eng. Chem. Res.* **2014**, *53*, 4821–4837. [CrossRef]
7. Wang, Z.; Yang, T.; Liu, Z.; Wang, S.; Gao, Y.; Wu, M. Mass transfer in a rotating packed bed: A critical review. *Chem. Eng. Process.-Process Intensif.* **2019**, *139*, 78–94. [CrossRef]
8. Karmakar, S.; Bhowal, A.; Das, P. A comparative study of liquid-liquid extraction in different rotating bed contactors. *Chem. Eng. Process.-Process Intensif.* **2018**, *132*, 187–193. [CrossRef]
9. Liang, Z.; Wei, T.; Xie, J.; Li, H.; Liu, H. Direct conversion of terminal alkenes to aldehydes via ozonolysis reaction in rotating zigzag bed. *J. Iran. Chem. Soc.* **2020**, *17*, 2379–2384. [CrossRef]
10. Li, Y.; Lu, Y.; Liu, X.; Wang, G.; Nie, Y.; Ji, J. Mass-transfer characteristics in a rotating zigzag bed as a Higee device. *Sep. Purif. Technol.* **2017**, *186*, 156–165. [CrossRef]
11. Li, Y.; Yu, Y.; Xu, Z.; Li, X.; Liu, X.; Ji, J. Rotating zigzag bed as trayed Higee and its power consumption. *Asia-Pac. J. Chem. Eng.* **2013**, *8*, 494–506. [CrossRef]

12. Liu, Z.; Esmaeili, A.; Zhang, H.; Xiao, H.; Yun, J.; Shao, L. Carbon dioxide absorption with aqueous amine solutions promoted by piperazine and 1-methylpiperazine in a rotating zigzag bed. *Fuel* **2021**, *302*, 121165. [CrossRef]
13. Zhang, L.; Wang, J.; Xiang, Y.; Zeng, X.; Chen, J. Absorption of carbon dioxide with ionic liquid in a rotating packed bed contactor: Mass transfer study. *Ind. Eng. Chem. Res.* **2011**, *50*, 6957–6964. [CrossRef]
14. Yi, F.; Zou, H.; Chu, G.; Shao, L.; Chen, J. Modeling and experimental studies on absorption of CO_2 by benfield solution in rotating packed bed. *Chem. Eng. J.* **2009**, *145*, 377–384. [CrossRef]
15. Sun, B.; Wang, X.; Chen, J.; Chu, G.; Chen, J.; Shao, L. Simultaneous absorption of CO_2 and NH_3 into water in a rotating packed bed. *Ind. Eng. Chem. Res.* **2009**, *48*, 11175–11180. [CrossRef]
16. Wang, D.; Liu, T.; Ma, L.; Wang, F.; Shao, L. Modeling and experimental studies on ozone absorption into phenolic solution in a rotating packed bed. *Ind. Eng. Chem. Res.* **2019**, *58*, 7052–7062. [CrossRef]
17. Zhao, Z.; Zhang, X.; Li, G.; Chu, G.; Sun, B.; Zou, H.; Arowo, M.; Shao, L. Mass transfer characteristics in a rotor-stator reactor. *Chem. Eng. Technol.* **2017**, *40*, 1078–1083. [CrossRef]
18. Qian, Z.; Xu, L.; Cao, H.; Guo, K. Modeling study on absorption of CO_2 by aqueous solutions of n-methyldiethanolamine in rotating packed bed. *Ind. Eng. Chem. Res.* **2009**, *48*, 9261–9267. [CrossRef]
19. Zhang, L.; Wang, J.; Liu, Z.; Lu, Y.; Chu, G.; Wang, W.; Chen, J. Efficient capture of carbon dioxide with novel mass-transfer intensification device using ionic liquids. *AlChE J.* **2013**, *59*, 2957–2965. [CrossRef]
20. Luo, Y.; Chu, G.; Zou, H.; Wang, F.; Xiang, Y.; Shao, L.; Chen, J. Mass transfer studies in a rotating packed bed with novel rotors: Chemisorption of CO_2. *Ind. Eng. Chem. Res.* **2012**, *51*, 9164–9172. [CrossRef]
21. Tsai, C.; Chen, Y. Effective interfacial area and liquid-side mass transfer coefficients in a rotating bed equipped with baffles. *Sep. Purif. Technol.* **2015**, *144*, 139–145. [CrossRef]
22. Li, Y.; Si, J.; Arowo, M.; Liu, Z.; Sun, B.; Song, Y.; Chu, G.; Shao, L. Experimental investigation of effective gas-liquid specific interfacial area in a rotor-stator reactor. *Chem. Eng. Process.-Process Intensif.* **2020**, *148*, 107801. [CrossRef]
23. Wang, H.; Li, Y.; Ji, J. Experimental study on gas flow field in a rotating zigzag bed. *Chin. J. Process Eng.* **2010**, *10*, 56–59. (In Chinese)
24. Li, Y.; Lu, Y.; Wang, G.; Nie, Y.; Ying, H.; Ji, J.; Liu, X. Liquid entrainment and flooding in a rotating zigzag bed. *Ind. Eng. Chem. Res.* **2015**, *54*, 2554–2563. [CrossRef]
25. Lu, Y.; Li, Y.; Yu, Y.; Liu, X.; Ji, J. Study on liquid hold-up of rotating zigzag bed. *Chin. J. Process Eng.* **2014**, *14*, 568–572. (In Chinese)
26. Cao, Z. Investigation into the Absorption in a Cyclone Absorber of Spraying Liquid from Side Wall. Ph.D. Thesis, South Yangtze University, Wuxi, China, 2008. (In Chinese).
27. Guo, F.; Zheng, C.; Guo, K.; Feng, Y.; Gardner, N. Hydrodynamics and mass transfer in cross-flow rotating packed bed. *Chem. Eng. Sci.* **1997**, *52*, 3853–3859. [CrossRef]
28. Liu, G.; Ma, L.; Xing, Z. *Handbook of Chemical Property Chart*; Chemical Industry Press: Beijing, China, 2002; pp. 47–163. (In Chinese)
29. Dutcher, C.; Wexler, A.; Clegg, S.L. Surface tensions of inorganic multicomponent aqueous electrolyte solutions and melts. *J. Phys. Chem. A* **2010**, *114*, 12216–12230. [CrossRef] [PubMed]
30. Sheng, M.; Xie, C.; Sun, B.; Luo, Y.; Zhang, L.; Chu, G.; Zou, H.; Chen, J. Effective mass transfer area measurement using a CO_2–NaOH system: Impact of different sources of kinetics models and physical properties. *Ind. Eng. Chem. Res.* **2019**, *58*, 11082–11092. [CrossRef]
31. Pohorecki, R.; Moniuk, W. Kinetics of reaction between carbon dioxide and hydroxyl ions in aqueous electrolyte solutions. *Chem. Eng. Sci.* **1988**, *43*, 1677–1684. [CrossRef]
32. Rajan, S.; Kumar, M.; Ansari, M.J.; Rao, D.; Kaistha, N. Limiting gas liquid flows and mass transfer in a novel rotating packed bed (HiGee). *Ind. Eng. Chem. Res.* **2011**, *50*, 986–997. [CrossRef]
33. Javed, K.; Mahmud, T.; Purba, E. The CO_2 capture performance of a high-intensity vortex spray scrubber. *Chem. Eng. J.* **2010**, *162*, 448–456. [CrossRef]

Article

Scaling up the Process of Catalytic Decomposition of Chlorinated Hydrocarbons with the Formation of Carbon Nanostructures

Chen Wang, Yury I. Bauman, Ilya V. Mishakov, Vladimir O. Stoyanovskii, Ekaterina V. Shelepova and Aleksey A. Vedyagin *

Department of Materials Science and Functional Materials, Boreskov Institute of Catalysis SB RAS, 630090 Novosibirsk, Russia; chen.wang0726@gmail.com (C.W.); bauman@catalysis.ru (Y.I.B.); mishakov@catalysis.ru (I.V.M.); stoyn@catalysis.ru (V.O.S.); shev@catalysis.ru (E.V.S.)
* Correspondence: vedyagin@catalysis.ru

Abstract: Catalytic processing of organochlorine wastes is considered an eco-friendly technology. Moreover, it allows us to obtain a value-added product—nanostructured carbon materials. However, the realization of this process is complicated by the aggressiveness of the reaction medium due to the presence of active chlorine species. The present research is focused on the characteristics of the carbon product obtained over the Ni-Pd catalyst containing 5 wt% of palladium in various quartz reactors: from a lab-scale reactor equipped with McBain balance to scaled-up reactors producing hundreds of grams. 1,2-dichloroethane was used as a model chlorine-substituted organic compound. The characterization of the materials was performed using scanning and transmission electron microscopies, Raman spectroscopy, and low-temperature nitrogen adsorption. Depending on the reactor type, the carbon yield varied from 14.0 to 24.2 g/g(cat). The resulting carbon nanofibers possess a segmented structure with disordered packaging of the graphene layers. It is shown that the carbon deposits are also different in density, structure, and morphology, depending on the type of reactor. Thus, the specific surface area changed from 405 to 262 and 286 m^2/g for the products from reactor #1, #2, and #3, correspondingly. The main condition providing the growth of a fluffy carbon product is found to be its ability to grow in any direction. If the reactor walls limit the carbon growing process, the carbon product is represented by very dense fibers that can finally crack the reactor.

Keywords: catalytic chemical vapor deposition; 1,2-dichloroethane; Ni-Pd alloy; metal dusting; carbon nanofibers; characterization

Citation: Wang, C.; Bauman, Y.I.; Mishakov, I.V.; Stoyanovskii, V.O.; Shelepova, E.V.; Vedyagin, A.A. Scaling up the Process of Catalytic Decomposition of Chlorinated Hydrocarbons with the Formation of Carbon Nanostructures. *Processes* **2022**, *10*, 506. https://doi.org/10.3390/pr10030506

Academic Editors: Vincenzo Russo, Elio Santacesaria and Riccardo Tesser

Received: 23 January 2022
Accepted: 1 March 2022
Published: 3 March 2022

Publisher's Note: MDPI stays neutral with regard to jurisdictional claims in published maps and institutional affiliations.

1. Introduction

Due to a growing world production volume of such chlorine-substituted hydrocarbons as dichloroethane, vinylidene chloride, vinyl chloride, etc., the utilization of harmful organochlorine wastes is of great importance [1]. Most of these wastes come from the production of vinyl chloride [2], which is a starting monomer for the manufacturing of polyvinyl chloride—a highly demanded material in the modern polymer industry. Thus, the production of one ton of vinyl chloride is accompanied by the appearance of nearly 50 kg of organochlorine wastes, represented by a complex mixture of chlorinated derivatives of ethane and ethylene. All these substances are xenobiotics, i.e., they are foreign to the body or to an ecological system and therefore exert highly toxic effects on living forms, including humans [3–5].

Most conventional utilization approaches are not applicable for chlorine-containing wastes. For instance, the open burning of chlorinated substances leads to the formation of even more hazardous compounds [6,7]. The landfill of such wastes also causes ecological disasters [8]. Therefore, the catalytic processing of chlorine-substituted organics seems to be the most prospective. In this case, two scenarios can be considered: hydrogen-assisted dechlorination with the formation of corresponding unsubstituted hydrocarbons and complete decomposition with the formation of a solid carbon product [9]. The present research deals with the second scenario. As is known, the catalytic decomposition of hydrocarbons, including chlorine-substituted ones, proceeds over iron subgroup metals via the so-called carbide cycle mechanism [10]. Among these metals, nickel possesses higher resistance to chlorination, since the formation of metal chlorides completely deactivates the catalyst. The addition of odd hydrogen into the reaction mixture allows cleaning the nickel surface from the chemisorbed chlorine species, thus stipulating the pulse regime of the catalytic process. The solid carbon product of this process is represented by carbon nanofibers (CNF) with a unique segmented structure [11,12]. It should be noted that the potential of the practical application of CNFs intensively grows every year [13,14]. In recent years, such materials have been notably attractive for application in various electrochemical processes [15–20].

Despite the high resistance of Ni to the chlorine action, pure nickel catalysts undergo deactivation caused by the formation of amorphous carbon blocking the active surface of nickel particles. Doping nickel with other metals (Cr, Co, Cu, Mo, W, Pd, Pt) results in the enhanced activity and elongated stability of Ni-based catalysts [11,21,22]. It should be emphasized that both bulk alloys and specially prepared porous alloy systems can serve as precursors of self-organizing catalysts [11,23]. In this case, the initial alloy samples undergo rapid disintegration under the action of an aggressive chlorine-containing medium. This phenomenon is well-known in the literature under the name of metal dusting (MD) [24,25]. In industry, the MD process is negatively associated with the destruction of metal reactors and pipelines [26–29]. However, in recent decades, this phenomenon has been more often considered an alternative method of catalyst preparation for CNF production [30,31]. At the same time, the chemical aggressiveness of the reaction gas mixture restricts the application of metal reactors for the target decomposition of organochlorine wastes. Therefore, all heated elements of the installations used for this purpose are usually made of quartz.

Until now, the pilot-scale utilization of chlorine-containing compounds deals with their incineration or catalytic oxidation [32–36]. At the same time, their gas-phase conversion into nanostructured carbon is mainly utilized at the fundamental level. Therefore, the present work aims to scale up the process of catalytic decomposition of chlorinated hydrocarbons with the formation of the nanostructured carbon product. In general, the upscaling of any process proceeds through several stages, including the detailed analysis of kinetics and thermodynamics [37–39]. In materials science, when the target product is obtained using the chemical vapor deposition technique, the main challenge to be solved in upscaling is to maintain the uniformity and quality of the material [40].

In the present work, 1,2-dichloroethane (DCE) was used as a representative of the mentioned class of organochlorine compounds. The Ni-Pd alloy with palladium content of 5 wt% was chosen as a catalyst providing high enough efficiency [23,41]. Three different reactor types were examined. Attention was mainly paid to the characteristics of the solid carbon product being produced. The carbon deposits were studied by scanning and transmission electron microscopies, low-temperature adsorption of nitrogen, and Raman spectroscopy.

2. Materials and Methods

2.1. Synthesis of the Ni-Pd Catalyst

An Ni-Pd alloy containing 5 wt% of palladium was synthesized by a co-precipitation technique as described elsewhere [42]. Precursor salts (K_2PdCl_4 and $Ni(NO_3)_2 \cdot 6H_2O$) were taken in a certain ratio and dissolved in deionized water (100 mL). The obtained joint solution was mixed dropwise with an aqueous solution of $NaHCO_3$ (0.1 M) at a temperature of 70 °C and was vigorously stirred. The pH was kept within a range of 7.0–9.0 by adding the precipitator ($NaHCO_3$). The sediment was centrifuged, washed with distilled water, and dried at 105 °C for 12 h. After the drying, the sample was treated in a hydrogen flow at 800 °C for 30 min and cooled down in a helium flow.

2.2. Scaling up the Catalytic Decomposition Process

The catalytic experiments were performed using 1,2-dichloroethane (DCE) as a model organochlorine compound. Three reactor types were used to study the scaling up of the DCE decomposition process. The principal schemes of the experimental installations are shown in Figure 1. In all three cases, the temperature of the process was 600 °C, and the duration was 2 h. The inlet gas flows (argon and hydrogen) were regulated using digital flow mass controllers (Figure 1, position 2). The outlet reaction mixture containing aggressive HCl vapors was passed through an alkali trap (Figure 1, position 14).

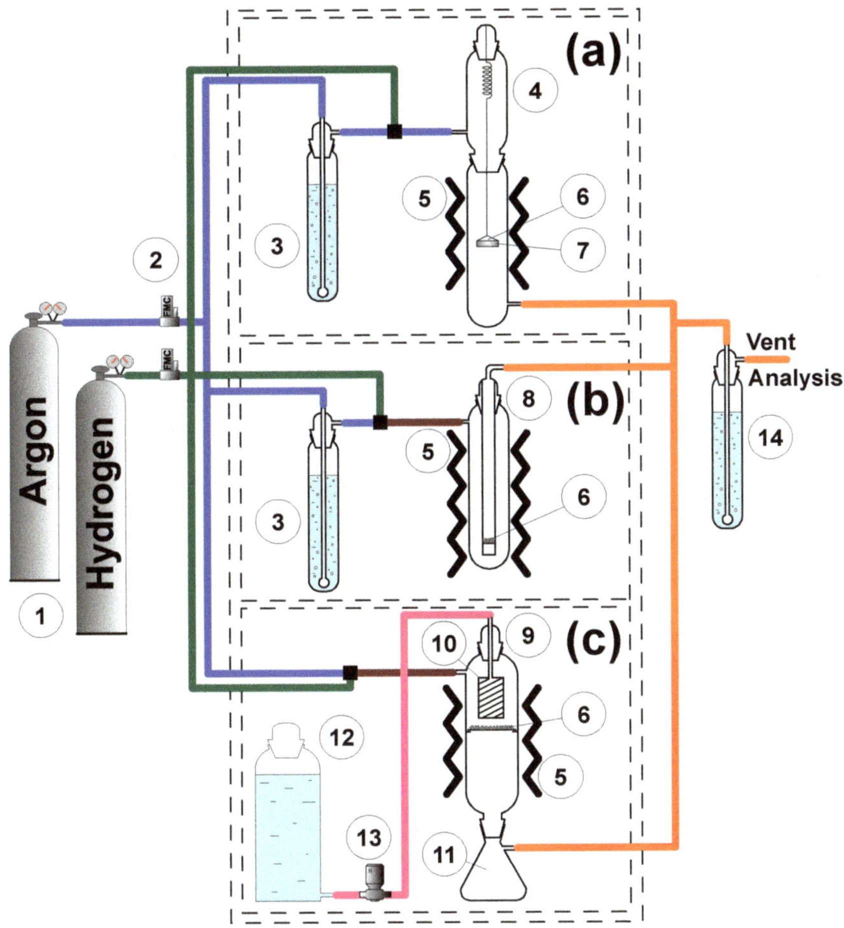

Figure 1. Principal schemes of the experimental installations equipped with reactor #1 (**a**), reactor #2 (**b**), and reactor #3 (**c**) and used for scaling up the process of catalytic decomposition of chlorinated hydrocarbons: 1—gas cylinders; 2—flow mass controllers; 3—saturator with DCE; 4—quartz reactor with McBain balance; 5—heating zone; 6—catalyst's sample; 7—foamed quartz basket; 8—lab flow-through quartz reactor; 9—scaled-up quartz reactor; 10—evaporator; 11—carbon collector; 12—vessel with DCE; 13—pump; 14—alkali trap.

The first reactor type was a tubular quartz reactor equipped with McBain balance (Figure 1a). This lab-scale reactor allowed us to measure the weight of the sample during the experiment, thus following the kinetics of the carbon deposition process. In this case, the specimen (2 mg) of NiPd alloy was placed inside a basket made of foamed quartz (Figure 1, position 7). The basket with the sample was fixed inside the reactor within the heating zone using quartz spring and quartz thread. The coefficient of a stretch of the spring was 12.68 mm/g. The stretching of the spring was monitored using a cathetometer with an accuracy of 0.01 mm. Before the experiment, the reactor was heated to 600 °C with a ramping rate of 10 °C/min in a flow of pure argon (9 L/h). Then, the reactor was fed with hydrogen (6 L/h) to reduce the surface oxide layer. After these pretreatment procedures, an argon flow (9 L/h) was passed through the saturator with DCE (Figure 1, position 3), mixed with a hydrogen flow (6 L/h), and directed to the reactor. The obtained composition of the reaction mixture was 7.5 vol% DCE, 37.5 vol% H_2, and 55 vol% Ar. Note that the reaction flow direction was downward. The sample weight was controlled every 2 min, and the time dependence of the weight gain was plotted. After 2 h, the experiment was stopped, and the reactor was cooled down to room temperature in an argon flow.

The second reactor type was a quartz tube with a quartz filter (Figure 1b). The tube was inserted into a quartz vessel with external heating elements. A specimen (100 mg) of NiPd alloy was located on the quartz filter of the reactor. All the pretreatment procedures and reaction conditions were similar to those in the previous case. In this case, the reaction flow direction was upward.

The third reactor type was an enlarged quartz tubular reactor (Figure 1c). The specimen (1 g) of NiPd alloy was placed on the quartz plate located below the evaporator. All the pretreatment procedures and reaction conditions were similar to those in the first case. The only exception was that DCE was fed into the reactor in a liquid state by the pump (Figure 1, position 13). Passing the evaporator (Figure 1, position 10), DCE vapors were mixed with argon and hydrogen flows, providing a reaction mixture composition of 7.5 vol% DCE, 37.5 vol% H_2, and 55 vol% Ar. In this case, the reaction flow direction was downward. The accumulated carbon product was unloaded via a collector (Figure 1, position 11) installed at the bottom of the reactor. The unloading of the carbon product from the plate occurred spontaneously when the plate was overfilled with the deposited carbon.

In all cases, the unloaded carbon product was weighted. The carbon yield (Y_C) was calculated as:

$$Y_C = \frac{m_C - m_{cat}}{m_{cat}} \tag{1}$$

where m_C is the carbon product weight (g), and m_{cat} is the catalyst's specimen (g). The bulk density was estimated as:

$$\varrho = \frac{V_C}{m_C} \tag{2}$$

where V_C is the carbon product volume (cm^3).

2.3. Characterization of the Carbon Product

The secondary structure of the carbon product was investigated by scanning electron microscopy (SEM) using a JSM-6460 microscope (JEOL Ltd., Tokyo, Japan). The microscope gives magnifications in a range from ×8 to ×300,000.

The studies of the primary structure and morphology of the carbon nanomaterials were performed by transmission electron microscopy using a JEM-2010 microscope (JEOL Ltd., Tokyo, Japan), working at an accelerating voltage of 200 kV and possessing a lattice resolution of 0.14 nm. The samples were ultrasonically dispersed and deposited on the carbon support over the copper grids.

The pore volume (V_{pore}) and specific surface area (SSA) of the carbon product were measured by means of low-temperature nitrogen adsorption. The adsorption/desorption isotherms were obtained at 77 K using an ASAP-2400 instrument (Micromeritics, Norcross, GA, USA).

The Raman spectra were recorded using a Horiba Jobin Yvon HR800 spectrometer (Horiba Ltd., Kyoto, Japan). An Nd:YAG laser line at 532 nm was used for excitation.

3. Results and Discussion

1,2-dichloroethane (also known as ethylene dichloride) is a colorless liquid with a boiling point of 84 °C. Therefore, in order to feed the reactor with DCE, the saturation of carrier gas with DCE vapors or direct feeding with subsequent evaporation of liquid DCE can be used.

It should also be noted that all catalytic experiments were performed at a temperature of 600 °C. Generally speaking, the reaction conditions, including the temperature, significantly affect the efficiency of the process [9,43]. At temperatures below 450 °C, the process is prevented due to the formation of nickel chloride phase on the surface of nickel particles [9]. On the contrary, at relatively high temperatures, the catalyst's surface can be blocked by the deposition of amorphous carbon. The optimal reaction temperature was found to be 600 °C, which corresponds to the highest yield of carbon product [42].

The first reaction scheme (Figure 1a) is suitable for fundamental studies of the DCE decomposition process. The use of McBain balances installed inside the reactor allows one to monitor the sample weight during the process, thus following the kinetics of carbon deposition. In this case, DCE was fed into the reactor through the saturation route. This scheme was utilized in a number of recently published papers [9,11,21–23]. The main disadvantages of this scheme should be noted. First of all, only a limited amount of the catalyst (a few mg) that can be loaded inside the basket. Secondly, the experiment has a limited duration since the carbon product overfills the basket at higher carbon yield values. Finally, the efficiency of the reaction gas contact with the catalyst is estimated not to exceed 10–15%. The other part of the reaction gas flows along the reactor walls without contact with the sample inside the basket.

Figure 2 demonstrates the carbon accumulation curve recorded using reactor #1 with McBain balance. Typically, the process is characterized by the presence of an induction period when no noticeable changes in the sample weight are seen. During this period, the carbon corrosion of the metal surface, the formation of microsized domains, and the appearance of the active particles catalyzing the growth of carbon fibers take place [23,44]. In the case of chlorine-containing medium, the induction period is significantly shortened if compared with other carburizing atmospheres [24,31] and does not exceed tens of minutes. After the induction period, the stage of intensive carbon deposition goes on (Figure 2). As one can notice from the graph, the experiment was stopped after 2 h from the beginning. The quartz basket with the accumulated carbon product is shown in Figure 3a. The basket is almost overfilled with the product. The unloaded carbon sample was weighted, and the carbon yield was estimated to be 19.3 g/g (cat) (see Table 1). The bulk density of this sample was found to be 0.038 cm^3/g.

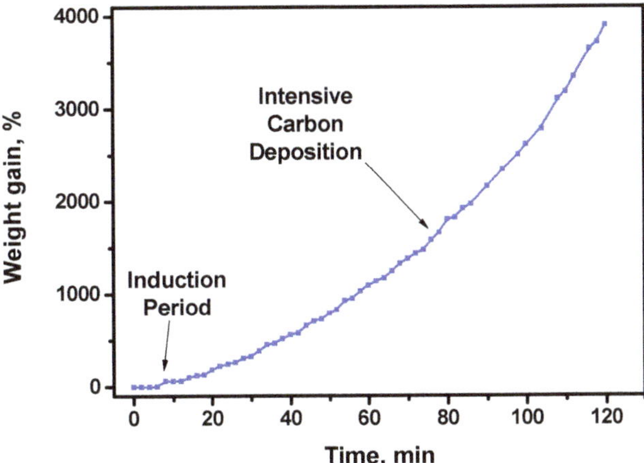

Figure 2. Kinetics of the carbon deposition over Ni-Pd catalyst at 600 °C for 2 h (reactor #1).

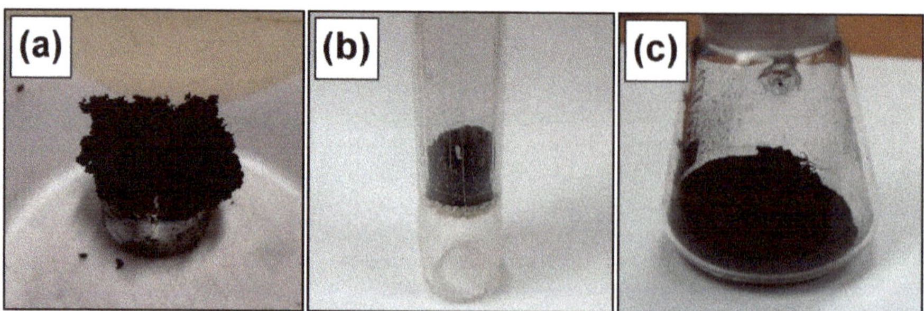

Figure 3. Photographs of the carbon products accumulated using various reactors: (**a**) reactor #1; (**b**) reactor #2; (**c**) reactor #3.

Table 1. Characteristics of the carbon products accumulated over Ni-Pd catalyst at 600 °C for 2 h.

Reactor	Carbon Yield, g/g (Cat)	Density, g/cm^3	SSA, m^2/g	V_{pore}, cm^3/g	I_D/I_G	L_a, Å
#1	19.3	0.038	405	0.87	1.79	15.9
#2	14.0	0.057	262	0.54	2.27	17.9
#3	24.2	0.036	286	0.67	2.34	18.2

In the second reactor scheme (Figure 1b), the quartz tube reactor with a quartz filter was used. The DCE-containing reaction mixture was also formed via the saturation route. The reaction gas flow goes upward through the filter and interacts with the uploaded sample. The efficiency of gas contact with the alloy particles was close to 100%, which is one of the advantages of this scheme. The second advantage is the more efficient preheating of the reaction mixture before its entrance into the reaction zone. Moreover, a large amount of the catalyst can be uploaded into the reactor, which is also a positive feature. On the other hand, the reaction volume is limited by the reactor walls. As is known, the growth of filamentous carbon proceeds in all directions. Therefore, at low gas flow rates, the carbon product growing in both the axial and radial directions halts the reactor near the filter. The gas flow rates used in the presented experiments were optimized in order to achieve the fluidized bed regime. In this case, each catalytic particle with growing carbon filaments is in motion. Such process parameters allow filling about 80% of the reactor volume. Herein, the process duration was also 2 h. Figure 3b demonstrates the reactor with the accumulated product. Despite the more efficient contact of the gas phase with the catalyst, the carbon yield was just 14.0 g/g(cat). Another feature of the obtained material is the highest bulk density—0.057 cm^3/g (Table 1). Such a low carbon yield can be explained by the observed compaction of the carbon product due to the realization of the fluidized bed regime. The formation of the dense carbon agglomerates complicates the diffusion of the reaction gas to the active metal particles.

The third reactor scheme is the next step in scaling up the process. Herein, DCE in a liquid state was fed into the evaporator installed directly inside the reactor above the main reaction zone. Note that the close location of the evaporator to the plate with the catalyst also provides nearly 100% efficiency of their contact. The reaction mixture flows in the downward direction. The use of the quartz plate to hold the sample eliminates the limitation effects of the reactor walls. The carbon product overfilling the plate falls down to the collector at the bottom of the reactor. The collector with the carbon product is shown in Figure 3c. The carbon yield after 2 h of the experiment reaches the highest value of 24.2 g/g (cat), whereas the density is lowest (Table 1).

Table 1 also presents the low-temperature nitrogen adsorption data for all the accumulated carbon products. The product from reactor #1 possesses the highest values of SSA and pore volume. In the case of reactor #2, the SSA value diminishes by more than 1.5 times, from 405 to 262 m^2/g. The pore volume also decreases by 1.6 times. These measurements correlate well with the observed changes in the bulk density. The carbon product accumulated in reactor #3 stands in the middle position between these samples, despite having the lowest density. As one can note, the textural characteristics of the nanostructured carbon product are affected by the reactor type. At the same time, Figure 4 indicates that the low-temperature nitrogen adsorption/desorption isotherms for all three carbon products are very similar to each other. In accordance with the International Union of Pure and Applied Chemistry (IUPAC) classification [45], all isotherms belong to Type IV with the H3 hysteresis loop.

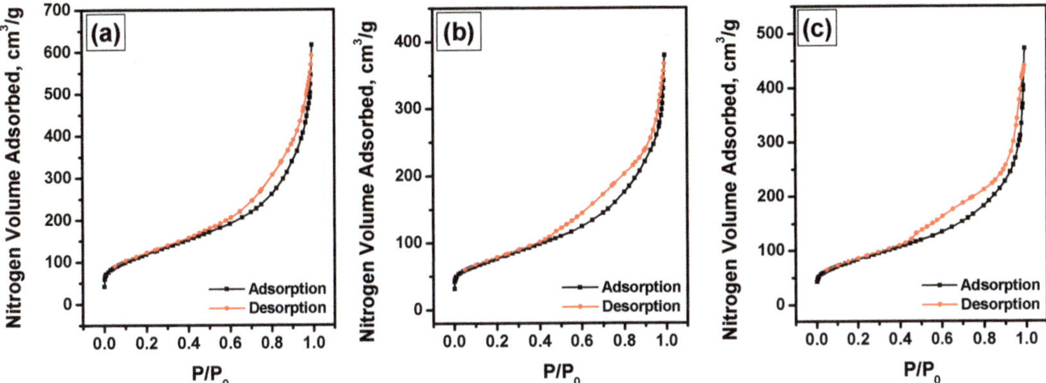

Figure 4. Low-temperature nitrogen adsorption/desorption isotherms of the carbon product obtained in different reactors: (**a**) reactor #1; (**b**) reactor #2; (**c**) reactor #3.

The precise characterization of the three obtained carbon samples was performed by means of electron microscopy and Raman spectrometry. SEM images of the carbon products from different reactors are shown in Figure 5. In general, all the samples are represented by bunches of carbon filaments. However, at a close examination, the secondary structures of the samples are different. The CNF#1 sample (from reactor #1) looks very fluffy and is composed of thin carbon filaments (Figure 5a). The metal particles catalyzing their growth can be seen. As shown in Figure 5b, the fibers have a regular segmented structure. One active metal particle can catalyze the simultaneous growth of a few fibers in various directions. On the contrary, the CNF#2 sample consists of dense lumps composed of carbon filaments (Figure 5c). The fibers are compact and well-shaped (Figure 5d). No segmented structuring is observed in this case. The third carbon product (sample CNF#3) is represented by the thinned bundles of carbon fibers (Figure 5e). Each filament possesses a segmented structure, which is not as regular as in the case of sample CNF#1 (Figure 5f). Such irregularity in the structure can be connected to a pulse character of DCE evaporation. It is important to note that, in all cases, the diameters of the representative filaments are very close to each other—about 0.8 μm.

The primary structure of the carbon filaments was examined by TEM. The corresponding images are shown in Figure 6. For sample CNF#1 (Figure 6(a1–a4)), the feather-like morphology of the filaments is easily seen. The fibers can be different in size, but they are similar in the primary structure. In the case of sample CNF#2, dense and highly ordered filaments along with areas of amorphous carbon are observed (Figure 6(b1,b2)). At the same time, some non-uniformity in the density of filaments can be supposed (Figure 6(b3,b4)). The filaments of the CNF#3 sample belong to two types (Figure 6(c1–c4)). The first type is relatively large fibers with an irregular defected structure. They partly echo the fibers of the sample CNF#1. The second type is much thinner fibers with a discrete structure. Their presence explains the lower SSA value for this sample if compared with the sample CNF#1 (Table 1).

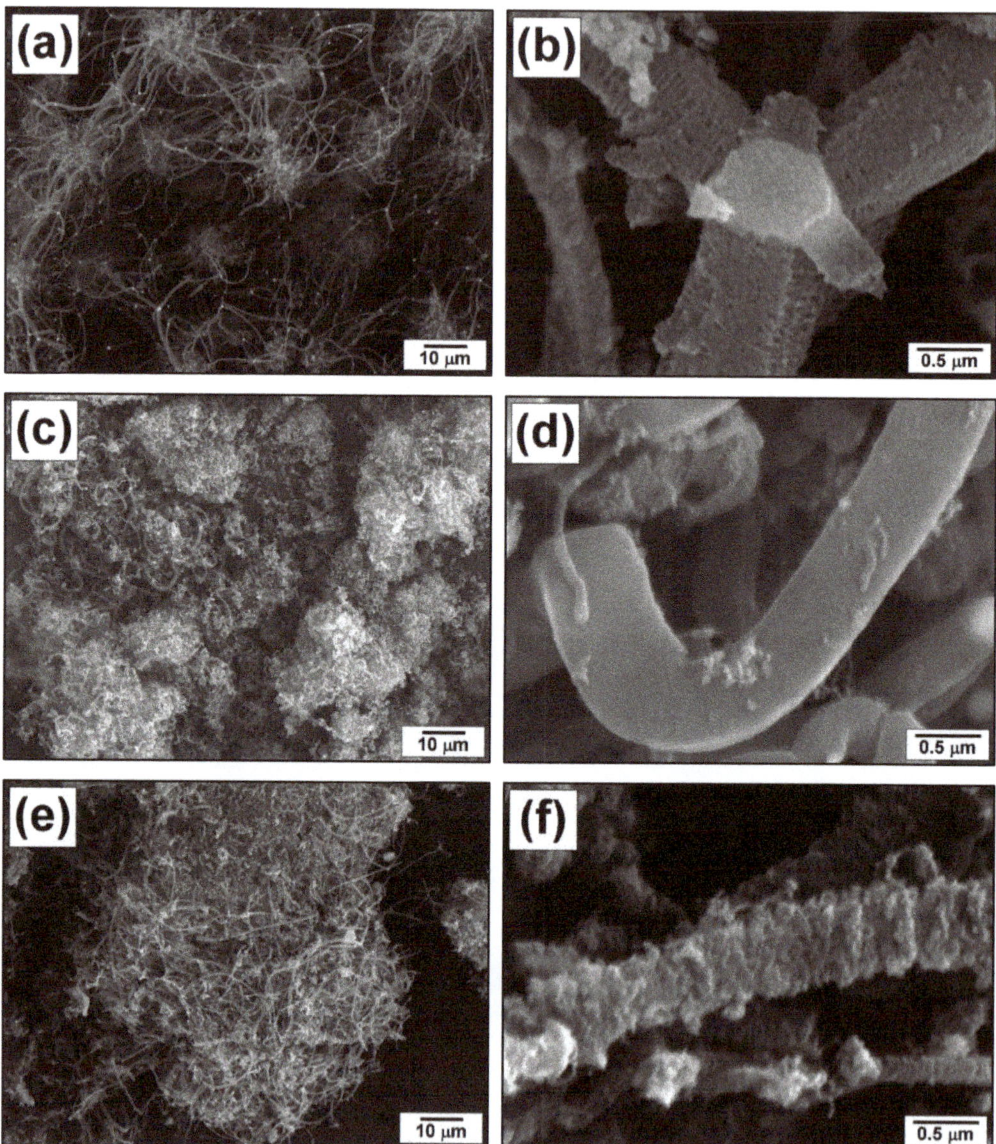

Figure 5. SEM images of the carbon nanostructures produced in different reactors: (**a**,**b**) reactor #1; (**c**,**d**) reactor #2; (**e**,**f**) reactor #3.

As opposed to microscopy techniques, Raman spectrometry gives average information regarding the bulk structure of carbon materials [46]. The Raman spectra for the accumulated carbon products are compared in Figure 7a–c. All the spectra exhibit the D band at ~1340 cm^{-1} and the G band at ~1590 cm^{-1}. The first band is connected with a breathing mode A$_{1g}$ [47,48], whereas the second band corresponds to the allowed vibrations E$_{2g}$ of the hexagonal lattice of graphite [49].

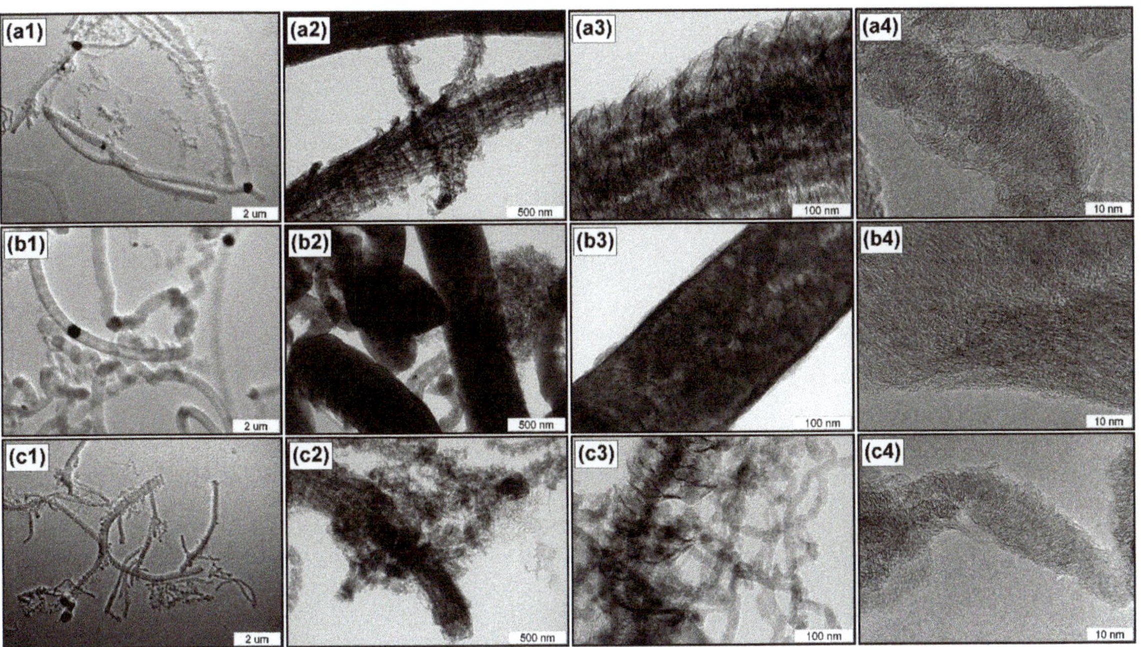

Figure 6. TEM images of the carbon nanostructures produced in different reactors: (**a1–a4**) reactor #1; (**b1–b4**) reactor #2; (**c1–c4**) reactor #3.

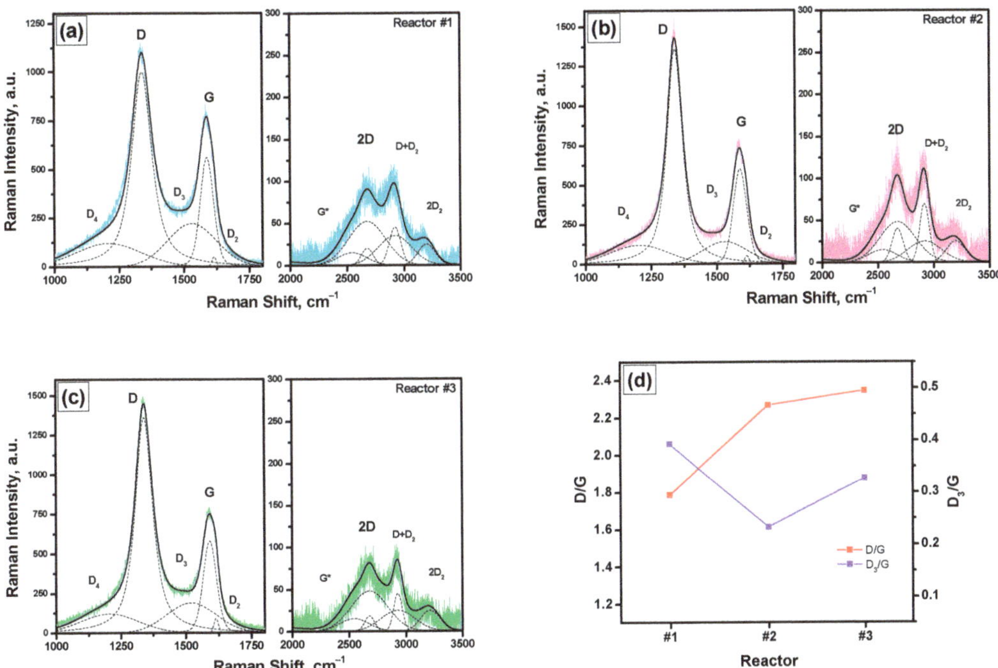

Figure 7. Raman spectra of the carbon nanostructures produced in different reactors: (**a**) reactor #1; (**b**) reactor #2; (**c**) reactor #3. Comparison of the D/G and D_3/G ratios for the studied samples (**d**).

The D_2 bands at ~1614 cm^{-1} are close to the bands at 1620 cm^{-1}, corresponding to the disordered graphitic lattice (surface graphene layers, E_{2g} symmetry) [50]. The bands D_3 at ~1520 cm^{-1} and D_4 at ~1200 cm^{-1} are observed for all samples. These bands can be assigned to amorphous carbon and disordered graphitic lattice (A_{1g} symmetry) or polyenes, which are typical for soot and related carbonaceous materials [51]. Their average relative intensities are $I_{D3/G}$ ~ 0.3 and $I_{D4/G}$ ~ 0.2. For the curve fitting of the second-order bands, the standard set of the bands 2D, D + D_2, $2D_2$, and G* ~ D_4+D was used. Along with this, two sets of the bands 2D and D + D_2 with significantly different half-width were used for the approximation of the experimental spectra. All this can correlate with the inhomogeneity of the carbon fibers or characterize the process of their growth. The typical sizes of the cluster diameter (in-plane correlation length) L_a calculated in accordance with the equation $I_D/I_G = C' (\lambda){\cdot}L_a{}^2$ proposed by Ferrari and Robertson [48] lie in a range of 16–18 Å. The intensity ratios I_D/I_G and L_a parameters for the samples are presented in Table 1. Although the changes in this parameter are not so crucial, the samples can be ranked in the following order with regard to the defectiveness of the bulk structure: CNF#3 < CNF#2 < CNF#1. As one can see in Figure 7d, along with the changes in the I_D/I_G ratio, the portion of amorphous carbon in the samples (I_{D3}/I_G) is influenced by the reactor type as well.

The main process parameters are summarized in Table 2. In order to obtain the numerical data, the Reynolds numbers and the residence time values were calculated. The Reynolds number is the main parameter defining the fluid dynamic properties of the gas flow. It was calculated using the following equation:

$$Re = \frac{u{\cdot}d}{v} \tag{3}$$

where u is a linear velocity (m·s^{-1}), d is a linear dimension defining the flow changes in the system (m), and v is a kinematic viscosity (m^2·s^{-1}). The value of the kinematic viscosity of argon at 600 °C of 8.7×10^{-5} m^2·s^{-1} was used for the calculations. The tube diameter was used as the linear dimension d.

Table 2. Process parameters for the considered reactor types.

Parameter	Reactor #1	Reactor #2	Reactor #3
Flow direction	downward	upward	downward
Specimen, mg	2	100	1000
Residence time, s	0.0003	0.016	0.16
Reynolds number	2.9	6.9	4.0
Flow regime	laminar	laminar	laminar

As is known, the flow regime is laminar when the Reynolds number is below the critical value. For the case of a gas flow in a cylindrical tube, the critical value is equal to 2000 [52]. Therefore, for all three cases, the flow regime is laminar.

The residence time was calculated as a ratio of the catalyst volume to the volume flow rate. The volume of the used Ni-Pd catalyst was estimated for the catalyst density $\rho = 1.5$ g·cm^{-3}. The upscaling of the process increases this parameter by 50 and 500 times for reactor #2 and reactor #3, correspondingly.

4. Conclusions

Chlorinated plastics are the third-most widely produced polymer materials. Besides their direct impact on the environment, the huge amounts of organochlorine wastes from their production are accumulated in depositories without appropriate processing. The use of conventional industrial equipment for the utilization of such wastes is restricted due to their high aggressiveness. The present research aimed to make a step forward in solving this problem by scaling up the process of catalytic decomposition of chlorinated organic compounds from the current fundamental level to the semi-pilot scale. All parts of the used reactor installations were made of quartz. 1,2-dichloroethane, used as a model chlorine-containing compound, was converted into nanostructured carbon materials, which are considered a value-added product. It was demonstrated that scaling up the process noticeably affects the characteristics of the deposited carbon. An increase in the bulk density of the filamentous carbon is accompanied by a drop in the specific surface area and pore volume of the materials. The variation of the reactor scheme with the reaction volume limited by the reactor walls gives the product with worst characteristics. Contrastingly, the scheme with a quartz plate used to hold the catalyst along with the liquid feeding of DCE into the reaction zone through the evaporator gives the product close in characteristics to the carbon nanostructures produced in a lab-scale reactor. It is important to note that the pump feeding of the liquid organochlorine substrate has an additional advantage. The real organochlorine wastes are represented by a complex mixture of various chlorine-substituted compounds. The use of the saturation approach is restricted due to the difference in boiling points of these chemicals. Therefore, the gas flow will be enriched with lighter components, and the gas phase composition will vary in time. In the case of pump feeding, all the components come to the reactor simultaneously at a constant concentration. The performed research has demonstrated that the process of DCE decomposition can be scaled up, thus producing a 500-times higher amount of carbon product with appropriate structural properties.

Author Contributions: Conceptualization, A.A.V.; methodology, A.A.V., I.V.M. and Y.I.B.; investigation, C.W., Y.I.B. and V.O.S.; formal analysis, E.V.S.; writing—original draft preparation, A.A.V.; writing—review and editing, A.A.V.; funding acquisition, I.V.M. All authors have read and agreed to the published version of the manuscript.

Funding: This research was funded by the Russian Ministry of Science and Higher Education (project number AAAA-A21-121011390054-1).

Institutional Review Board Statement: Not applicable.

Informed Consent Statement: Not applicable.

Data Availability Statement: The data presented in this study are available on request from the corresponding author.

Acknowledgments: Characterization of the samples was performed using the equipment of the Center of Collective Use "National Center of Catalysts Research".

Conflicts of Interest: The authors declare no conflict of interest. The funders had no role in the design of the study; in the collection, analyses, or interpretation of data; in the writing of the manuscript, or in the decision to publish the results.

References

1. Treger:, Y.; Flid, M. State of the Art and Problems of Organochlorine Synthesis. In *Chemistry beyond Chlorine*; Springer International Publishing AG: Cham, Switzerland, 2016; pp. 533–555. [CrossRef]
2. Garside, M. Global Production Capacity of Vinyl Chloride Monomer 2018 & 2023. Available online: https://www.statista.com/statistics/1063677/global-vinyl-chloride-monomer-production-capacity/ (accessed on 21 January 2022).
3. LeBaron, M.J.; Hotchkiss, J.A.; Zhang, F.; Koehler, M.W.; Boverhof, D.R. Investigation of potential early key events and mode of action for 1,2-dichloroethane-induced mammary tumors in female rats. *J. Appl. Toxicol.* **2021**, *41*, 362–374. [CrossRef] [PubMed]
4. Doan, T.Q.; Berntsen, H.F.; Verhaegen, S.; Ropstad, E.; Connolly, L.; Igout, A.; Muller, M.; Scippo, M.L. A mixture of persistent organic pollutants relevant for human exposure inhibits the transactivation activity of the aryl hydrocarbon receptor in vitro. *Environ. Pollut.* **2019**, *254*, 113098. [CrossRef] [PubMed]
5. Nunes, L.M. Organochlorine Compounds in Beached Plastics and Marine Organisms. *Front. Environ. Sci.* **2022**, *9*, 784317. [CrossRef]
6. Zhang, M.; Buekens, A.; Li, X. Open burning as a source of dioxins. *Crit. Rev. Environ. Sci. Technol.* **2017**, *47*, 543–620. [CrossRef]
7. Estrellan, C.R.; Iino, F. Toxic emissions from open burning. *Chemosphere* **2010**, *80*, 193–207. [CrossRef]
8. Li, J.; Xu, L.; Zhou, Y.; Yin, G.; Wu, Y.; Yuan, G.-L.; Du, X. Short-chain chlorinated paraffins in soils indicate landfills as local sources in the Tibetan Plateau. *Chemosphere* **2021**, *263*, 128341. [CrossRef]

9.   Mishakov, I.V.; Vedyagin, A.A.; Bauman, Y.I.; Potylitsyna, A.R.; Kadtsyna, A.S.; Chesnokov, V.V.; Nalivaiko, A.Y.; Gromov, A.A.; Buyanov, R.A. Two Scenarios of Dechlorination of the Chlorinated Hydrocarbons over Nickel-Alumina Catalyst. *Catalysts* **2020**, *10*, 1446. [CrossRef]

10.   Mishakov, I.V.; Buyanov, R.A.; Zaikovskii, V.I.; Strel'tsov, I.A.; Vedyagin, A.A. Catalytic synthesis of nanosized feathery carbon structures via the carbide cycle mechanism. *Kinet. Catal.* **2008**, *49*, 868–872. [CrossRef]

11.   Bauman, Y.I.; Rudneva, Y.V.; Mishakov, I.V.; Plyusnin, P.E.; Shubin, Y.V.; Korneev, D.V.; Stoyanovskii, V.O.; Vedyagin, A.A.; Buyanov, R.A. Effect of Mo on the catalytic activity of Ni-based self-organizing catalysts for processing of dichloroethane into segmented carbon nanomaterials. *Heliyon* **2019**, *5*, e02428. [CrossRef]

12.   Bauman, Y.I.; Mishakov, I.V.; Rudneva, Y.V.; Popov, A.A.; Rieder, D.; Korneev, D.V.; Serkova, A.N.; Shubin, Y.V.; Vedyagin, A.A. Catalytic synthesis of segmented carbon filaments via decomposition of chlorinated hydrocarbons on Ni-Pt alloys. *Catal. Today* **2020**, *348*, 102–110. [CrossRef]

13.   Gopinath, K.P.; Vo, D.-V.N.; Gnana Prakash, D.; Adithya Joseph, A.; Viswanathan, S.; Arun, J. Environmental applications of carbon-based materials: A review. *Environ. Chem. Lett.* **2020**, *19*, 557–582. [CrossRef]

14.   Bauman, Y.I.; Netskina, O.V.; Mukha, S.A.; Mishakov, I.V.; Shubin, Y.V.; Stoyanovskii, V.O.; Nalivaiko, A.Y.; Vedyagin, A.A.; Gromov, A.A. Adsorption of 1,2-Dichlorobenzene on a Carbon Nanomaterial Prepared by Decomposition of 1,2-Dichloroethane on Nickel Alloys. *Russ. J. Appl. Chem.* **2021**, *93*, 1873–1882. [CrossRef]

15.   Kamedulski, P.; Lukaszewicz, J.P.; Witczak, L.; Szroeder, P.; Ziolkowski, P. The Importance of Structural Factors for the Electrochemical Performance of Graphene/Carbon Nanotube/Melamine Powders towards the Catalytic Activity of Oxygen Reduction Reaction. *Materials* **2021**, *14*, 2448. [CrossRef] [PubMed]

16.   Barhoum, A.; Favre, T.; Sayegh, S.; Tanos, F.; Coy, E.; Iatsunskyi, I.; Razzouk, A.; Cretin, M.; Bechelany, M. 3D Self-Supported Nitrogen-Doped Carbon Nanofiber Electrodes Incorporated Co/CoOx Nanoparticles: Application to Dyes Degradation by Electro-Fenton-Based Process. *Nanomaterials* **2021**, *11*, 2686. [CrossRef]

17.   Sun, J.; Ge, Q.; Guo, L.; Yang, Z. Nitrogen doped carbon fibers derived from carbonization of electrospun polyacrylonitrile as efficient metal-free HER electrocatalyst. *Int. J. Hydrogy Energy* **2020**, *45*, 4035–4042. [CrossRef]

18.   Filik, H.; Avan, A.A. Review on applications of carbon nanomaterials for simultaneous electrochemical sensing of environmental contaminant dihydroxybenzene isomers. *Arab. J. Chem.* **2020**, *13*, 6092–6105. [CrossRef]

19.   Liu, J.; Ji, Y.-G.; Qiao, B.; Zhao, F.; Gao, H.; Chen, P.; An, Z.; Chen, X.; Chen, Y. N,S Co-Doped Carbon Nanofibers Derived from Bacterial Cellulose/Poly(Methylene blue) Hybrids: Efficient Electrocatalyst for Oxygen Reduction Reaction. *Catalysts* **2018**, *8*, 269. [CrossRef]

20.   Buan, M.E.M.; Muthuswamy, N.; Walmsley, J.C.; Chen, D.; Rønning, M. Nitrogen-doped carbon nanofibers on expanded graphite as oxygen reduction electrocatalysts. *Carbon* **2016**, *101*, 191–202. [CrossRef]

21.   Veselov, G.B.; Karnaukhov, T.M.; Bauman, Y.I.; Mishakov, I.V.; Vedyagin, A.A. Sol-Gel-Prepared Ni-Mo-Mg-O System for Catalytic Transformation of Chlorinated Organic Wastes into Nanostructured Carbon. *Materials* **2020**, *13*, 4404. [CrossRef]

22.   Bauman, Y.I.; Mishakov, I.V.; Vedyagin, A.A.; Rudnev, A.V.; Plyusnin, P.E.; Shubin, Y.V.; Buyanov, R.A. Promoting Effect of Co, Cu, Cr and Fe on Activity of Ni-Based Alloys in Catalytic Processing of Chlorinated Hydrocarbons. *Top. Catal.* **2016**, *60*, 171–177. [CrossRef]

23.   Bauman, Y.I.; Mishakov, I.V.; Rudneva, Y.V.; Plyusnin, P.E.; Shubin, Y.V.; Korneev, D.V.; Vedyagin, A.A. Formation of Active Sites of Carbon Nanofibers Growth in Self-Organizing Ni–Pd Catalyst during Hydrogen-Assisted Decomposition of 1,2-Dichloroethane. *Ind. Eng. Chem. Res.* **2019**, *58*, 685–694. [CrossRef]

24.   Wu, Q.; Zhang, J.; Young, D.J. Metal dusting behaviour of several nickel- and cobalt-base alloys in CO-H_2-H_2O atmosphere. *Mater. Corros.* **2011**, *62*, 521–530. [CrossRef]

25.   Grabke, H.J. Metal dusting. *Mater. Corros.* **2003**, *54*, 736–746. [CrossRef]

26.   Al-Meshari, A.; van Zyl, G.; Al-Musharraf, M. Metal Dusting of Process Gas Heater Convection Tubes. *J. Fail. Anal. Preven.* **2017**, *17*, 363–369. [CrossRef]

27.   Guo, X.; Vullum, P.E.; Venvik, H.J. Inhibition of metal dusting corrosion on Fe-based alloy by combined near surface severe plastic deformation (NS-SPD) and thermochemical treatment. *Corros. Sci.* **2021**, *190*, 109702. [CrossRef]

28.   Mathieu, S.; Le Pivaingt, L.; Ferry, O.; Vilasi, M.; Stuppfler, A.; Guichard, J.L.; Vande Put, A.; Monceau, D. Investigation of the metal dusting attack on the temperature range 500−700 °C using X-ray tomography. *Corros. Sci.* **2021**, *192*, 109863. [CrossRef]

29.   Bentria, E.T.; Akande, S.O.; Ramesh, A.; Laycock, N.; Hamer, W.; Normand, M.; Becquart, C.; Bouhali, O.; El-Mellouhi, F. Insights on the effect of water content in carburizing gas mixtures on the metal dusting corrosion of iron. *Appl. Surf. Sci.* **2022**, *579*, 152138. [CrossRef]

30.   Sridhar, D.; Omanovic, S.; Meunier, J.-L. Direct growth of carbon nanofiber forest on nickel foam without any external catalyst. *Diam. Relat. Mater.* **2018**, *81*, 70–76. [CrossRef]

31.   Romero, P.; Oro, R.; Campos, M.; Torralba, J.M.; Guzman de Villoria, R. Simultaneous synthesis of vertically aligned carbon nanotubes and amorphous carbon thin films on stainless steel. *Carbon* **2015**, *82*, 31–38. [CrossRef]

32.   Corella, J.; Toledo, J.M. Testing total oxidation catalysts for gas cleanup in waste incineration at pilot scale. *Ind. Eng. Chem. Res.* **2002**, *41*, 1171–1181. [CrossRef]

33.   Yim, S.D.; Koh, D.J.; Nam, I.S. A pilot plant study for catalytic decomposition of PCDDs/PCDFs over supported chromium oxide catalysts. *Catal. Today* **2002**, *75*, 269–276. [CrossRef]

34. Everaert, K.; Baeyens, J. Catalytic combustion of volatile organic compounds. *J. Hazard. Mater.* **2004**, *109*, 113–139. [CrossRef] [PubMed]
35. Yang, Y.; Yu, G.; Deng, S.B.; Wang, S.W.; Xu, Z.Z.; Huang, J.; Wang, B. Catalytic oxidation of hexachlorobenzene in simulated gas on V_2O_5-WO_3/TiO_2 catalyst. *Chem. Eng. J.* **2012**, *192*, 284–291. [CrossRef]
36. Kajiwara, N.; Noma, Y.; Matsukami, H.; Tamiya, M.; Koyama, T.; Terai, T.; Koiwa, M.; Sakai, S. Environmentally sound destruction of hexachlorobutadiene during waste incineration in commercial- and pilot-scale rotary kilns. *J. Environ. Chem. Eng.* **2019**, *7*, 103464. [CrossRef]
37. Rostrup-Nielsen, J. Reaction kinetics and scale-up of catalytic processes. *J. Mol. Catal. A-Chem.* **2000**, *163*, 157–162. [CrossRef]
38. Rönnholm, M.R.; Warna, J.; Salmi, T. Scale-up of catalytic three-phase reactors from first principles. In *European Symposium on Computer-Aided Process Engineering-14, 37th European Symposium of the Working Party on Computer-Aided Process Engineering*; Elsevier: Amsterdam, The Netherlands, 2004; pp. 277–282.
39. Piccinno, F.; Hischier, R.; Seeger, S.; Som, C. From laboratory to industrial scale: A scale-up framework for chemical processes in life cycle assessment studies. *J. Clean. Product.* **2016**, *135*, 1085–1097. [CrossRef]
40. Van der Werf, C.H.M.; Hardeman, A.J.; van Veenendaal, P.A.T.T.; van Veen, M.K.; Rath, J.K.; Schropp, R.E.I. Investigation of scaling-up issues in hot-wire CVD of polycrystalline silicon. *Thin Solid Film.* **2003**, *427*, 41–45. [CrossRef]
41. Rudneva, Y.V.; Shubin, Y.V.; Plyusnin, P.E.; Bauman, Y.I.; Mishakov, I.V.; Korenev, S.V.; Vedyagin, A.A. Preparation of highly dispersed Ni1-xPdx alloys for the decomposition of chlorinated hydrocarbons. *J. Alloys Compd.* **2019**, *782*, 716–722. [CrossRef]
42. Bauman, Y.I.; Shorstkaya, Y.V.; Mishakov, I.V.; Plyusnin, P.E.; Shubin, Y.V.; Korneev, D.V.; Stoyanovskii, V.O.; Vedyagin, A.A. Catalytic conversion of 1,2-dichloroethane over Ni-Pd system into filamentous carbon material. *Catal. Today* **2017**, *293*, 23–32. [CrossRef]
43. Wang, C.; Bauman, Y.; Mishakov, I.; Vedyagin, A.A. Features of the Carbon Nanofibers Growth over Ni-Pd Catalyst Depending on the Reaction Conditions. *Mater. Sci. Forum* **2019**, *950*, 144–148. [CrossRef]
44. Mishakov, I.V.; Bauman, Y.I.; Korneev, D.V.; Vedyagin, A.A. Metal Dusting as a Route to Produce Active Catalyst for Processing Chlorinated Hydrocarbons into Carbon Nanomaterials. *Top. Catal.* **2013**, *56*, 1026–1032. [CrossRef]
45. Sing, K.S.W. Reporting physisorption data for gas/solid systems with special reference to the determination of surface area and porosity (Recommendations 1984). *Pure Appl. Chem.* **1985**, *57*, 603–619. [CrossRef]
46. Dresselhaus, M.S.; Dresselhaus, G.; Saito, R.; Jorio, A. Raman spectroscopy of carbon nanotubes. *Phys. Rep.* **2005**, *409*, 47–99. [CrossRef]
47. Tuinstra, F.; Koenig, J.L. Raman Spectrum of Graphite. *J. Chem. Phys.* **1970**, *53*, 1126–1130. [CrossRef]
48. Ferrari, A.C.; Robertson, J. Interpretation of Raman spectra of disordered and amorphous carbon. *Phys. Rev. B* **2000**, *61*, 14095–14107. [CrossRef]
49. Nemanich, R.J.; Solin, S.A. First- and second-order Raman scattering from finite-size crystals of graphite. *Phys. Rev. B* **1979**, *20*, 392–401. [CrossRef]
50. Wang, Y.; Alsmeyer, D.C.; McCreery, R.L. Raman spectroscopy of carbon materials: Structural basis of observed spectra. *Chem. Mater.* **1990**, *2*, 557–563. [CrossRef]
51. Sadezky, A.; Muckenhuber, H.; Grothe, H.; Niessner, R.; Pöschl, U. Raman microspectroscopy of soot and related carbonaceous materials: Spectral analysis and structural information. *Carbon* **2005**, *43*, 1731–1742. [CrossRef]
52. Salmi, T.O.; Mikkola, J.-P.; Warna, J.P. *Chemical Reaction Engineering and Reaction Technology*; CRC Press: Boca Raton, FL, USA, 2008; p. 615.

Article

Study of Static and Dynamic Behavior of a Membrane Reactor for Hydrogen Production

Rubayyi T. Alqahtani [1,*], Abdelhamid Ajbar [2], Samir Kumar Bhowmik [3] and Rabab Ali Alghamdi [1]

[1] Department of Mathematics and Statistics, College of Science, Imam Mohammad Ibn Saud Islamic University (IMSIU), Riyadh 11432, Saudi Arabia; raalghamdi@imamu.edu.sa
[2] Department of Chemical Engineering, King Saud University, Riyadh 11421, Saudi Arabia; aajbar@ksu.edu.sa
[3] Department of Mathematics, University of Dhaka, Dhaka 1000, Bangladesh; bhowmiksk@du.ac.bd
* Correspondence: rtalqahtani@imamu.edu.sa

Abstract: The paper investigates the stability and bifurcation phenomena that can occur in membrane reactors for the production of hydrogen by ammonia decomposition. A simplified mixed model of the membrane reactor is studied and two expressions of hydrogen permeation are investigated. The effect of the model design and operating parameters on the existence of steady state multiplicity is discussed. In this regard, it is shown that the adsorption-inhibition effect caused by the competitive adsorption of ammonia can lead to the occurrence of multiple steady states in the model. The steady state multiplicity exists for a wide range of feed ammonia concentration and reactor residence time. The effect of the adsorption constant, the membrane surface area and its permeability on the steady state multiplicity is delineated. The analysis also shows that no Hopf bifurcation can occur in the studied model.

Keywords: membrane reactor; hydrogen; ammonia; permeation; multiple steady states

Citation: Alqahtani, R.T.; Ajbar, A.; Bhowmik, S.K.; Alghamdi, R.A. Study of Static and Dynamic Behavior of a Membrane Reactor for Hydrogen Production. *Processes* **2021**, *9*, 2275. https://doi.org/10.3390/pr9122275

Academic Editors: Elio Santacesaria, Riccardo Tesser and Vincenzo Russo

Received: 20 October 2021
Accepted: 15 December 2021
Published: 18 December 2021

Publisher's Note: MDPI stays neutral with regard to jurisdictional claims in published maps and institutional affiliations.

1. Introduction

Research in the use of renewable energy is being actively pursued in order to mitigate global climate challenges and preserve sustainability. In this regard, hydrogen is a promising clean energy that can be utilized in many applications in the chemical industry [1] and in the production of such systems as membrane fuel cells that can be used in electronic devices and transportation [2,3]. However, the use of hydrogen either in gaseous or liquid form is known to be hampered by difficulties in storage and transportation [4]. In order to overcome these logistic problems, active research is being carried out in the use of other energy carriers that can provide on site hydrogen production. Methanol and ammonia are two candidates for this purpose due to their high energy densities and limited storage problems. However, methanol reforming is known to produce greenhouse gases in the form of carbon oxides. Ammonia, on the other hand, is carbon free and can produce via catalytic or thermal cracking only hydrogen and nitrogen. Moreover, ammonia has a well-established storage and distribution infrastructure making it an economically feasible large scale energy carrier [4–6].

There has been active research in recent years in the use of catalytic membrane reactors for cracking ammonia to produce pure hydrogen [7–15]. de Nooijer et al. [10], for instance, studied the thermal stability and performance of Pd-Ag films used in membrane reactors for hydrogen production via the reforming of different feedstocks. Petriev et al. [11] carried out experimental modification of the surface of Pd-Ag films and showed that hydrogen permeability rate depends not only on the surface area but also on the structure of the membrane modifying layer. Lytkina et al. [12] carried out a comparative study of methanol steam reforming in conventional and membrane reactors. It was found that the use of a membrane reactor allowed the production of ultra pure hydrogen (i.e., CO-free) that can be directly fed to a polymer electrolyte membrane fuel cell. Barba et al. [13] investigated the merits of a Reformer and Membrane Module configuration which allowed to alternate between steam reforming (SR) reactions and hydrogen separation in the membrane module in order to overcome the thermodynamic limitations of the conventional (SR). Cechetto et al. [14] carried out an experimental decomposition of ammonia in a Pd-based membrane reactor over a Ru-based catalyst and compared its results with those of a conventional packed bed reactor. Membrane reactors were found to enjoy clear advantages over traditional reactors since the reaction and separation tasks are carried out in one single unit which results in a compact device that is more economical over conventional reactors. The use of membrane reactors offers an additional advantage in the case of ammonia decomposition to hydrogen. Compared to steam reforming of hydrocarbons, the ammonia decomposition achieve almost complete equilibrium conversion at temperatures close to 400 °C. However, at low temperatures the rate of hydrogen production is inhibited by hydrogen itself [14]. Using a membrane reactor could overcome both thermodynamic and kinetic limitations because the removal of hydrogen as it is produced would shift the equilibrium towards higher conversions. Operating at lower temperatures represents also an advantage in terms of energy requirements of the process.

While remarkable achievements have been accomplished to improve catalysts for ammonia decomposition, theoretical studies on using ammonia as feedstock for the production of hydrogen are limited [15,16]. Murmura et al. [17] carried out recently a numerical analysis of a model of membrane reactor for the production of pure hydrogen from ammonia. The authors concluded that a membrane reactor allowed the operation at low temperature, the selective removal of hydrogen and the enhancement of the reaction rate.

One area that deserves theoretical investigation is the stability behavior of such reactors. Unlike stirred tank reactors [18] or bioreactors [19], the theoretical analysis of stability of catalytic membrane reactors in general and the production of hydrogen in particular did not receive much attention in the literature. Reactors, including membrane ones, can present operational problems that can manifest themselves in form of steady state input, output multiplicities and undesired oscillations. Input multiplicities occur when different values of an input variable produce the same value of the output variable. Output multiplicities occur when the same value of an input variable produces different values of the output variable. The hysteresis phenomenon is the most common form of output multiplicity and is associated with the existence of a region of unstable behavior. Besides steady state multiplicity (input and output multiplicities), a nonlinear model can exhibit periodic or non-periodic oscillations [18,19]. Solutions that form a closed curve for some model parameters' values are called limit cycle (or periodic orbit). This is associated with bifurcation from equilibria to limit cycles. This bifurcation is commonly called a Hopf bifurcation [18]. All such nonlinear phenomena (steady state multiplicity or oscillations) are generally detrimental to the operation of the reactor. An early detection of such operating regions in reactors would be useful since this would allow the removal or at least the reduction of these operational problems in the early stage of process design and would allow the selection of operating parameters that avoid such behavior. Recently, Murmura et al. [20] investigated the bistability in membrane reactors for the ammonia cracking to hydrogen. The authors studied both mixed reactor model and fixed bed reactor model and concluded that bistability is possible in such models.

Inspired by these results, the aim of the present work is to study in more detail the stability of a model of membrane reactor for the production of hydrogen from ammonia for two different expressions of hydrogen permeation rate, and to examine the effect of operating and design parameters on the existence of such behavior. The rest of the paper is structured as follows: We present the model in Section 2. In Section 3 we analyze the boundedness of the model solutions. In Section 4 we investigate the stability of the model and in Section 5 we present numerical simulations for the different expressions of hydrogen permeation rate.

2. Process Model

Ammonia NH_3 catalytic decomposition is assumed to follow the reaction

$$2NH_3 \Longleftrightarrow 3H_2 + N_2.$$

(1)

We consider an isothermal membrane reactor with pure ammonia feed. The membrane reactor is represented by the following set of ordinary differential equations resulting from mass balances on ammonia (denoted A) and hydrogen (denoted H).

$$\frac{dC_A}{dt} = -2r + \frac{C_{A0} - C_A}{\theta},$$

(2)

$$\frac{dC_H}{dt} = -\frac{C_H}{\theta} + 3r - \frac{Ja_m}{F\theta},$$

(3)

t is time, C_A and C_H are the concentrations of ammonia and hydrogen respectively, F the volumetric flow rate, θ the reactor residence time (volume over volumetric flow rate), r the rate of reaction, a_m the membrane area and J the hydrogen flux through the selective membrane. For the reaction rate (r), studies have shown that at high temperatures, above 500 °C, the reaction rate depends only on the partial pressure of ammonia [21], whereas at temperatures lower than 500 °C and high hydrogen partial pressures, the reaction rate is inhibited by hydrogen and can be described by the following relation [21,22]:

$$r = k \left(\frac{P_A{}^2}{P_H{}^3} \right)^{0.25}.$$

(4)

where P_A and P_H are partial pressures of ammonia and hydrogen respectively and k is the reaction rate constant.

As to hydrogen permeation, assumed to be through dense palladium membranes, it is commonly described by Sieverts' law [23]:

$$J = \phi(\sqrt{p_H^r} - \sqrt{p_H^p}),$$

(5)

where p_H^r and p_H^p are partial pressures of hydrogen in the retentate and permeate sides, respectively, and ϕ is the permeance of the membrane whose value depends on the membrane characteristics and on the operating temperature. Sieverts Law is widely applied in analyzing the steady state hydrogen permeation through Pd-based membranes. Alraeesi and Gardner [24], for instance, have assessed the validity of this law and modelled the effects of pressure and temperature on the accuracy of this power law.

Neglecting hydrogen partial pressure in the permeate side, and assuming an ideal gas law, J becomes:

$$J = \phi\sqrt{RT}\sqrt{C_H},$$

(6)

where T is the temperature and R is the ideal gas constant. Hydrogen permeation is also known to be inhibited by the adsorption of ammonia on the membrane. A number of expressions were proposed in literature [25,26] to account for this inhibition. We will consider in this paper the following expression

$$J = \phi \frac{\sqrt{RT}\sqrt{C_H}}{1 + K_{ad}C_A},$$

(7)

where K_{ad} is the adsorption constant of ammonia.

3. Model Analysis

Before we analyze the stability of the model, we first need to see whether the proposed model is mathematically correct and that it can represent real situations. This means that we need to check whether the solutions (C_A, C_H) of the model remain bounded. We consider in the analysis that follows a general dependence of rate (r) and flux (J) on C_A and C_H. To facilitate our analysis we reformulate the model by defining

$$(C_A, C_H) := (x_1, x_2) = x \in \Omega, \tag{8}$$

and

$$f(x_1, x_2) = (f_1(x_1, x_2), f_2(x_1, x_2)), \tag{9}$$

where

$$f_1(x_1, x_2) = -2\, r(x_1, x_2) - \frac{(x_1 - C_{A0})}{\theta}, \tag{10}$$

and

$$f_2(x_1, x_2) = -\frac{a_m}{F\theta} J(x_1, x_2) + 3\, r(x_1, x_2) - \frac{x_2}{\theta}. \tag{11}$$

Ω is the domain of interest for the model and is defined as follow:

$$\Omega := \{x \in (0, \infty) \times (0, \infty)\}. \tag{12}$$

Then the system of differential equations can be written as

$$\dot{x} = f(x), \; x(0) = x_0 > 0. \tag{13}$$

Theorem 1. *The nonlinear autonomous model* (13) *has a bounded unique solution over* Ω *if* $x(0) \in \Omega$.

Proof. Here both functions f_1 and f_2 are continuous and Lipchitz over Ω which guarantee the uniqueness of solutions. f is also a differentiable function over Ω, and we have:

$$|f_1| \leq c_1 + c_2 \|x\| \leq c_1 + c_2 \|x\|, \tag{14}$$

and

$$|f_2| \leq c_3 |x_2| \leq c_3 \|x\|. \tag{15}$$

Therefore

$$\|f\|_\infty \leq c_1 + c\|x\|, \tag{16}$$

where c_i, $(i = 1, 3)$ depend on the parameters and $c = max\{c_2, c_3\}$. The above inequality shows a linear growth of the nonlinearity f which guarantees the uniqueness of solutions of (13) which is bounded [27,28]. $\square$

4. Local Stability

In this section the stability of the steady state solutions are studied. We consider a general dependence of flux J on C_A and C_H but we consider the rate equation as defined by Equation (4), and written as function of concentrations,

$$r(C_A, C_H) = k \left(\frac{C_A{}^2}{k_1 C_H{}^3} \right)^{0.25} \text{ with } k_1 = RT. \tag{17}$$

The Jacobian matrix of the system is

$$J = \begin{bmatrix} j_{11} & j_{12} \\ j_{21} & j_{22} \end{bmatrix}, \tag{18}$$

where,

$$j_{11} = -\theta^{-1} - 0.50\, \frac{2kC_A}{k_1\, C_H{}^3} \left(\frac{C_A{}^2}{k_1\, C_H{}^3} \right)^{-0.75}, \tag{19}$$

$$j_{12} = 0.75\, \frac{2kC_A{}^2}{k_1\, C_H{}^4} \left(\frac{C_A{}^2}{k_1\, C_H{}^3} \right)^{-0.75}, \tag{20}$$

$$j_{21} = 0.50\, \frac{3\,kC_A}{k_1\, C_H{}^3} \left(\frac{C_A{}^2}{k_1\, C_H{}^3} \right)^{-0.75} - \frac{\left(\frac{\partial}{\partial C_A} J(C_A, C_H) \right) a_m}{F\theta}, \tag{21}$$

$$j_{22} = -\theta^{-1} - 0.75\, \frac{3\,kC_A{}^2}{k_1\, C_H{}^4} \left(\frac{C_A{}^2}{k_1\, C_H{}^3} \right)^{-0.75} - \frac{\left(\frac{\partial}{\partial C_H} J(C_A, C_H) \right) a_m}{F\theta}. \tag{22}$$

The trace and determinant of the matrix are:

$$Trace(J) = -(T_1 + T_2), \tag{23}$$

$$T_1 = \frac{2 + \frac{\left(\frac{\partial}{\partial C_H} J(C_A, C_H) \right) a_m}{F}}{\theta}, \tag{24}$$

$$T_2 = \frac{(0.5 \times 2C_H + 0.75 \times 3\, C_A) kC_A}{k_1\, C_H{}^4} \left(\frac{C_A{}^2}{k_1\, C_H{}^3} \right)^{-3/4}, \tag{25}$$

and

$$Det(J) = \left[\left(\frac{C_A{}^2}{k_1\, C_H{}^3} \right)^{-3/4} \right] A,$$

$$A = 0.75\, \frac{2kC_A{}^2 a_m \frac{\partial}{\partial C_A} J(C_A, C_H)}{\theta\, k_1\, C_H{}^4 F}$$

$$+ 0.5\, \frac{\left(\frac{\partial}{\partial C_H} J(C_A, C_H) \right) a_m}{k_1\, C_H{}^3 \theta^2 F} \left(2.0 \left(\frac{C_A{}^2}{k_1\, C_H{}^3} \right)^{3/4} k_1\, C_H{}^3 + 1.0 \times 2kC_A\, \theta \right)$$

$$+ \frac{1}{k_1\, C_H{}^4 \theta^2} \left(\left(\frac{C_A{}^2}{k_1\, C_H{}^3} \right)^{3/4} k_1\, C_H{}^4 + 0.75 \times 3\, kC_A{}^2 \theta + 0.5 \times 2kC_A\, \theta\, C_H \right). \tag{26}$$

Case 1

For first case when J is given by

$$J = \phi\, \sqrt{RT} \sqrt{C_H}, \tag{27}$$

we have,

$$\frac{dJ}{dC_H} = \frac{\phi\, \sqrt{RT}}{2\sqrt{C_H}}, \tag{28}$$

$$\frac{dJ}{dC_A} = 0. \tag{29}$$

In this case the trace is negative and the determinant is positive. Thus the steady state solution is always stable.

Case 2

For the second case when J is given by

$$J = \frac{\phi \sqrt{RT} \sqrt{C_H}}{1 + K_{ad} C_A},$$ (30)

we have,

$$\frac{dJ}{dC_H} = \frac{\phi \sqrt{RT}}{2\sqrt{C_H}(K_{ad} C_A + 1)},$$ (31)

$$\frac{dJ}{dC_A} = -\frac{\phi \sqrt{RT} \sqrt{C_H} K_{ad}}{(K_{ad} C_A + 1)^2}.$$ (32)

In this case the trace is always negative but the determinant can be either positive or negative. Thus the equilibria can be either stable or unstable. However, a Hopf point can not exist in the model for both expressions of J since the trace is always strictly negative. We conclude therefore that adsorption inhibition is necessary for the occurrence of static multiplicity but oscillatory behavior is ruled out for the chosen expressions of permeation.

5. Numerical Simulations

The nominal values of the model parameters are shown in Table 1. There are two design parameters for the model: The permeance (ϕ) whose value depends on the membrane characteristics and on the operating temperature, and the membrane area a_m. The values of the design parameters were taken from experimentally validated studies [8,19]. As to the operating parameters, the pressure was selected to be 5×10^5 Pa in accordance with the work in [8,29].

Table 1. Nominal Values of Model parameters.

Parameter	Value	Source
C_{A0}	89.36 (mol/m^3)	
k	50 (kgH$_2$· Pa/m^3· s)	[19]
P	5×10^5 (Pa)	[8,29]
T	400 °C	[8,29]
ϕ	2.5×10^{-5} (kgH$_2$/s·m^2·Pa$^{0.5}$)	[19]
a_m	25×10^{-4} (m^2)	[8]
θ	1.46 (s)	[8]

For the operating temperature of 400 °C, it was selected in accordance with considerations mentioned earlier in model description, that operating at low temperatures lower than 500 °C will shift the equilibrium through the selective removal of hydrogen as well as favoring the reaction kinetics by removing the inhibiting component.

The nominal value of feed concentration shown in Table 1 was calculated from the given pressure and temperature conditions i.e., $C_{A0} = \frac{P}{RT}$. Since in practice the ammonia decomposition reaction is generally carried out at pressures between 10^5 Pa and 5×10^5 Pa and in the 300–500 °C temperature range, we considered in the simulations a range of feed concentration from 15 mol/m^3 to 105 mol/m^3. The simulations and the plots were extended sometimes above this range in order to show the range outside the multiplicity region. The nominal value of residence time (i.e., ratio of volume to volumetric flow rate) was taken to be 1.46 s which corresponded to a flow rate of 5.0×10^{-6} m^3/s given that the volume of the reactor adopted from [8] is 7.3×10^{-6} m^3. The residence time was changed by varying the feed flow rate in line with the experimental values in [8]. The maximum value of residence time was 42.9 s and corresponded to a small value of flow rate

1.7×10^{-7} m^3/s. The smallest value of residence time was, on the other hand, 0.55 s and corresponded to a flow rate of 1.33×10^{-5} m^3/s. While large values of residence times are not practical from a productivity point of view, the graphs nevertheless were extended to make sure that the model does not predict any other nonlinear phenomena at these values.

The numerical analysis was carried out using elementary continuation techniques with the help of the software AUTO [30]. The AUTO continuation package can trace out the entire steady state branches, locate the static limit points and continue these points in two parameters. Compared to dynamic simulation (i.e., time traces), continuation methods have the advantage of locating both stable and unstable solutions.

Figure 1 shows the continuity diagram for the first expression of the permeation rate (Equation (6)). As it can be seen the diagram features a single stable branch as the concentration of ammonia decreases with the residence time, while the concentration of hydrogen exhibits a maximum of C_H = 41.96 mol/m^3 at θ = 13.3 s before decreasing. Figure 2 shows the continuity diagram when the feed concentration of ammonia is chosen as the bifurcation parameter. Again one stable branch characterizes the behavior of the model. As expected, the concentration of ammonia increases with its feed concentration while the same can be said about the hydrogen concentration which increases almost linearly. Figure 3 shows the bifurcation diagram for the second expression of the permeation rate (Equation (7)). A multiplicity in the form of hysteresis can be seen in the digram. The region of instability extends from θ = 155.2 s to θ = 187.9 s. The hydrogen concentration can be seen to go through a maximum of C_H = 117.42 mol/m^3 at θ = 98.10 s that occurs outside the hysteresis region. Within the multiplicity region, abrupt changes in the feed conditions can push the operation of the membrane reactor from high conversion to low conversion conditions. The same multiplicity can be seen to occur when the ammonia concentration C_{A0} is varied (Figure 4). As C_{A0} increases, both ammonia and hydrogen concentrations increase but a region of hysteresis can be seen to exist between C_{A0} = 89.4 mol/m^3 (P = 5×10^5 Pa) and C_{A0} = 108.1 mol/m^3 (P = 6.05×10^5 Pa). This region can limit the performance of the reactor as the safe operation of the reactor dictates (barring the use of any feedback control) the operation away from the hysteresis region, that is to operate at the lower residence time (i.e., larger flow rates) since operating at larger residence times (lower flow rates) will decrease the productivity.

Next we examine the effect of some operating and design parameters of the membrane reactor on the existence of such multiplicity. Figure 5 shows the loci of the static limit points in the parameter space (θ, C_{A0}). Each curve represents the locus of the static limit point of Figures 3 and 4. It can be seen that steady state multiplicity does not exist when C_{A0} and θ are below the cusp point (A, θ = 50.1 s and C_{A0} = 31.2 mol/m^3). However, if either θ or C_{A0} is increased, the range of steady state multiplicity increases. The effect of the adsorption constant on the multiplicity is shown in Figure 6a,b. It can be seen that for the nominal value of feed concentration C_{A0}, the instability exists only when the adsorption constant is below the cusp (k_{ad} = 192.2 m^3/mol) that occurs at θ = 83.7 s. This means that, notwithstanding issues affecting productivity, the region of multiplicity can be avoided by operating the reactor at residence times lower than the cusp point. The range of multiplicity (in term of residence time) increases as the value of k_{ad} decreases. On the other hand, for a given residence time, the region of multiplicity can be seen to exist for any value of feed concentration and can be seen to increase (in term of feed concentration) with the increase in the adsorption constant. Figure 6a,b showed therefore a conflicting effect of the adsorption constant. The selection of values of operating parameters (residence time θ) and feed concentration (C_{A0}) is to be made judiciously in order avoid or at least minimize the instability introduced by the adsorption constant.

The effect of the surface area of the membrane is shown in Figure 7a,b. The multiplicity in term of residence time occurs only when the surface area is larger than a critical value, a_m = 0.00167 m^2, and the region increases as the surface area is increased. In term of C_{A0}, Figure 6b shows that the multiplicity also exists only for values of a_m larger than certain value, and increases in range as the surface area is increased. Finally, Figure 8 shows the effect of membrane permeance. The diagram is similar to Figure 7 and similar conclusions can be drawn concerning the range of hysteresis in term of θ and C_{A0}. An important question to ask is whether the range of parameters in which the multiplicity exists can be found in practice. We have seen that when the feed concentration was the bifurcation parameter, the multiplicity region seemed to occur within practical values such as the ones reported in [8]. The values, on the other hand, of residence time at which instability occurred are larger than what would be met in practical operations. However, a note should be made about the ammonia adsorption constant. It is the only parameter for which reliable data are still not available [20], therefore the diagrams could change as more accurate values are obtained. Figure 6a has effectively shown the important effect of the adsorption constant. In any case, experimental investigation of these instability phenomena is still needed to confirm our results as well as those reported in similar previous studies [20,31].

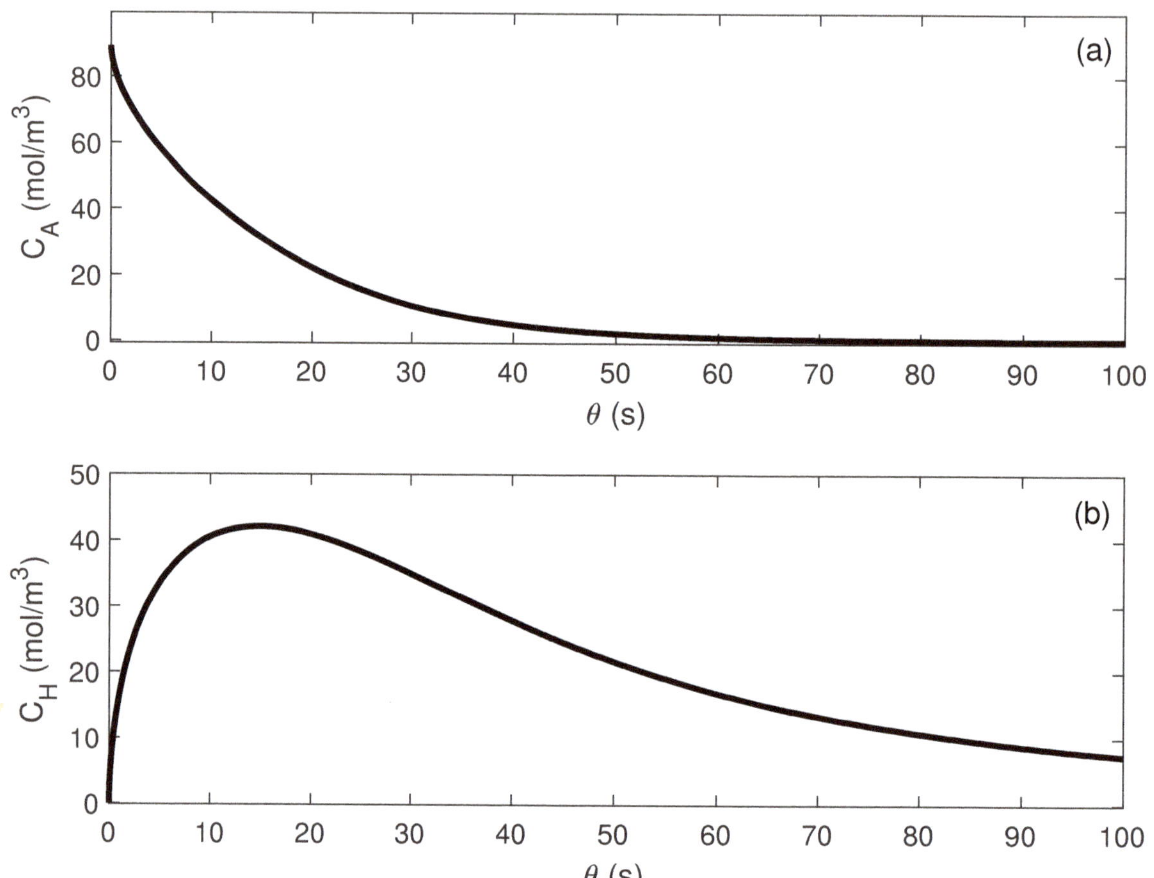

Figure 1. Continuity diagram for the expression of permeation rate given by Equation (6): (**a**) Variations of ammonia concentration with residence time; (**b**) Variations of hydrogen concentration with residence time. Solid line (-) stable branch.

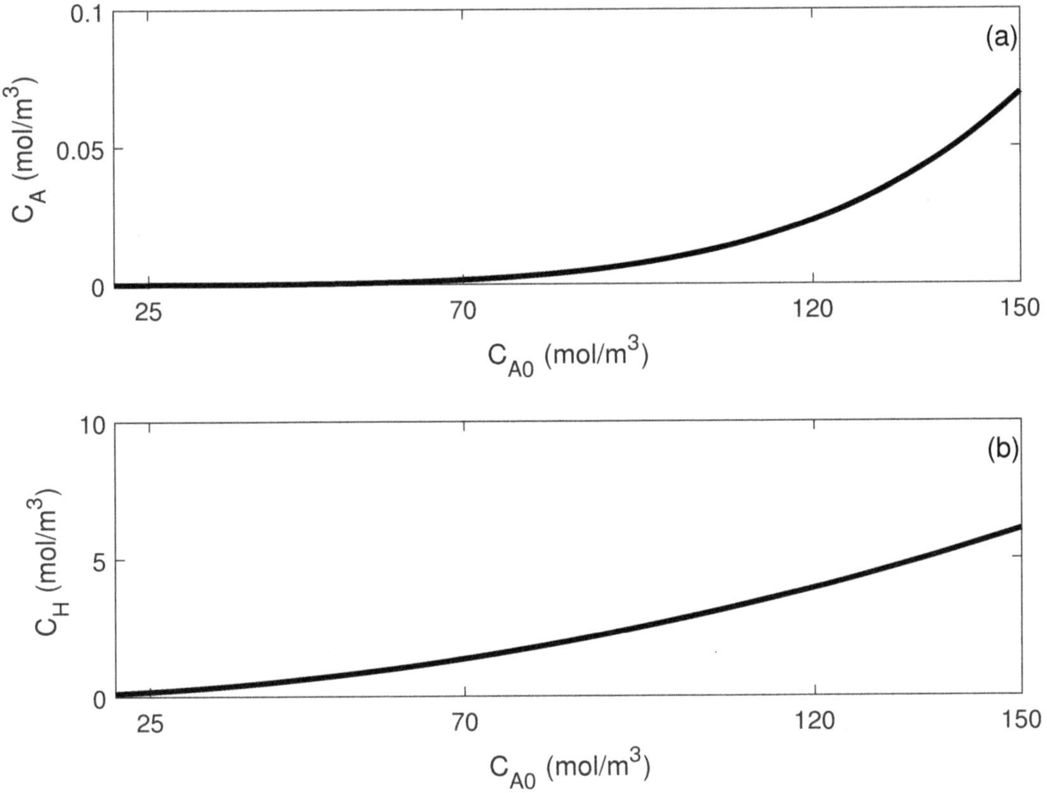

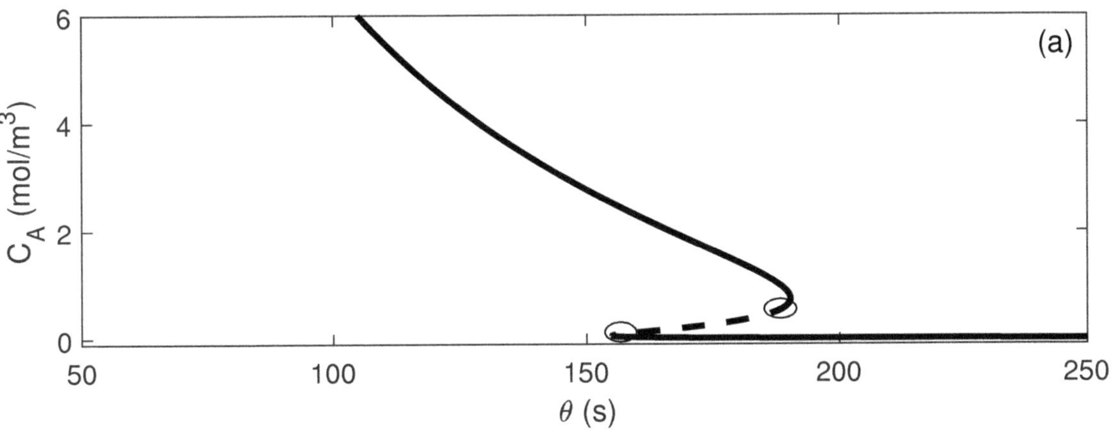

Figure 2. Continuity diagram for the expression of permeation rate given by Equation (6): (**a**) Variations of ammonia concentration with feed ammonia concentration; (**b**) Variations of hydrogen concentration with feed ammonia concentration. Solid line (-) stable branch.

Figure 3. *Cont.*

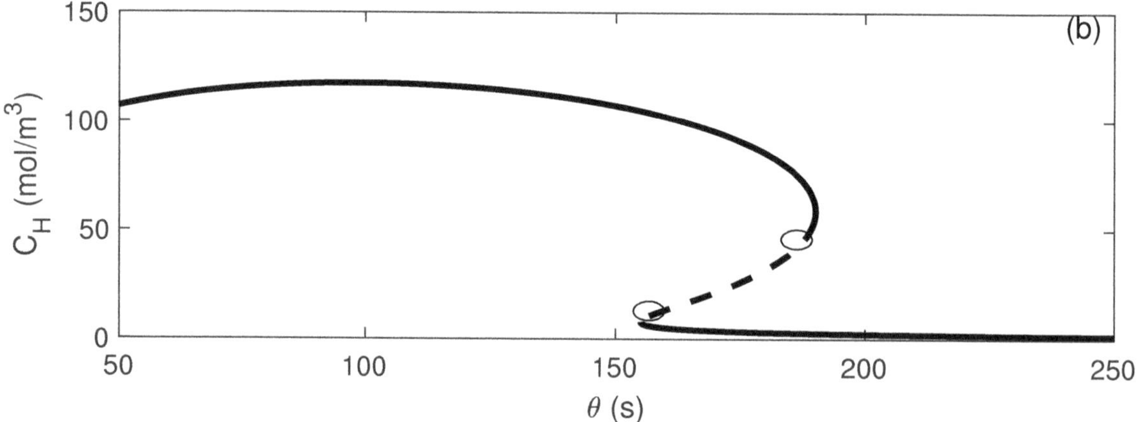

Figure 3. Continuity diagram for the expression of permeation rate given by Equation (7): (**a**) Variations of ammonia concentration with residence time; (**b**) Variations of hydrogen concentration with residence time. Solid line (-) stable branch; dashed line (–) unstable branch; circle (o) static limit point.

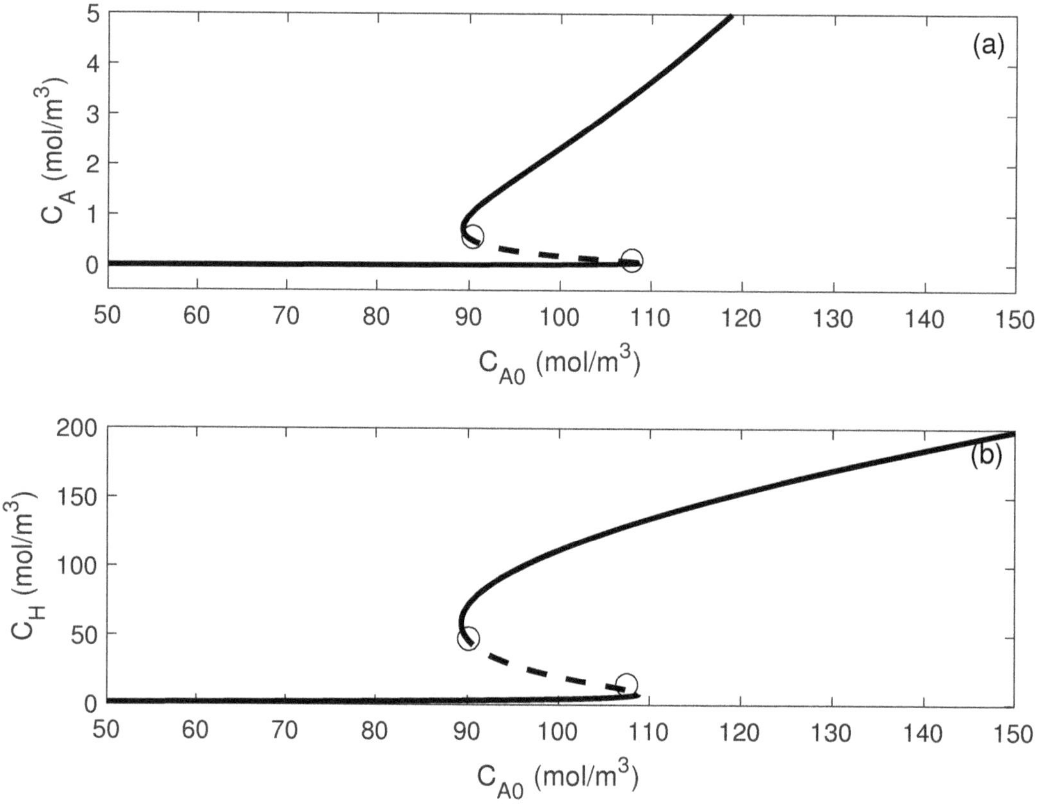

Figure 4. Continuity diagram for the expression of permeation rate given by Equation (7): (**a**) Variations of ammonia concentration with feed ammonia concentration; (**b**) Variations of hydrogen concentration with feed ammonia concentration. Solid line (-) stable branch; dashed line (–) unstable branch; circle (o) static limit point.

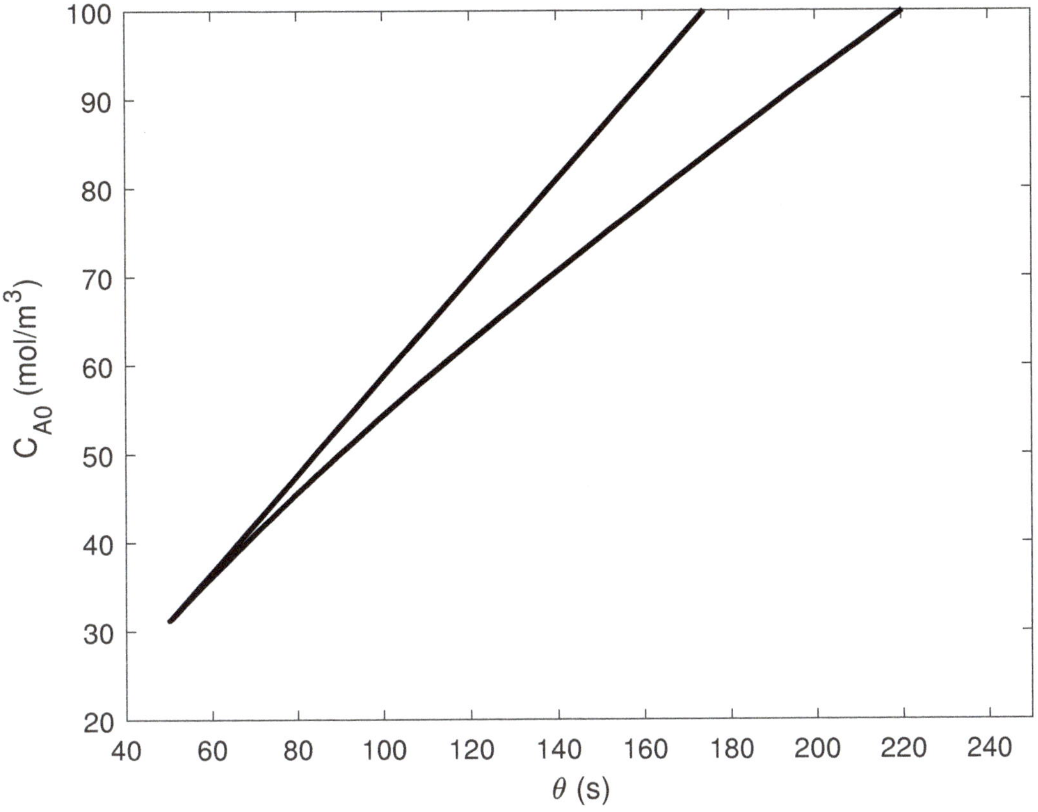

Figure 5. Two-parameter continuation showing the loci of static limit points of Figures 3 and 4 in the parameter space (θ, C_{A0}).

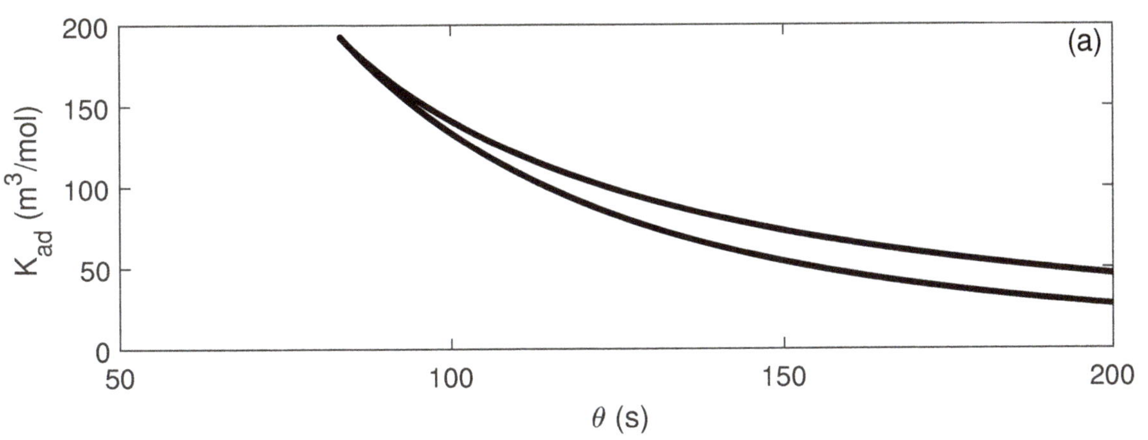

Figure 6. *Cont.*

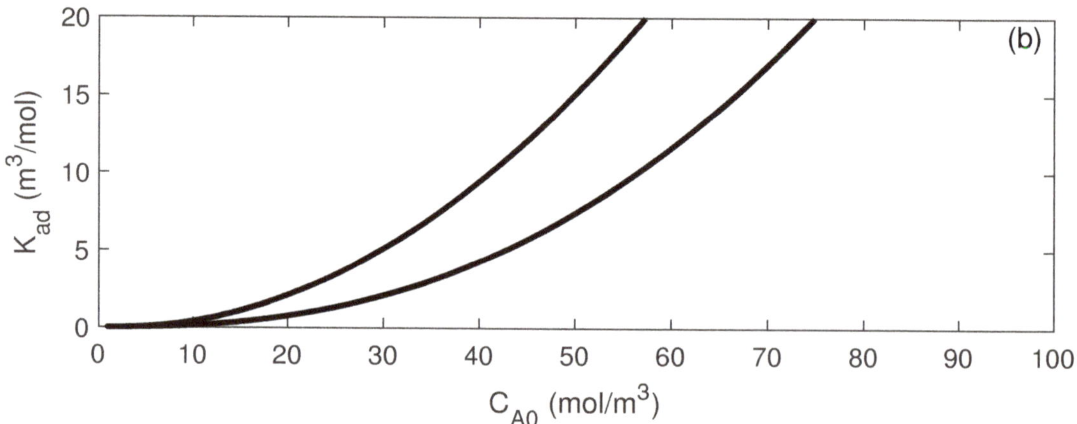

Figure 6. Effect of adsorption constant on the locus of the static limit points of Figures 3 and 4: (**a**) parameter space (θ, K_{ad}); (**b**) parameter space (C_{A0}, K_{ad}).

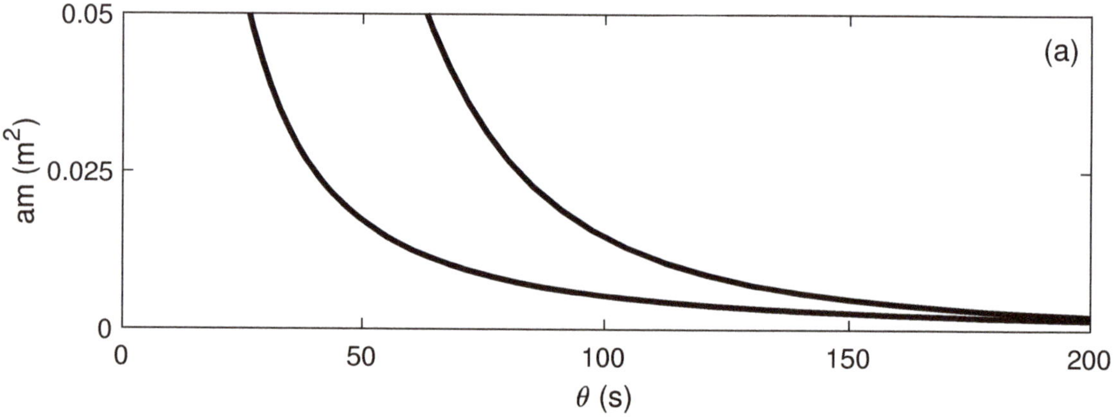

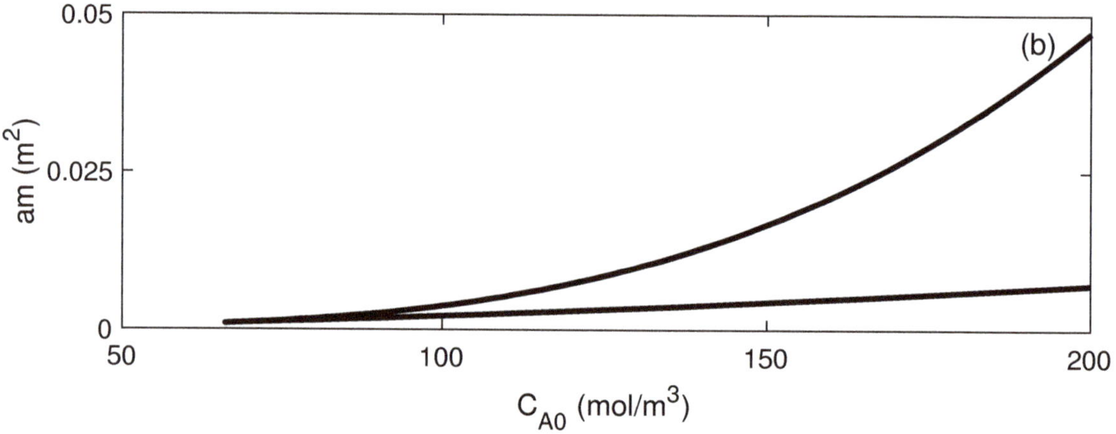

Figure 7. Effect of area of membrane on the locus of the static limit points of Figures 3 and 4: (**a**) parameter space (θ, a_m); (**b**) parameter space (C_{A0}, a_m).

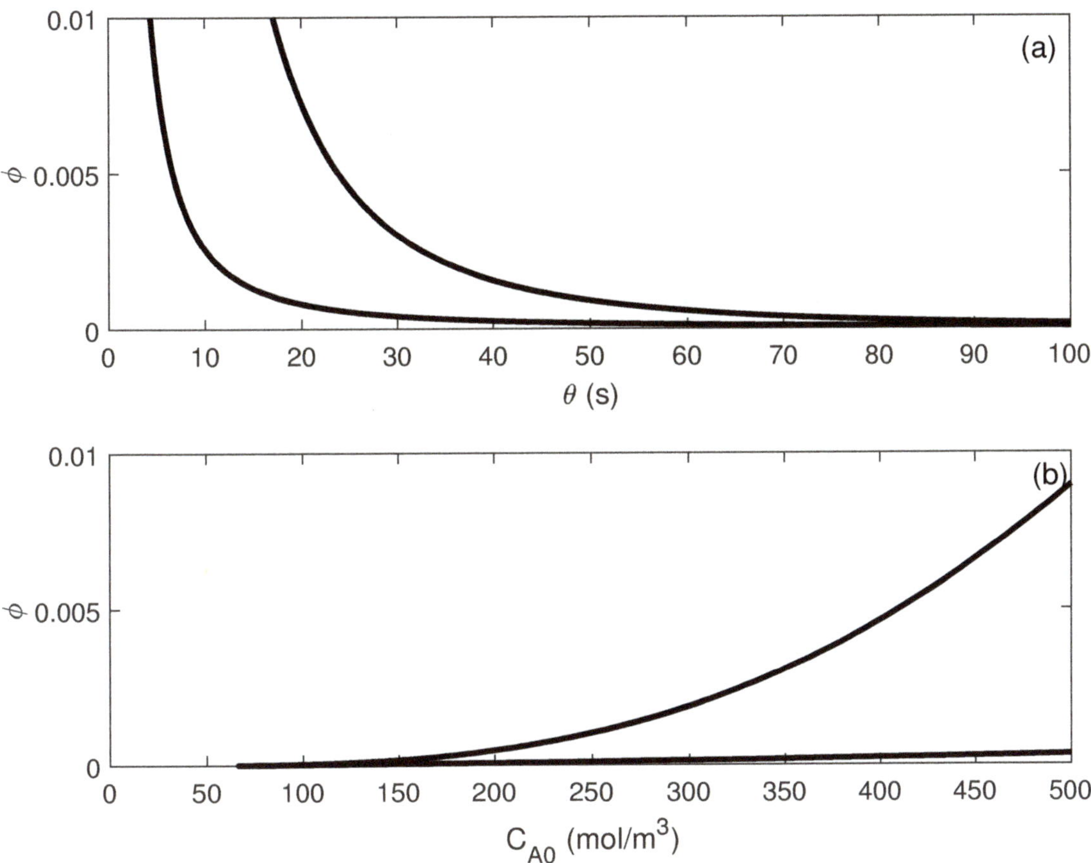

Figure 8. Effect of membrane permeance on the locus of the static limit points of Figures 3 and 4: (**a**) parameter space (θ, ϕ); (**b**) parameter space (C_{A0}, ϕ).

6. Conclusions

The paper investigated the occurrence of steady state multiplicity in a membrane reactor for the cracking of ammonia into hydrogen. Using a simple mixed model for the reactor the analysis was carried out for two expressions of hydrogen permeation rates. An adsorption-inhibition form was found to be necessary for the occurrence of static multiplicity but no oscillatory behavior i.e., Hopf bifurcations was found in the studied expressions of hydrogen flux.

Instability in the form of hysteresis was found to exist for a wide range of operating conditions (residence time and feed ammonia concentration). From a practical point of view, the region of hysteresis is detrimental to the operation of the reactor and should be avoided. In this regard, for given residence time (feed flow rate) lowering feed ammonia concentration decreases the range of instability. On the other hand, for given feed ammonia concentration, operating at smaller residence time (high flow rate) decreases this multiplicity region. Larger values of membrane areas or its permeance were found to accentuate the region of multiplicity.

For given value of feed ammonia concentration, static multiplicity is expected only when the value of adsorption constant is smaller than a critical value. On the contrary, for given flow rate the range of static multiplicity, in term of feed ammonia concentration is expected to accentuate when the value of adsorption constant is large. A careful choice should be made for the residence time (i.e., flow rate) and feed concentration (i.e., pressure) in order to avoid the instability region while maintaining a reasonably high level of reactor productivity.

It could be well argued that the mixed model studied in this paper is too simple to represent the actual operation of the membrane reactor which should be modeled as a distributed parameter system (represented by partial differential equations). A rigorous modeling could include a heterogeneous fixed bed membrane reactor such as the one developed in [15] for the production of ultra-pure hydrogen where different reactor configurations were compared. Gomez-Garci et al. [32] also proposed valuable systematic design guidelines for the design and analysis of the membrane reactor for ammonia decomposition. Recently, Murmura et al. [33] also provided a review and critical analysis of the modeling of fixed bed membrane reactors for hydrogen from steam reforming reactions. Nevertheless, the merit of this study is that it is one of few studies, along with [20], that have tackled the instability that could be found in the operation of membrane reactors for ammonia decomposition to hydrogen. We think that the trends obtained using the simple mixed model can be useful to shed light at least qualitatively on the complex stability behavior expected in such reactors.

Author Contributions: Conceptualization, R.T.A. and A.A.; methodology, R.T.A., A.A. and S.K.B.; writing—original draft preparation, R.T.A., A.A. and S.K.B.; writing—review and editing, R.T.A., A.A., S.K.B. and R.A.A. All authors have read and agreed to the published version of the manuscript.

Funding: This research was supported by the Deanship of Scientific Research, Imam Mohammad Ibn Saud Islamic University, Saudi Arabia, Grant No. (20-13-12-015).

Institutional Review Board Statement: Not applicable.

Informed Consent Statement: Not applicable.

Conflicts of Interest: The authors declare no conflict of interest. The funders had no role in the design of the study; in the collection, analyses, or interpretation of data; in the writing of the manuscript, or in the decision to publish the results.

References

1. Abdin, Z.; Zafaranloo, A.; Rafee, A.; Merida, W.; Lipinski, W.; Khalilpour, K.R. Hydrogen as an energy vector. *Renew. Sustain. Energy Rev.* **2020**, *120*, 109620. [CrossRef]
2. Jo, Y.S.; Cha, J.; Lee, C.H.; Jeong, H.; Yoon, C.H.; Nam, S.W.; Han, J. A viable membrane reactor option for sustainable hydrogen production from ammonia. *J. Power Sources* **2018**, *400*, 518–526. [CrossRef]
3. Rambhujun, N.; Salman, M.S.; Wang, T.; Pratthana, C.; Sapkota, P.; Costalin, M.; Lai, Q.; Aguey-Zinsou, K.F. Renewable hydrogen for the chemical industry. *MRS Energy Sustain.* **2020**, *7*, E33. [CrossRef]
4. Kojima, Y. Hydrogen storage materials for hydrogen and energy carriers. *Int. J. Hydrog. Energy* **2019**, *44*, 18179–18192. [CrossRef]
5. Lu, N.; Xie, D. Novel Membrane Reactor Concepts for Hydrogen Production from Hydrocarbons: A Review. *Int. J. Chem. React.* **2016**, *14*, 1–31. [CrossRef]
6. Hasan, M.H.; Mahlia, T.M.I.; Mofijur, M.; Rizwanul Fattah, I.M.; Handayani, F.; Ong, H.C.; Silitonga, A.S. A Comprehensive Review on the Recent Development of Ammonia as a Renewable Energy Carrier. *Energies* **2021**, *14*, 3732. [CrossRef]
7. Itoh, N.; Kikuchi, Y.; Furusawa, T.; Sato, T. Tube-wall catalytic membrane reactor for hydrogen production by low-temperature ammonia decomposition. *Int. J. Hydrog.* **2021**, *46*, 20257–20265. [CrossRef]
8. Zhang, Z.; Liguori, S.; Fuerst, T.F.; Way, J.D.; Wolden, C.A. Efficient Ammonia Decomposition in a Catalytic Membrane Reactor To Enable Hydrogen Storage and Utilization. *ACS Sustain. Chem. Eng.* **2019**, *7*, 5975–5985. [CrossRef]
9. Lamb, K.; Dolan, M.D.; Kennedy, D.F. Ammonia for hydrogen storage; A review of catalytic ammonia decomposition and hydrogen separation and purification. *Int. J. Hydrog.* **2019**, *44*, 3580–3593. [CrossRef]
10. Nooijer, N.D.; Arratibel Plazaola, A.; Melendez Rey, J.; Fernandez, E.; Pacheco Tanaka, D.A.; Sint Annaland, M.; Gallucci, F. Long-Term Stability of Thin-Film Pd-Based Supported Membranes. *Processes* **2019**, *7*, 106. [CrossRef]
11. Petriev, I.; Pushankina, P.; Bolotin, S.; Lutsenko, I.; Kukueva, E.; Baryshev, M. The influence of modifying nanoflower and nanostar type Pd coatings on low temperature hydrogen permeability through Pd-containing membranes. *J. Membr. Sci.* **2021**, *620*, 118894. [CrossRef]
12. Lytkina, A.A.; Orekhova, N.V.; Ermilova, M.M.; Petriev, I.S.; Baryshev, M.G.; Yaroslavtsev, A.B. Ru-Rh based catalysts for hydrogen production via methanol steam reforming in conventional and membrane reactors. *Int. J. Hydrog. Energ.* **2019**, *44*, 13310–13322. [CrossRef]

13. Barba, D.; Capocelli, M.; De Falco, M.; Franchi, G.; Piemonte, V. Mass Transfer Coefficient in Multi-Stage Reformer/Membrane Modules for Hydrogen Production. *Membranes* **2018**, *8*, 109. [CrossRef]
14. Cechetto, V.; Di Felice, L.; Medrano, J.A.; Makhloufi, C.; Zuniga, J.; Gallucci, F. H2 production via ammonia decomposition in a catalytic membrane reactor. *Fuel Process. Technol.* **2021**, *216*, 106772. [CrossRef]
15. Abashar, M.E.E. Ultra-clean hydrogen production by ammonia decomposition. *J. King Saud Univ. Eng. Sci.* **2018**, *30*, 2–11. [CrossRef]
16. Hla, S.S.; Dolan, M.D. CFD modelling of a membrane reactor for hydrogen production from ammonia. *IOP Conf. Ser. Mater. Sci. Eng.* **2018**, *297*, 012027.
17. Murmura, M.A.; Annesini, M.C. Numerical Analysis of the Performance of Membrane Reactors for NH3 Decomposition. *Chem. Eng. Trans.* **2021**, *86*, 829–834.
18. Alhumaizi, K.; Aris, R. *Surveying a Dynamical System: A Study of the Gray-Scott Reaction in a Two-Phase Reactor*; Chapman and Hall/CRC: London, UK, 1996.
19. Ajbar, A.; Alhumaizi, K. *Dynamics of the Chemostat: A Bifurcation Theory Approach*; CRC Press Taylor and Francis: London, UK, 2011.
20. Murmura, M.A.; Annesini, M.C.; Sheintuch, M. Bistability in membrane reactors due to membrane inhibition by competitive adsorption of reactants. *Chem. Eng. J.* **2018**, *334*, 1594–1604. [CrossRef]
21. Chellappa, A.S.; Fischer, C.M.; Thomson, W.J. Ammonia decomposition kinetics over Ni-Pt/Al2O3 for PEM fuel cell applications. *Appl. Catal. A* **2002**, *227*, 231–240. [CrossRef]
22. Zhang, J.; Xu, H.; Li, W. Kinetic study of NH3 decomposition over Ni nanoparticles: The role of La promoter, structure sensitivity and compensation effect. *Appl. Catal. A* **2005**, *296*, 257–267. [CrossRef]
23. Ward, T.; Dao, T. Model of hydrogen permeation behavior in palladium membranes. *J. Membr. Sci.* **1999**, *153*, 211–231. [CrossRef]
24. Alraeesi, A.; Gardner, T. Assessment of Sieverts Law Assumptions and n Values in Palladium Membranes: Experimental and Theoretical Analyses. *Membranes* **2021**, *11*, 778. [CrossRef]
25. Israni, S.; Harold, M. Methanol steam reforming in Pd-Ag membrane reactors: Effects of reaction system species on transmembrane hydrogen flux. *Ind. Eng. Chem. Res.* **2010**, *49*, 10242–20150. [CrossRef]
26. Catalano, J.; Baschetti, M.G.; Sarti, G. Hydrogen permeation in palladium based membranes in the presence of carbon monoxide. *J. Membr. Sci.* **2010**, *362*, 221–233. [CrossRef]
27. Coddington, E.A.; Levinson, N.; *Theory of Ordinary Differential Equations*; McGraw-Hill Book Company, Inc.: New York, NY, USA, 1955.
28. Serhani, M.; Gouzé, J.; Raissi, N. Dynamical study and robustness for a nonlinear wastewater treatment model. *Nonlinear Anal. Real World Appl.* **2010**, *12*, 487–500. [CrossRef]
29. Murmura, M.A.; Annesini, M.C.; Cerbelli, S. Designing the optimal geometry of a membrane reactor for hydrogen production from a pre-reformed gas mixture based on the extent of the reaction boundary layer. *Chem. Eng. Process* **2017**, *120*, 148–160. [CrossRef]
30. Doedel, E.J. *AUTO: Continuation and Bifurcation Software for Ordinary Differential Equations*; Concordia University Press: Montreal, QC, Canada, 2007.
31. Chen, Z.; Elnashaie, S. Bifurcation Behavior and Efficient Pure Hydrogen Production for Fuel Cells Using a Novel Autothermic Membrane Circulating Fluidized-Bed (CFB) Reformer: Sequential Debottlenecking and the Contribution of John Grace. *Ind. Eng. Chem. Res.* **2004**, *43*, 5449–5459. [CrossRef]
32. Gómez-García, M.Á., Dobrosz-Gómez, J.I.; Fontalvo, J.; Rynkowski, J.M. Membrane reactor design guidelines for ammonia decomposition. *Catal. Today* **2012**, *191*, 165e8. [CrossRef]
33. Murmura, M.A.; Cerbelli, S.; Annesini, M.C. Modeling Fixed Bed Membrane Reactors for Hydrogen Production through Steam Reforming Reactions: A Critical Analysis. *Membranes* **2018**, *8*, 34. [CrossRef]

Review

The Perspective of Using the System Ethanol-Ethyl Acetate in a Liquid Organic Hydrogen Carrier (LOHC) Cycle

Elio Santacesaria [1,*], **Riccardo Tesser** [2,3], **Sara Fulignati** [3,4] **and Anna Maria Raspolli Galletti** [3,4]

1. Eurochem Engineering Ltd., 20139 Milan, Italy
2. NICL–Department of Chemical Science, University of Naples Federico II, 80126 Naples, Italy
3. Consorzio Interuniversitario Reattività Chimica e Catalisi (CIRCC), 70121 Bari, Italy
4. Department of Chemistry and Industrial Chemistry, University of Pisa, 56124 Pisa, Italy
* Correspondence: elio.santacesaria@eurochemengineering.com

Abstract: Starting from bioethanol it is possible, by using an appropriate catalyst, to produce ethyl acetate in a single reaction step and pure hydrogen as a by-product. Two molecules of hydrogen can be obtained for each molecule of ethyl acetate produced. The mentioned reaction is reversible, therefore, it is possible to hydrogenate ethyl acetate to reobtain ethanol, so closing the chemical cycle of a Liquid Organic Hydrogen Carrier (LOHC) process. In other words, bioethanol can be conveniently used as a hydrogen carrier. Many papers have been published in the literature dealing with both the ethanol dehydrogenation and the ethyl acetate hydrogenation to ethanol so demonstrating the feasibility of this process. In this review all the aspects of the entire LOHC cycle are considered and discussed. We examined in particular: the most convenient catalysts for the two main reactions, the best operative conditions, the kinetics of all the reactions involved in the process, the scaling up of both ethanol dehydrogenation and ethyl acetate hydrogenation from the laboratory to industrial plant, the techno-economic aspects of the process and the perspective for improvements. In particular, the use of bioethanol in a LOHC process has three main advantages: (1) the hydrogen carrier is a renewable resource; (2) ethanol and ethyl acetate are both green products benign for both the environment and human safety; (3) the processes of hydrogenation and dehydrogenation occur in relatively mild operative conditions of temperature and pressure and with high energetic efficiency. The main disadvantage with respect to other more conventional LOHC systems is the relatively low hydrogen storage density.

Keywords: ethanol; ethyl acetate; hydrogen; LOHC

Citation: Santacesaria, E.; Tesser, R.; Fulignati, S.; Raspolli Galletti, A.M. The Perspective of Using the System Ethanol-Ethyl Acetate in a Liquid Organic Hydrogen Carrier (LOHC) Cycle. *Processes* **2023**, *11*, 785. https://doi.org/10.3390/pr11030785

Academic Editor: Davide Dionisi

Received: 1 February 2023
Revised: 1 March 2023
Accepted: 3 March 2023
Published: 7 March 2023

1. Introduction

Bioethanol is a renewable, biodegradable, non-toxic raw material nowadays produced in an ever-larger amount as the base for a possible green transition to energy production. Its global production is expected to grow up to 41.4 billion liters by 2025, with a progressive reduction in price [1]. It is well known that bioethanol can selectively be dehydrogenated to ethyl acetate under mild conditions, using copper-based catalysts [2–11].

$$2C_2H_5OH \rightleftarrows CH_3COOC_2H_5 + 2H_2 \qquad (1)$$

Pure hydrogen exempt from CO and directly usable in fuel cells can be obtained from ethanol through this route. Moreover, the reaction is reversible and ethyl acetate can then be hydrogenated to ethanol closing the chemical cycle. On this basis, in the present work we suggest the possibility to use ethanol or derivatives as hydrogen vectors (LOHC, Liquid Organic Hydrogen Carrier) operating in a new original way. As known, hydrogen is a powerful vector of energy, but it is difficult to transport [12]. In fact, its volumetric energy density is very low due to its very scarce density under standard temperature and pressure conditions (0.0824 kg/m^3 under ideal gas conditions). This low density causes a very low volumetric energy content of hydrogen, i.e., 0.01 MJ/L H_2 for the gas at ambient conditions and 8.5 MJ/L H_2 for liquefied H_2. One possible solution is to store hydrogen inside a liquid molecule that can easily be transported and that can release it when needed. In this regard, many different LOHCs have been proposed in the literature such as methylcyclohexane, decaline, dodecahydro-N-ethyl carbazole, and other molecules containing aromatic rings. A more detailed list of possible LOHC molecules is reported in some recently published reviews [13–15]. All these LOHCs have the advantage of a high hydrogen density but also many disadvantages such as: (*i*) aromatic molecules are harmful, not renewable, and not biodegradable; (*ii*) their dehydrogenation occurs at relatively high temperatures with high consumption of heat; (*iii*) catalysts promoting the reaction are normally based on precious noble metals (Pt, Pd) susceptible to deactivation or poisoning. Another proposed approach is the use of methanol, ethanol, and hydrocarbons to produce hydrogen by steam reforming but in this case, CO and CO_2 are obtained as by-products. CO_2 must be separated, and CO contaminates the produced hydrogen. Moreover, the process is irreversible and consumes the reactants causing CO_2 emissions [16]. On the other hand, several researchers have described the ethanol dehydrogenation to ethyl acetate by using different catalysts [2–11] and some patents have also been published on this reaction describing the employment of different catalysts and related operative conditions [17–19]. Moreover, an industrial process has been developed and patented by Davy Process Technology, a Johnson and Matthey company [20–22]. More recently, Eurochem Engineering developed and patented in 2011 an industrial process for ethanol dehydrogenation to ethyl acetate carried out in a packed bed tubular reactor containing a more efficient copper-copper chromite commercial catalyst [23]. Under the best conditions, an ethanol conversion of 50–60 mol% and selectivity to ethyl acetate higher than 97% were obtained. The reaction was conducted in a temperature range of 473–493 K and 20–25 atm of pressure with an ethanol contact time of about 90–100 g·h/mol. In that process, 1 mole of pure hydrogen was obtained as a co-product for each mole of converted ethanol. The obtained hydrogen can easily be separated by condensing ethanol and ethyl acetate. All the aspects of this process have been studied in a detailed way and the results have been already reported in some different publications [24–27]. The kinetics of the reaction has been accurately studied by performing many experimental runs in different operative conditions, in the perspective of a suitable scale-up towards the industrial plant and a dedicated paper has been published [28]. The dehydrogenation reaction is reversible, and this is the reason for the limited ethanol conversion obtainable in a single step. The reverse reaction is therefore feasible and different works have studied the hydrogenation of ethyl acetate to ethanol by using different copper-based catalysts [29–33]. All these works, respectively devoted to ethanol dehydrogenation and ethyl acetate hydrogenation, demonstrate the feasibility of the use of ethanol as a LOHC because it is possible to close the chemical cycle of dehydrogenation of ethanol and hydrogenation of ethyl acetate. However, until now no-one has suggested the use of ethanol as a LOHC except for the proposal reported in a work recently published by Tran et al. [34] and in the more recent one by Mevawala et al. [35]. Tran et al. proposed the use of ethanol as a LOHC considering the cycle dehydrogenation-hydrogenation promoted by a ruthenium complex in homogeneous phase. Although the homogeneous catalytic approach is not convenient with respect to the heterogeneous one, the work is important because it is a further demonstration of the feasibility of the idea to use ethanol as a hydrogen carrier. On the other hand, the paper of Mevawala et al. [35] studied the thermodynamic and environmental aspects of the ethanol–ethyl acetate cycle for hydrogen storage applications, without an in-depth analysis of the catalytic systems. Their modeling of the overall cycle determined an energy efficiency up to 88%, a significantly high value with respect to those of other LOHCs, mainly addressed to the very low endothermicity of the reaction. Moreover, a preliminary eco-balance evidenced that the use of ethanol was more sustainable from a carbon emission perspective when compared with fossil LOHCs. Therefore, starting from the existing literature and considering, as a reference, the performance of the already well-known copper-copper chromite catalyst in promoting ethanol dehydrogenation we would like to define the best operative conditions for both ethanol dehydrogenation and ethyl acetate hydrogenation, also with the aim to find alternative catalysts exempt of chromium with similar or superior performances. In the perspective of the scale-up, the kinetics of both the dehydrogenation and hydrogenation reactions in conventional reactors will be defined for a preliminary determination of the techno-economic aspects of a structured process with the aim to optimize the characteristics of three different application areas: energy-storage, energy-transport, and mobility application.

2. Methodology of Ethanol Dehydrogenation

2.1. Thermodynamics of the Occurring Reactions

The conversion of ethanol to ethyl acetate occurs according to the following stoichiometry:

$$2C_2H_5OH \rightleftarrows CH_3COOC_2H_5 + 2H_2 \tag{2}$$

$\Delta H°_{500\,K} = 33.64\ kJ/mol;\ \Delta G°_{500\,K} = -6.44\ kJ/mol$
From experimental observation, this reaction is the addition of two consecutive reactions that are:

$$(1)\ C_2H_5OH \rightleftarrows CH_3CHO + H_2 \tag{3}$$

$\Delta H°_{500\,K} = 71.30\ kJ/mol;\ \Delta G°_{500\,K} = 10.59\ kJ/mol$

$$(2)\ CH_3CHO + CH_3CH_2OH \rightleftarrows CH_3COOCH_2CH_3 + H_2 \tag{4}$$

$\Delta H°_{500\,K} = -37.66\ kJ/mol;\ \Delta G°_{500\,K} = -17.03\ kJ/mol$
It is opportune to observe that the overall reaction (2) is moderately endothermic, while reaction (3) is endothermic, and the successive reaction (4) is on the contrary exothermic. The equilibrium constants of the three reactions, at 500 K, are respectively:

$$K_{p2\text{-}500\,K} = 4.69;\ K_{p3\text{-}500\,K} = 0.080;\ K_{p4\text{-}500\,K} = 58.94$$

The equilibrium of reaction (3) is shifted to the left but as acetaldehyde is consumed by reaction (4) it proceeds to the right and the equilibrium of the overall reaction is moderately shifted to the right at 500 K. The overall reaction (2) is more favored at 600 K with the equilibrium constant being about 17. Thermodynamic calculations have been made on data reported by Stull et al. [36]. The choice of 500 K, as reference temperature, is a compromise between the thermodynamic and kinetic properties of the involved reactions, because higher temperatures favor the occurrence of undesired side reactions.

2.2. Catalysts Normally Employed for Promoting the Ethanol Dehydrogenation

Many different catalysts have been used for promoting ethanol dehydrogenation [2–11,17–28]. It is possible to distinguish two different classes of copper catalysts respectively containing copper/copper chromite, and copper metal supported and/or promoted by different oxides such as Al_2O_3, Cr_2O_3, ZnO, ZrO_2, and SiO_2. The presence of oxide compounds has the scope to improve the dispersion of Cu thus, as consequence, slowing down the catalyst deactivation due to the metal sintering. Some other catalysts containing Ni or Pd have also been tested with lower performances. Interesting is the use of ruthenium complexes acting in a homogeneous phase, at very low temperatures, and promoting both ethanol dehydrogenation and ethyl acetate hydrogenation. Many papers and reviews have been published on the subject and nowadays some conclusions can be drawn [37–39]. According to Finger et al. [37], the dehydrogenation of ethanol to acetaldehyde occurs on the Cu surface, while the coupling of acetaldehyde with ethanol occurs mainly at the interface Cu-metal oxide (ZnO, ZrO_2) where ethanol is adsorbed as alkoxide. According to Pang et al. [38], acetaldehyde is generated preferably on an unreduced Cu^+ site while H_2 is liberated on Cu^0. The ratio Cu^+/Cu^0 would be therefore important for the selectivity, and a high presence of Cu^+ favors the formation of acetaldehyde, while the contrary favors the formation of ethyl acetate. As previously reported, the

Cu dispersion is of paramount importance, the activity being normally proportional to the specific surface area of Cu. In this regard, the support has a great influence on catalyst activity, selectivity and stability. For example, a high acidity of the support is detrimental, favoring the dehydration path instead of the dehydrogenation one; thus, the oxides, being weak acids or basic, favor dehydrogenation. Oxides trapping the Cu particles favor stability, hindering the deactivation due to sintering. It seems that stability can be induced also by alloying Cu with a moderate quantity of Ni. The operative conditions are also very important, considering that the formation of acetaldehyde is favored by low pressure (<1 atm) and high temperature (>573 K), whilst the selectivity to ethyl acetate increases at a relatively high pressure (10–30 atm) and low temperature (<523 K). The optimal operative conditions will be chosen through a compromise that allows the achievement of the maximum ethyl acetate yield. Phung [12] recently published a review on copper-based catalysts for ethanol dehydrogenation examining, on the basis of the previous literature, the respective roles of: Cu loading, Cu dispersion, the particle size, the Cu support and related acidity or basicity and contact time.

2.3. A Reference Catalyst, Reaction Mechanism and Related Kinetic Model

In this work, the behavior of a copper-copper chromite catalyst could be used as a reference starting point for comparing kinetic data obtained by employing other catalysts because the kinetics of that catalyst are well known from different sources [28,40]. Different kinetic models have been proposed in the literature. For example, Tu et al. [40] considered the main overall reaction to be of the pseudo-first order; thus, the model suggested by these authors was oversimplified. Moreover, the tested copper chromium catalysts were deactivated with time and a deactivation kinetic law was proposed by the same authors. More recently, a kinetic model based on reliable reaction mechanisms has been published by Carotenuto et al. [28]. The model proposed by these authors considered the following reaction scheme:

$$CH_3CH_2OH \rightleftarrows CH_3CHO + H_2 \tag{5}$$

$$CH_3CHO + CH_3CH_2OH \rightleftarrows CH_3COOCH_2CH_3 + H_2 \tag{6}$$

$$2CH_3CHO \rightarrow \text{Other products} \tag{7}$$

Probably occurring according to the following reaction mechanisms:

$$C_2H_5OH + \sigma_0 \rightleftarrows \sigma_{EtOH} \tag{8}$$

$$\sigma_{EtOH} + \sigma_0 \rightarrow \sigma_{AcH} + \sigma_{H2} \text{ rate determining step} \tag{9}$$

$$\sigma_{AcH} \rightleftarrows \sigma_0 + CH_3CHO \tag{10}$$

$$\sigma_{H2} \rightarrow \sigma_0 + H_2 \tag{11}$$

σ_0 is the fraction of free active sites on the catalytic surface, while σ_i is the fraction of active sites occupied by adsorbed "i" molecules. The "rate determining step" should be the surface reaction between chemisorbed ethanol and a catalyst void site to form adsorbed acetaldehyde. For the second reaction the following mechanism was suggested:

$$C_2H_5OH + \sigma_0 \rightleftarrows \sigma_{EtOH} \tag{12}$$

$$CH_3CHO + \sigma_0 \rightleftarrows \sigma_{AcH} \tag{13}$$

$$\sigma_{EtOH} + \sigma_{AcH} \rightleftarrows \sigma_{EA} + \sigma_{H2} \text{ rate determining step} \tag{14}$$

$$\sigma_{EA} \rightleftarrows \sigma_0 + CH_3COOCH_2CH_3 \tag{15}$$

$$\sigma_{H2} \rightleftarrows \sigma_0 + H_2 \tag{16}$$

In this case, the rate determining step would be the reaction between adsorbed ethanol and adsorbed acetaldehyde. Based on the postulated reaction mechanisms, it is possible to write the following kinetic law expressions:

$$r_1 = \frac{k_1 P_{EtOH}\left(1 - \frac{1}{Ke_1}\frac{P_{AcH}P_{H_2}}{P_{EtOH}}\right)}{\left(1 + b_{EtOH}P_{EtOH} + b_{AcH}P_{AcH} + b_H P_H + b_{EA}P_{EA}\right)^2} \tag{17}$$

$$r_2 = \frac{k_2 P_{EtOH}P_{AcH}\left(1 - \frac{1}{Ke_2}\frac{P_{EA}P_{H_2}}{P_{EtOH}P_{AcH}}\right)}{\left(1 + b_{EtOH}P_{EtOH} + b_{AcH}P_{AcH} + b_H P_H + b_{EA}P_{EA}\right)^2} \tag{18}$$

$$r_3 = k_3 P^2{}_{AcH} \tag{19}$$

The reactions of acetaldehyde to other undesired products are normally reactions of acetaldehyde condensation as deeply investigated by Inui et al. [7] and Colley et al. [8]; many different undesired products could be obtained in the worst operative conditions as in the following scheme (Scheme 1):

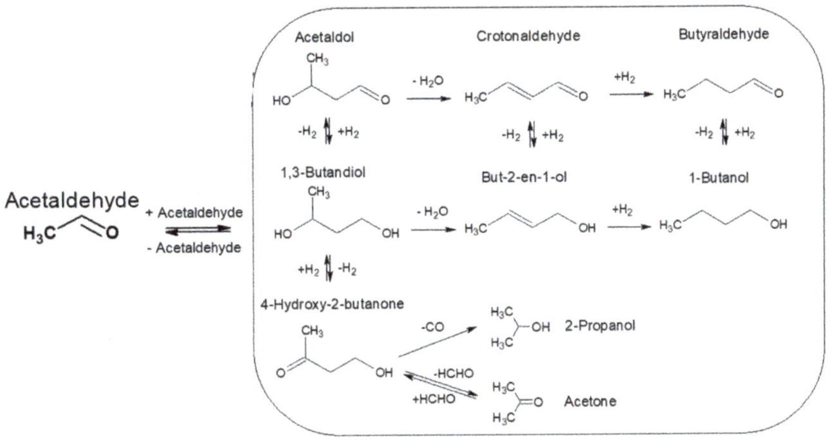

Scheme 1. Pathways of by-products formation from acetaldehyde condensation.

To avoid the formation of acetaldehyde condensation by-products it is clearly imperative to keep the acetaldehyde concentration in the system low and to achieve this result the catalyst selectivity becomes of paramount importance. In contrast, no-one observed an acetaldehyde decomposition reaction to CO by using copper-copper chromite catalysts.

In the kinetic approach followed by Carotenuto et al. [28], the reaction of acetaldehyde to other undesired side products was simplified to a pseudo-second-order reaction because it was characterized by a very low conversion. The best fitting parameters obtained by interpreting all the experimental runs performed are summarized in Table 1. The kinetic parameters reported in Table 1 seem reliable because the activation energies and the adsorption constants are reasonable values compatible with the postulated reaction mechanism. The very low value of k_3 and the negligible value of the corresponding activation energy is justified by: (*i*) the approximation introduced by arbitrarily assuming a pseudo-second order rate law; (*ii*) the fact that reaction (7) is not a single reaction but is an assembly of different reactions exclusively consuming acetaldehyde; (*iii*) the very low amount of by-products that determines high analytical errors.

Table 1. Optimized kinetic parameters obtained for a commercial copper-copper chromite catalyst, from [28], reported with the permission of Elsevier.

Kinetic Constants		Activation Energy (kJ/mol)
$k_{1\text{-}493\,K}$	$97.1 \pm 6.8 (\mathrm{mol}/(\mathrm{g_{cat}} \cdot \mathrm{h} \cdot \mathrm{atm}))$	151.67 ± 18.20
$k_{2\text{-}493\,K}$	$0.089 \pm 9.8 \times 10^{-3}\ (\mathrm{mol}/(\mathrm{g_{cat}} \cdot \mathrm{h} \cdot \mathrm{atm}^2))$	54.18 ± 2.72
$k_{3\text{-}493\,K}$	$0.0011 \pm 7.8 \times 10^{-4}\ (\mathrm{mol}/(\mathrm{g_{cat}} \cdot \mathrm{h} \cdot \mathrm{atm}^2))$	$6.69 \times 10^{-4} \pm 7.53 \times 10^{-5}$
Adsorption parameters (atm^{-1})		**Adsorption Enthalpy (kJ/mol)**
$b_{\mathrm{EtOH}\text{-}493\,K}$	10.4 ± 0.83	-106.82 ± 10.67
$b_{\mathrm{AcH}\text{-}493\,K}$	98.4 ± 12.79	-29.37 ± 1.46
$b_{\mathrm{EA}\text{-}493\,K}$	41.2 ± 4.94	-58.20 ± 0.59
$b_{\mathrm{H}\text{-}493\,K}$	$2.5 \times 10^{-4} \pm 3.5 \times 10^{-5}$	-55.81 ± 6.15

This reaction system is singular because, looking at the reaction scheme, we can observe that ethanol transformation to ethyl acetate passes through the formation of acetaldehyde, an endothermic reaction ($\Delta H \approx 71$ kJ/mol), while the successive reaction is moderately exothermic ($\Delta H \approx -40$ kJ/mol). This means that using a tubular reactor it is opportune to separate it in different stages (at least 2) differently heated if the system should be maintained approximately isotherm. In the first part it is necessary to furnish heat, whilst in the second one the cooling of the reactor is required. On the other hand, if a unique adiabatic reactor is adopted and a hot reagents stream (e.g., 508 K) is fed, it is possible to observe initially a decrease in the temperature followed by a progressive moderate increase. For an isothermal tubular reactor, the following system of differential equations must be simultaneously solved:

$$\frac{dF_{EtOH}}{dZ} = -W_{cat}(r_1 + r_2) \tag{20}$$

$$\frac{dF_{AcH}}{dZ} = W_{cat}(r_1 - r_2 - 2r_3) \tag{21}$$

$$\frac{dF_{AcOEt}}{dZ} = W_{cat}r_2 \tag{22}$$

$$\frac{dF_{H2}}{dZ} = W_{cat}(r_1 + r_2) \tag{23}$$

where F_i is the molar flow rate of component i, $Z = L/L_{bed}$ is the dimensionless reactor length, W_{cat} is the loaded catalyst weight and r_j are the rates of the j-reactions referred to the unit of catalyst weight. Figure 1 shows the conversion of ethanol and selectivities to respectively ethyl acetate and acetaldehyde as a function of the contact time W/F (g·h/mol), at 493 K and 20 atm, calculated with the described kinetic model and related parameters.

A similar kinetic approach has been published by Men'shchikov et al. [41], who interpreted runs performed on a catalyst composed of copper-zinc-chromium supported on alumina (CuO-ZnO-Cr_2O_3-Al_2O_3) in the temperature range 503–563 K and pressure of 10–20 atm. The runs were carried out in a continuous fixed-bed reactor and interpreted with a Langmuir–Hinshelwood kinetic model. The conversion to ethyl acetate changed with the temperature from 10 to 47%, acetaldehyde was produced in a small amount from 0.7 to 3%, while increasing the temperature greatly increased the presence of other by-products from 0.3% to 7%. Optimal kinetic parameters of the model were determined and interpreted for all the kinetic runs showing an error of $\pm 20\%$. Starting from similar results, it would be possible to do a scale-up and formulate the best arrangement of the plant equipped with a conventional reactor for producing hydrogen and ethyl acetate.

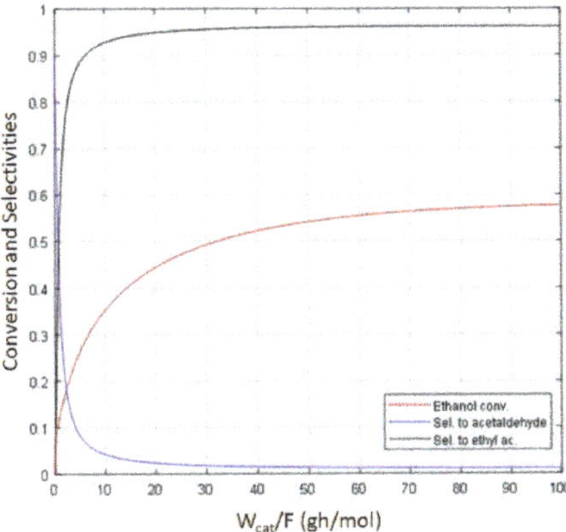

Figure 1. Kinetic behavior of copper-copper-chromite catalyst. Ethanol conversion and selectivities to respectively ethyl acetate and acetaldehyde, for different contact times at 493 K and 20 atm. Curves have been obtained by solving the system of differential Equations (10)–(13) with the parameters of Table 1.

2.4. Laboratory and Pilot Chemical Plants

The ethanol dehydrogenation kinetic data, related to the previous section, have been collected in a laboratory chemical plant, as schematized in Figure 2 [42].

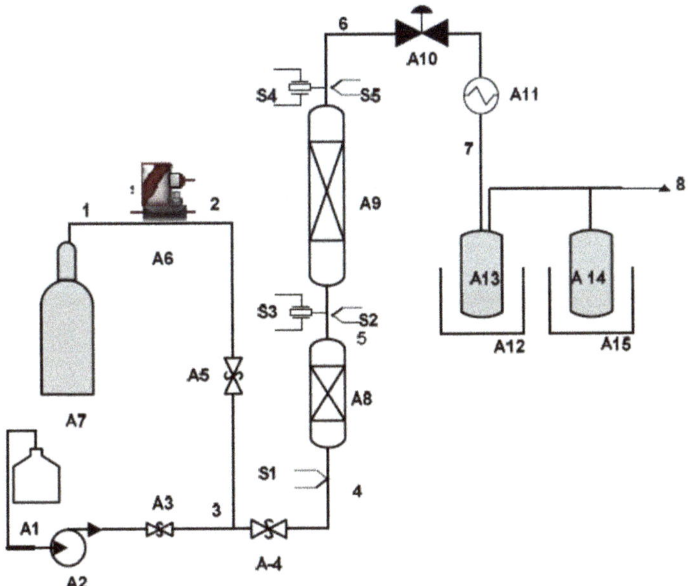

Figure 2. A1—Ethanol tank; A2—HPLC pump; A3, A5—Check valves; A4—relief valve; A6—Flow mass controller, A7—Cylinder containing a mixture of hydrogen 5% and nitrogen; A8—pre-heater to vaporize ethanol; A9—Stainless steel packed bed tubular reactor. A10—Back pressure regulator; A11—Heat exchanger; A13, A14—Product tanks raising; A12/A15—Liquid nitrogen dewars; S1/S2/S5—Temperature probes; S3/S4—Pressure transducers. The reactor A9 was heated by two independently thermoregulated heaters.

Recently, Semenov et al. [10] published a work describing the performance of a pilot plant working with a Cu-ZnO catalyst. A scheme of the employed pilot plant, containing

0.5 L of catalyst, is shown in Figure 3. The authors worked in a temperature range of 503–573 K and 1–20 atm. The best obtained results were 63% of ethanol conversion per pass and 94% of selectivity to ethyl acetate, together with very small amounts of side products such as butyl alcohol, butyl acetate, and ethyl butyrate. Moreover, the catalyst remained stable during all the time investigated in a long-term test (1250 h on stream).

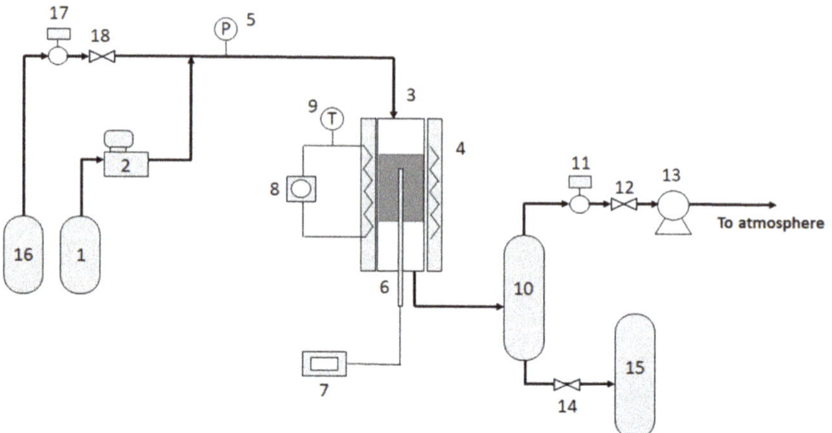

Figure 3. Scheme of a pilot plant, containing 0.5 L of catalyst, used by Semenov et al. to verify catalyst stability in runs prolonged for 1250 h. (1) Raw material container; (2) Dosing pump; (3) Tubular fixed bed reactor; (4) Heating furnace; (5) Pressure gauge; (6) Movable thermocouple; (7) Temperature reading; (8) and (9) Temperature regulation system; (10) Ethyl acetate, ethanol condenser; (11)–(13) Hydrogen flow measurement; (15) Container to collect liquid produced; (16) Hydrogen bottle; (17) Pressure reducer; (18) valve.

However, considering the change of enthalpy passing from the dehydrogenation of acetaldehyde (endothermic) to the successive reaction of ethyl acetate formation (exothermic) it would be opportune, as before mentioned, to consider the reactor subdivided into two or three zones differently heated, as in the scheme of Figure 4.

2.5. Scale-Up: Scheme of the Industrial Plant and Most Opportune Operative Conditions

As mentioned before, an industrial process has been realized and commercialized by Davy Company in South Africa some years ago. The catalyst used by Davy for the ethanol dehydrogenation was a commercial copper chromite catalyst. A simplified scheme of their proposed plant is the one reported in Figure 5 taken from their private communication [43].

The step of selective hydrogenation is related to the hydrogenation of by-products of acetaldehyde condensation such as methyl ethyl ketone and butyraldehyde to obtain the corresponding alcohols that are easier to separate from ethyl acetate. Catalysts employed for this hydrogenation step were supported ruthenium or nickel. The molar conversion-per-passage of ethanol dehydrogenation was kept at 30–40% and selectivity to ethyl acetate was 90–96%. The relatively low selectivity to ethyl acetate justified the necessity of a supplementary hydrogenation step for an easier separation of the by-products. The dehydrogenation catalyst was pre-activated at 473–493 K in a stream containing 5% of hydrogen in nitrogen. The activated catalyst was then used at about 14 atm of pressure and 493–518 K of the temperature range. More recently, Eurochem Engineering has studied the same reaction and proposed the following simplified scheme for an industrial plant [24,42] (see Figure 6). An ethanol conversion per passage of 50–60% can normally be adopted and 5% of hydrogen is recycled acting as carrier gas and catalyst stabilizer. At least two distillation columns are needed to break the ethanol-ethyl acetate azeotrope and the purge is opportune to avoid side-products accumulation. The process scheme is simplified with respect to the one proposed by Davy because the copper-copper chromite catalyst employed by Eurochem Engineering was more selective and the amount of produced by-products was lower, thus not requiring their following hydrogenation and separation from the recycle stream. As before mentioned, this reaction system is singular because, looking at the reaction scheme, we can observe that ethanol transformation to ethyl acetate passes through the formation of acetaldehyde and this is an endothermic reaction ($\Delta H \approx 71$ kJ/mol), while the succeeding reaction is moderately exothermic ($\Delta H \approx -40$ kJ/mol). This means that the operation on both laboratory or pilot scale can be conducted in two ways: by using an approximately isotherm jacketed tubular reactor or by dividing the tubular reactor into two/three stages differently heated if the system should be maintained approximately isotherm.

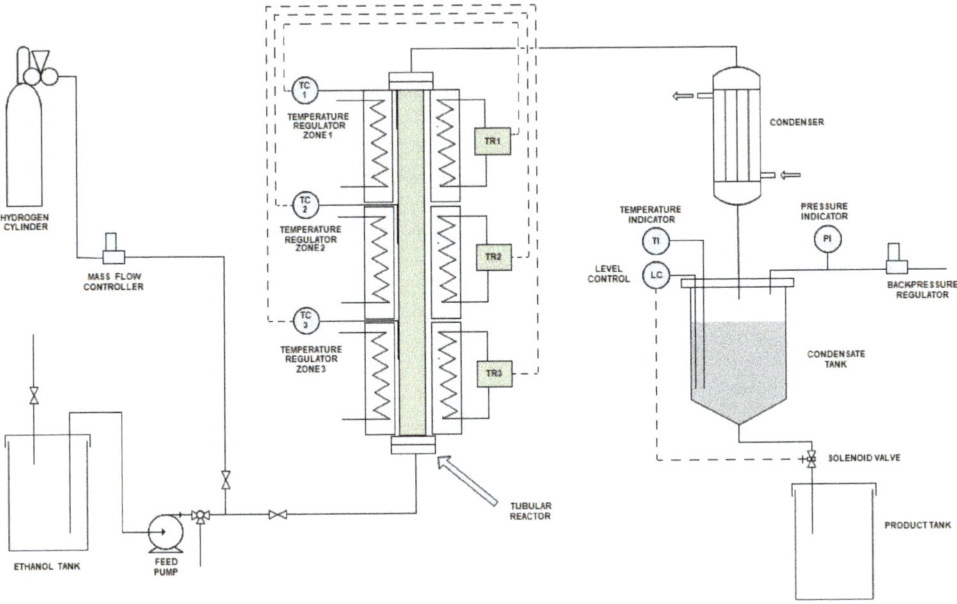

Figure 4. An example of the pilot plant suggested by the authors.

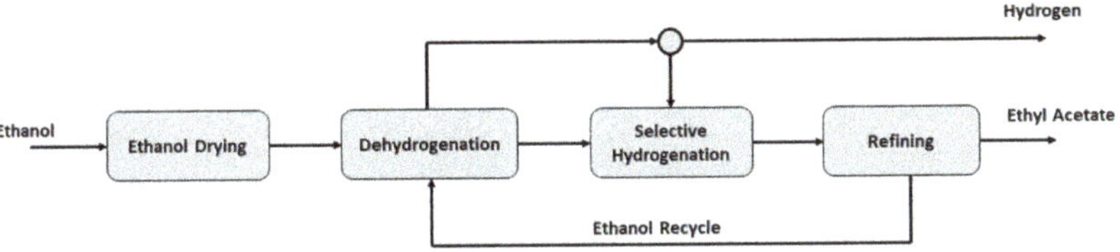

Figure 5. Simplified scheme of the DAVY process for ethanol dehydrogenation [43].

At the industrial scale level, it is opportune to operate with three or four adiabatic reactors in sequence with a heat exchanger between a reactor and the successive stages to repristinate the desired reaction temperature. A preliminary estimation [42], based on the previously described copper chromite catalyst and related kinetic model, has been made using ChemCad software by simulating a complete industrial plant producing 21 tons/h of ethyl acetate and 1 ton/h of hydrogen. The overall process was divided into four different sections:

Section (1) was devoted to fresh ethanol purification and recycling. In this section, ethanol was dehydrated because water is dangerous for the catalyst. Dehydration was realized by distillation in the presence of ethylene glycol as an entrainer.

Section (2) Catalytic reactors. Dehydrated ethanol is heated, vaporized, and pressurized before entering the reactors. Four adiabatic reactors, each of 10 m^3, were considered in sequence with inter-bed heating to compensate for the endothermicity of the reaction. The inlet temperature was 493 K for the first two reactors and 503 K for the other two.

Section (3) Hydrogen separation through two flash units. 5% of hydrogen is recycled acting as carrier gas and catalyst stabilizer.

Section (4) Separation of ethyl acetate from unreacted ethanol + by-products. To break the azeotrope ethanol-ethyl acetate a pressure swing distillation system has been considered by varying the pressure from 20 atm (the reaction pressure) to 1 atm. The separation scheme consists of a first distillation column working at 20 atm in which on the overhead the azeotrope ethyl acetate-ethanol was separated from pure ethyl acetate. The azeotropic mixture was then depressurized to 1 atm and fed to a second distillation column for further separation. Alternatively, the separation can be made by azeotropic distillation by using an effective agent such as ethyl ether, methyl formate, or cyclohexane [44] or some ionic liquids as entrainers [45].

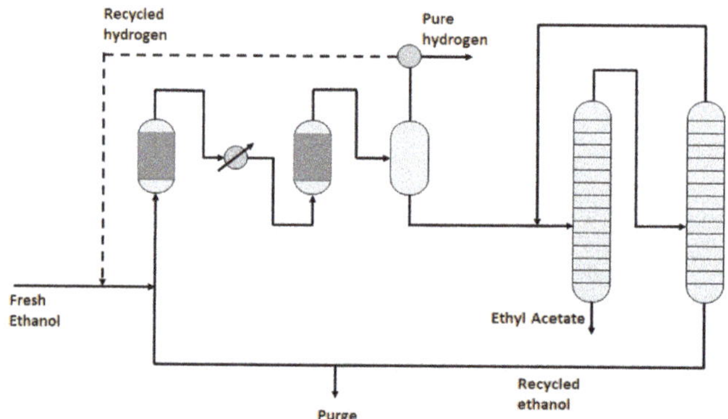

Figure 6. Simplified scheme of the Eurochem process for ethanol dehydrogenation. Employed catalyst CuCrO$_4$/CuO/Cu/BaCrO$_4$/Al$_2$O$_3$ [42].

2.6. Techno-Economic Analysis of the Process

A techno-economic analysis of this dehydrogenation process has recently been made by Khanhaeng et al. [16] with the scope of determining the cost of hydrogen produced through the steam reforming of ethanol and the ethanol dehydrogenation to acetaldehyde. The comparison has been performed by using Aspen PLUS V10. The hydrogen produced would be used for the synthesis of methanol by reacting hydrogen with CO$_2$. This comparison is not useful for our purpose, but the cost estimation for ethanol dehydrogenation could be a useful reference for this study. The scheme of the industrial process considered by the authors is shown in Figure 7.

The kinetics published by Carotenuto et al. [28] have been used by Khanhaeng et al. for the calculations related to ethanol dehydrogenation. As reported in Figure 7, ethanol preheated at 493 K was pumped to 20 atm (E-200) by using high-pressure steam before entering in the fixed bed reactor (R-200). The conditions inside the reactor were maintained at 20 atm and 513 K. The reaction mixture (unreacted ethanol, ethyl acetate, and hydrogen) was cooled (E-202) at 308 K by cooling water. Unreacted ethanol and ethyl acetate were separated from hydrogen in a flash vessel (V-200). The hydrogen collected had a purity of 98.87%. The mixture of unreacted ethanol-ethyl acetate collected from the bottom stage was separated by extractive distillation using dimethylsulfoxide (DMSO) as suggested by Zhang et al. [46]. The extractive distillation occurred in a distillation column with 48 theoretical trays (T-200). 99.64% of ethyl acetate was separated with a purity of 99.5%. Then, the mixture of ethanol–DMSO was separated in a simple distillation column with 10 theoretical trays and the unreacted ethanol was then recycled. The authors evaluated the overall cost of the process consisting of Fixed Capital Investment + Cost of Manufacturing. The Fixed Capital Investment for a plant producing 1665 tons/year of hydrogen was estimated to be 4,784,141 USD, while the Cost of Manufacturing was calculated to be 6,919,315 USD/year. This cost also includes the value of the employed raw materials (ethanol 3,192,398 USD/year and DMSO 1924 USD/year) that must be calculated just for undergoing the first LOHC cycle, while, after the successive LOHC cycles the same ethanol is reused. Hence the effective cost of manufacturing would be: 6,919,315 − 3,192,398 − 1924 = 3,724,993 USD/year corresponding to a cost of 2237 USD/ton of hydrogen produced. The cost must be increased by considering the Fixed Capital Investment + Cost of Raw Materials for the start-up: 4,784,141 + 3,192,398 +1924 = 7,978,463 USD. If these expenses are amortized in 10 years we have a charge of 794,846 USD/year which is 479 USD/ton of hydrogen produced. Therefore, the estimation of the hydrogen cost will be 2237 + 479 = 2716 USD/ton of hydrogen produced. This means a cost of energy of about 0.0684 USD/kWh. However, we must point out that some energy must be supplied to the system for sustaining: (*i*) ethanol heating from room temperature to the reaction temperature, including the ethanol vaporization heat; (*ii*) the heat necessary for supplying the heat of the endothermic reaction; (*iii*) the energy necessary for the separations; (*iv*) the energy for ethanol recycle. This energy can be supplied by burning the purge that mainly contains ethanol with small amounts of acetaldehyde and very small amounts of other organic by-products. In this case, a moderate excess of ethanol must be employed in the process. An alternative could be the use of waste heat where available.

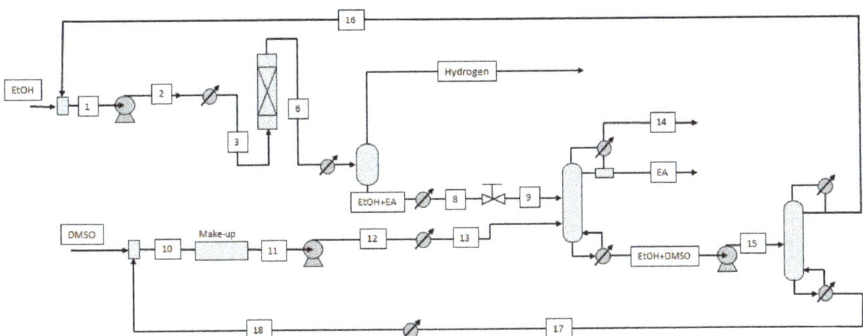

Figure 7. Scheme of an industrial plant of ethanol dehydrogenation to ethyl acetate. Plant capacity was calibrated for producing 1665 tons/year of hydrogen with an ethanol feed of 38,700 tons/year. Re-elaborated from [16] and reported with the permission of Elsevier.

The different described behaviors suggest that there is a large margin for improvement in defining the best operative modality either increasing the catalyst activity or improving the operative conduct. We can now summarize some of the advantages of this process:

- Only ethanol is required as feedstock. Then, if ethanol derives from a biological source the process is based on a renewable resource.
- No problems of corrosion are envisaged during the plant operation. This allows using lower grades of material in the equipment fabrication and the plant costs are therefore reduced.
- Pure hydrogen is obtained exempt from CO and CO_2 and is therefore directly usable in fuel cells.
- High purity of the product ethyl acetate (>99%), which can be directly sent to the second hydrogen reloading step without significant action.

In conclusion, starting from the described kinetic models, adapted to any newly developed catalyst, a more detailed process design can be obtained using ChemCAD or ASPEN. By imposing the desired productivity, a process optimization study can be realized with the aim of reducing the cost of plant realization and the cost of hydrogen production.

2.7. Process Intensification and Adoption of Membrane Reactors

An attempt to introduce process intensification has recently been made by the Startup Co. Greenyung [47–52], which published different patents proposing an innovative ethanol dehydrogenation process performed in a reactive distillation column, according to the following scheme (Figure 8).

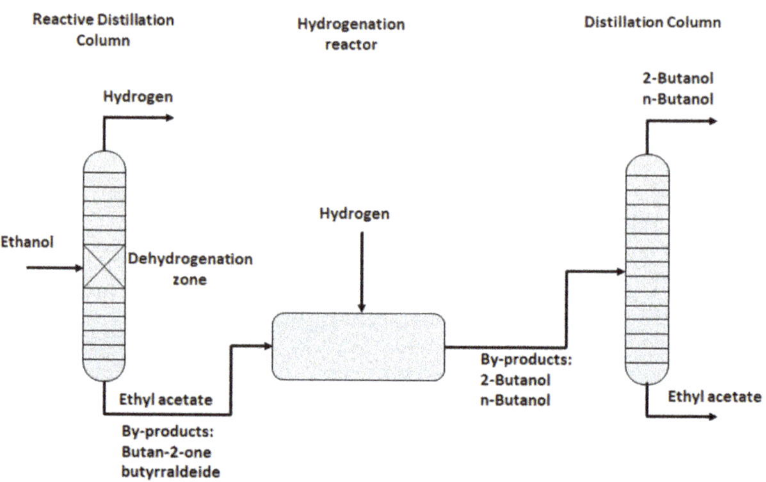

Figure 8. Simplified scheme of the Greenyung Process.

In this case, ethanol is fed in a reactive distillation column, in which a dehydrogenation zone is present where the main reactions occur together with the associated side reactions. The employed catalyst was a physical mixture of CuO/Al_2O_3 and ZnO/Al_2O_3. Two streams exit from the reactor: *(i)* the produced hydrogen, which exits from the top of the column; *(ii)* the by-products, which are then hydrogenated because some side products are hard to separate from ethyl acetate. This means that the catalyst employed by Greenyung was not selective enough, but more details about this process and some data on costs are reported in a report of the Pennsylvania University available online [53] and on the published patents [47–52]. However, the idea of process intensification combining the dehydrogenation reaction with the hydrogen separation can be considered as a profitable opportunity. Another very promising possibility is the adoption of a membrane reactor as can be argued from the published literature [54–59]. The introduction of a membrane reactor allows the shifting of the equilibria of reactions (3) and (4) to the right by subtracting hydrogen from the chemical environment, so increasing the ethanol conversion at lower temperatures, without any loss in selectivity. Moreover, the use of a membrane reactor greatly simplifies the subsequent separation process that in this case occurs contemporarily with the reaction [13]. The dehydrogenation reaction of ethanol to acetaldehyde in membrane reactors has been studied by many authors [54–57] using particularly Pd and Pd-Ag membranes. More specifically the technology has been applied also to the ethanol dehydrogenation to ethyl acetate [58,59] using Pd, Pd-Ag and Pd-Zn membranes. As the dehydrogenation of ethanol to acetaldehyde is much faster than the ethanol-acetaldehyde coupling reaction, the residence time must be opportunely regulated to obtain the desired product. Moreover, in the presence of an opportune membrane, generated hydrogen can be selectively separated in situ from the reaction mixture and the reaction can be driven to completion. Moreover, reaction and hydrogen separation steps occur inside the same reactor, thus avoiding a further separation unit.

3. Ethyl Acetate Hydrogenation

3.1. Thermodynamics of the Occurring Reactions

Ethyl acetate hydrogenation is the reverse reaction of ethanol dehydrogenation. Therefore, we can write for the overall reaction:

$$CH_3COOC_2H_5 + 2H_2 \rightleftarrows 2C_2H_5OH \tag{24}$$

$\Delta H^\circ_{500\,K} = -33.64$ kJ/mol; $\Delta G^\circ_{500\,K} = 6.44$ kJ/mol $K_{p\text{-}500} = 0.21$
This reaction very probably occurs in two steps that are:

$$(1)\ CH_3COOCH_2CH_3 + H_2 \rightleftarrows CH_3CHO + CH_3CH_2OH \tag{25}$$

$\Delta H^\circ_{500\,K} = 37.66$ kJ/mol; $\Delta G^\circ_{500\,K} = 17.03$ kJ/mol

$$(2)\ CH_3CHO + H_2 \rightleftarrows CH_3CH_2OH \tag{26}$$

$\Delta H^\circ_{500\,K} = -71.30$ kJ/mol; $\Delta G^\circ_{500\,K} = -10.59$ kJ/mol
We have then $K_{p1\text{-}500} = 0.0166$ and $K_{p2\text{-}500} = 12.76$.
Calculations have been made on thermodynamic data reported by Stull et al. [36].

3.2. Catalysts Normally Employed for Promoting the Ethyl Acetate Hydrogenation Reaction

The interest in ethyl acetate hydrogenation to ethanol is due to the possibility to use ethanol as an alternative fuel and many papers have been published on the subject. Adkins and Folker [60] firstly described the reaction of esters hydrogenation promoted by a copper-chromium catalyst, being copper with chromium additive often used for fatty esters hydrogenation [61]. In fact, copper catalysts are highly active for C=O bond hydrogenation but much less active for C-C bond cleavage and this explains the choice of copper-based catalysts for ester hydrogenation. More recently, as chromium poses environmental problems, research is focused on the development of chromium-free catalysts. For this purpose, the addition to copper of other metal oxides, such as ZrO_2, Fe_2O_3, or ZnO, has been investigated with the aim of promoting the activity of copper in the hydrogenation of esters to alcohols [62–66]. These works are of great interest for the LOHC process because very active catalysts in the hydrogenation of esters are probably also active in the reverse reaction of alcohol dehydrogenation. However, the hydrogenation of ethyl acetate to ethanol has received poor attention in the past and according to the few published works before 2014 a conversion of 40% of ethyl acetate with an ethanol selectivity of 80% has been reported [32]. In this article, a bimetallic Cu-Zn/SiO_2 catalyst has been employed. The catalyst was obtained by calcination of the synthesized compound $CuZn(OH)_4(H_2SiO_3)\cdot 2.4H_2O$ at 473 K in air followed by a reduction step with hydrogen at 573 K. The Cu/Zn molar ratio of 1:1 was the optimal one leading to ethyl acetate conversion of 82% and ethanol selectivity of 94%. In another work [31], a Zn-promoted Cu-Al_2O_3 catalyst has been positively tested in a fixed bed reactor carried out at 523 K and 20 atm. The catalyst was prepared by impregnating alumina with a mixture of copper nitrate and zinc nitrate in a solution by incipient wetness. By opportunely regulating calcination and reduction conditions the best results were ethyl acetate conversion of 71.5% and ethanol selectivity of 95%. An activity test was prolonged for 90 h without any apparent catalyst deactivation. Interesting results have been obtained also by Di et al. [30] by using two different synthesized Cu/SiO_2 catalysts. In the first case, an acidic copper nitrate solution was added to a basic sodium silicate solution obtaining a precipitate that was then calcined and reduced, thus providing the Cu/SiO_2 catalyst. The other one was prepared by using the urea hydrolysis deposition-precipitation method. Catalytic runs were performed in a fixed bed reactor operating in the temperature range of 483–553 K at 30 atm using 1 g of catalyst and a LHSV = 1.24 h^{-1}. The reactor was fed with a molar ratio H_2/EA = 29. The best data obtained were 99.7% of ethyl acetate conversion and 99.07% of selectivity to ethanol. These data seem to be optimistic, probably because of the high H_2/EA ratio. Good results have been obtained by Huang et al. [67] by using a

Cu/SiO$_2$ catalyst for hydrogenating methyl acetate to methanol + ethanol, obtaining a conversion of 94% and a selectivity of 94%. The catalyst was prepared by the precipitation-gel method by reacting copper nitrate with NaOH and then supporting the gel on silica. On the other hand, Lu et al. [33] prepared by coprecipitation and tested three different catalysts of the type Cu/ZnO/MO$_x$, where MO$_x$ could be the supports SiO$_2$, Al$_2$O$_3$, and ZrO$_2$. The best catalyst results were from Cu/ZnO/Al$_2$O$_3$ that, in a fixed bed reactor (filled with 3 g of catalyst) kept at 553 K and 20 atm, gave a conversion of 80% with a selectivity of 95%. A value of the activation energy equal to 115 kJ/mol was estimated for this catalyst. This work can be reinterpreted for a kinetic approach to this reaction. A confirmation of the good performances of the CuZn-SiO$_2$ catalyst has been furnished by Zhao et al. [68] who studied the hydrogenation of methyl acetate on the same catalyst. In conclusion, a catalyst active in the hydrogenation of ethyl acetate and probably also in the reverse reaction of ethanol dehydrogenation is a copper-based catalyst with copper extremely dispersed and hindered from sintering by the presence of other oxides. In this regard, the role of zinc as a promoter seems very important, while the support cannot be acidic to avoid the formation of by-products from the reactions of acetaldehyde condensation.

3.3. Ethyl Acetate Hydrogenation, Kinetic Methodology in Conventional Reactors

Very few papers have considered the kinetics of the ethyl acetate hydrogenation to ethanol on copper-based catalysts [69,70]. The first one determined the intrinsic kinetics of ethyl acetate hydrogenolysis to ethanol over a Cu-Zn/Al$_2$O$_3$ catalyst operating in a tubular reactor in the temperature range of 453–503 K, pressure 10–15 atm, LHSV 0.7–1.9 h^{-1} and molar ratio H$_2$/EA = 20–50. A power law kinetic model was applied for interpreting the experimental data and the optimal kinetic parameters were determined by mathematical regression. The second paper studied the kinetic behavior in the ethyl acetate hydrogenation promoted by a Cu/ZrO$_2$ catalyst prepared by coprecipitation from copper nitrate and zirconium oxynitrate with NaOH as precipitating agent. The kinetics of hydrogenation were studied at atmospheric pressure in the temperature range 293–513 K. The hydrogenation reaction was followed by gradually increasing the temperature from 293 to 513 K, stopping the temperature level for a determined time for evaluating conversion and yield at intermediate temperatures. The conversion varied therefore from 8 to 45%. The first ethyl acetate hydrogenation reaction was always far from the equilibrium, while acetaldehyde quickly reached the thermodynamic equilibrium. As a matter of fact, acetaldehyde hydrogenation is 6000 times more active than ethyl acetate hydrogenation. Data was collected on conversion as a function of the residence time, and hydrogenation rates as a function of the ratio H$_2$/ester at different temperatures. The authors assumed the reaction mechanism suggested by Natal Santiago et al. [71], which is characterized by dissociative adsorption of ethyl acetate yielding acyl and alkoxy species:

$$CH_3COOCH_2CH_3 + 2^* \rightarrow CH_3CO* + CH_3CH_2O^* \tag{27}$$

The alkoxy fragment is rapidly hydrogenated to ethanol, whilst the acyl fragment is less reactive and, more slowly, can be partially hydrogenated to acetaldehyde or fully hydrogenated to ethanol. A moderate deactivation was observed due to coke deposition occurring at higher temperatures. This phenomenon is hindered by operating in the presence of an excess of hydrogen. A kinetic model is mentioned in the work by using, for the overall reaction, a power law kinetic expression. Reaction order results were 0.1–0.3 for hydrogen and −0.4 to 0.1 for ethyl acetate; the apparent activation energy was 74 kJ/mol.

An example of a laboratory-scale plant (Figure 9) is reported by the already cited Zhu et al. [31].

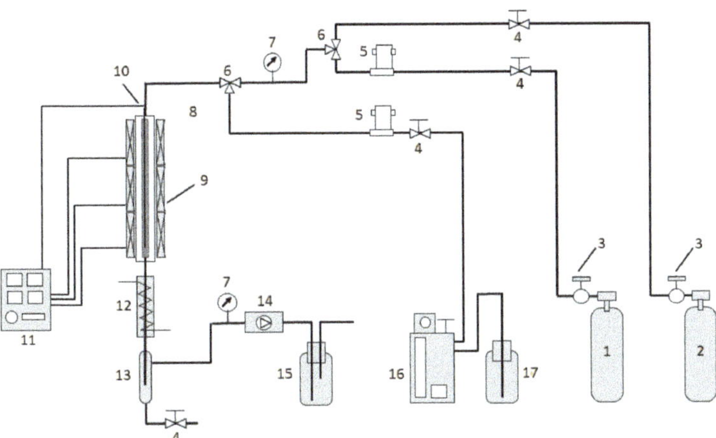

Figure 9. Fixed-bed reaction system. 1—H_2; 2—N_2; 3—Reducing valve; 4—Valve; 5—Mass flow meter; 6—Three way valve; 7—Pressure gauge; 8—Reactor; 9—Heater; 10—Thermocouple; 11—Temperature controller; 12—Condenser; 13—Gas-liquid separator; 14—Back pressure valve; 15—Gas washing bottle; 16—Micro syringe pump; 17—Raw materials. See ref. [31]. Reported with the permission of Elsevier.

Another example is the one reported by Jiang et al. (see Figure 10) [69].

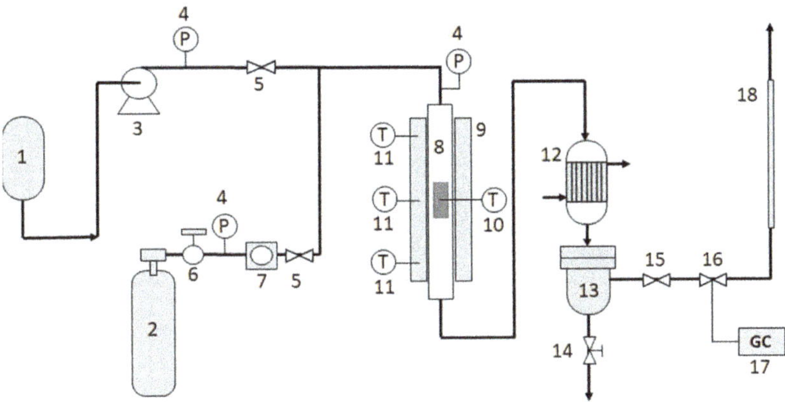

Figure 10. 1—Ethyl acetate reservoir; 2—Hydrogen bottle; 3—Pump; 4—Pressure gauge; 5—Valves; 6—Pressure reducer; 7—Non-return valve; 8—Tubular reactor; 9—Heating furnace; 10–11—Thermocouples and thermoregulators; 12—Liquid condenser; 13—Liquid product container; 14—Valve; 15,16—Valves; 17—Gas Chromatograph;18—Hydrogen to the vent. See Ref. [69].

Based on these data, it could be possible to elaborate a simplified scheme for the corresponding industrial plant.

3.4. The Industrial Ethanol-Ethylacetate LOHC Process

A work published very recently by Mevawala et al. [35] has specifically been devoted to the possibility of using the ethanol-ethyl acetate system as a LOHC by examining three different aspects: the thermodynamics of the chemical reaction, the energy balance of the process and the assessment of greenhouse gas emission. This work confirmed that the energy demand for dehydrogenation is small, and the authors calculated an energy efficiency of the system equal to 88%. The results obtained by the authors show that the ethanol-ethyl acetate system is very promising as a LOHC and is worth further study. In particular, the same authors furnished a complete Block Flow Diagram developed in Aspen Plus 10 of the ethanol-ethyl acetate LOHC process for a production of 500 kg/h of hydrogen.

4. Ethanol as a Possible LOHC in Comparison with Other Candidate Molecules

"Green" electricity provided by renewable sources of energy such as photovoltaic plants, wind turbines, hydroelectric plants, geothermal energy plants, and energy from biomass can be used within an electrolysis process to produce gaseous hydrogen from water. However, when the renewable sources of energy are in places where the energy demand is low, it is opportune to store the excess of produced energy. This can be done by producing hydrogen and storing it inside a LOHC through a hydrogenation reaction. The hydrogenated molecule must allow long-term storage, must be suitable for long-distance transport and must be ready, at the time of need, to unload hydrogen after the transport and storage in a place where the energy demand is high. Many different LOHC substances have been proposed until now such as: methylcyclohexane, 12H-N-ethylcarbazole, 18H dibenzyl toluene, naphthalene and others. The use of a LOHC is based on a two steps cycle: *(i)* loading hydrogen into the LOHC molecule through hydrogenation; *(ii)* recovering hydrogen through dehydrogenation. Normally, hydrogenation is an exothermic process, and the reaction heat can be used to keep the reaction pressure and temperature at the desired level. In contrast, dehydrogenation is an endothermic process occurring at low pressure and high temperatures. The hydrogen released by dehydrogenation requires, therefore, heat from an appropriate heat source to drive the dehydrogenation process. This occurs when LOHCs are molecules containing aromatic rings. The heat necessary during dehydrogenation can be furnished by burning part of the hydrogen produced, so reducing the efficiency of the overall process, or alternatively by building the plant near a waste heat source. Let us consider, for this purpose, the LOHC system methylcyclohexane-toluene as a reference for comparison [72]. The use of bioethanol as LOHC allows storing of 4.3% by weight of hydrogen, which means storage of 34 kg of H_2/m^3 of ethanol or, alternatively, 1.35 kWh/L of ethanol. Bioethanol is a free-flowing liquid that has the advantage of being non-toxic, renewable, biodegradable, and largely available as raw material at a lower and lower cost. Bioethanol is, therefore, a good LOHC candidate for the application of green energy storage according to the cycle of Figure 11.

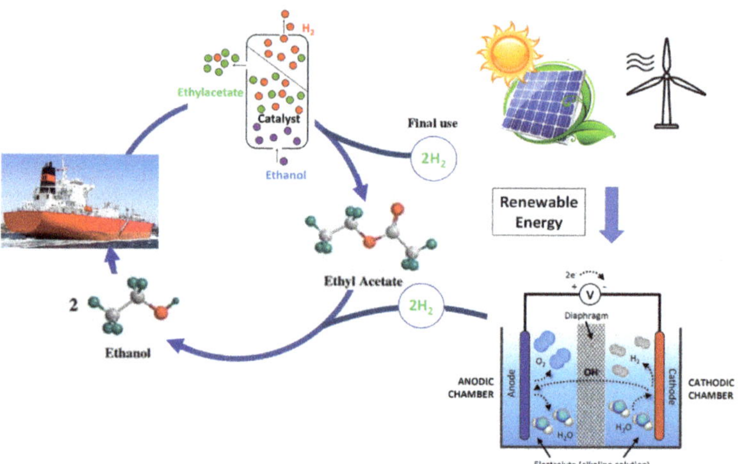

Figure 11. Overview of coupling ethanol dehydrogenation and ethyl acetate hydrogenation of ethyl acetate to ethanol with the scope of storing hydrogen inside the ethanol molecule and transporting it to a place of final use.

Ethyl acetate can also be considered a green product and, like ethanol, benign for both the environment and human safety. To compare the cost of using ethanol as LOHC with respect to other compounds, we can neglect the cost of transportation, this being similar for all the liquid organic compounds; therefore, the comparison will be made on the cost of the dehydrogenation-hydrogenation cyclic operations. Let us consider, for this purpose, one of the most studied LOHCs, namely, methylcyclohexane (MCH). The MCH-toluene reaction is promoted by catalysts of platinum or nickel supported on alumina, at 623–723 K, with a yield between 50 and 92%. The highest hydrogen yield (95% at 593 K) was obtained on a K-Pt/Al$_2$O$_3$ catalyst [15]. MCH has a hydrogen storage capacity of 6.11% by weight, which means density storage of 47.5 kg of H$_2$/m^3.

Despite the advantage in the storage capacity of MCH with respect to ethanol, the energy that must be consumed in the MCH dehydrogenation favors the use of ethanol. The overall reaction enthalpy for the reaction has already been seen:

$$2C_2H_5OH \rightleftarrows CH_3COOC_2H_5 + 2H_2 \tag{28}$$

$\Delta H^\circ_{500\,K} = 33.64$ kJ/mol.

This value is much lower than that of MCH and the reaction temperature range, between 493–543 K, is lower, too:

$$\text{Methylcyclohexane} \rightleftarrows \text{Toluene} + 3H_2 \tag{29}$$

$\Delta H^\circ_{500\,K} = 213.22$ kJ/mol.

Moreover, the drawback of the higher endothermicity of the dehydrogenation of MCH respect to that of ethanol is more evident considering the enthalpies calculated per moles of produced hydrogen. In fact, although the dehydrogenation of ethanol leads to the formation of two moles of hydrogen whilst three moles are produced from the dehydrogenation of MCH, the enthalpy per mole of produced hydrogen is markedly lower in the first case, being 16.82 kJ/mol instead of 71.11 kJ/mol, thus confirming that the dehydrogenation reaction of MCH requires a higher amount of heat than that of ethanol. The heat can be provided in several ways, e.g., from the combustion of compounds external to the process or, more advantageously, from the combustion of a part of the produced hydrogen. In fact, the enthalpy of hydrogen combustion is about 243 kJ/mol thus, being comparable with the enthalpy of the MCH-toluene reaction (213.22 kJ/mol), about one mole of hydrogen produced could be used to sustain the dehydrogenation reaction of MCH. However, this strategy lowers the effective hydrogen storage to about 5%, which becomes comparable with the level for ethanol. Hydrogen release from ethanol is achieved at lower temperatures and with significantly less energy consumption. Moreover, hydrogenation, in this case, is just moderately exothermic, while, for MCH we have a highly exothermic reaction with problems of heat transfer. Besides, MCH dehydrogenation, considering its high endothermicity, requires a more sophisticated reactor characterized by very high heat exchange efficiency. An opposite situation characterizes the hydrogenation reaction that being very exothermic requires a very efficient heat exchange system. This means additional costs for the plant realization.

The mentioned advantages of ethanol as LOHC can be further implemented by increasing the dehydrogenation reaction rate and the overall reaction yield. This can be done by improving the catalyst performance and by employing membrane reactors. The catalyst activity is strongly related to the specific surface area of dispersed copper while selectivity and stability depend on the copper chemical environment. The improvement of catalyst performance in terms of activity and selectivity is therefore of paramount importance for the overall efficiency of the process.

As previously mentioned, the introduction of a membrane reactor allows the shifting of the equilibrium to the right by subtracting hydrogen from the chemical environment, so increasing the ethanol conversion at lower temperatures, without any loss in selectivity. The introduction of a suitable membrane reactor requires a study oriented to the specific case to evaluate the best membrane to be used and the most opportune operative conditions. On the other hand, many studies have been published on hydrogen separation for similar purposes such as, for example, in the case of ethanol and methanol steam reforming [73,74].

5. Conclusions

The potential impact of the employment of ethanol as a hydrogen carrier through the chemical cycle dehydrogenation-hydrogenation can be summarized in the following points:

(a) The realization of an innovative and simple system for hydrogen storage and transportation.

(b) The use of inexpensive raw materials (bio-ethanol, ethyl acetate), which are abundant, renewable, biodegradable, non-toxic, and not dangerous for the environment.

(c) The synthesis of more active and selective dehydrogenation-hydrogenation catalysts and the development of techniques for their characterization and kinetic testing.

(d) The realization and modeling of membrane reactors for process intensification.

(e) Modeling of the overall cyclic system for different operative scales.

Significant progress in each of the mentioned points would render feasible and convenient the process by overcoming all the disadvantages in the use of the system ethanol-ethyl acetate as LOHC with respect to the more conventional systems.

Author Contributions: Conceptualization, E.S.; data curation, E.S. and R.T.; writing—original draft preparation, E.S.; writing—review and editing, E.S., R.T., S.F. and A.M.R.G.; supervision, E.S. and A.M.R.G. All authors have read and agreed to the published version of the manuscript.

Funding: This research received no external funding.

Data Availability Statement: No new data were created or analyzed in this study. Data sharing is not applicable to this article.

Acknowledgments: Eurochem Engineering is acknowledged for the financial support. Thanks are due also to Giuseppina Carotenuto for some useful information reported in her PhD thesis.

Conflicts of Interest: The authors declare no conflict of interest.

References

1. OECD/Food and Agriculture Organization of the United Nations. Biofuels. In *OECD-FAO Agricultural Outlook 2022–2031*; OECD Publishing: Paris, France, 2022; pp. 232–243. [CrossRef]

2. Iwasa, N.; Takezawa, N. Reforming of ethanol—Dehydrogenation to ethyl acetate and steam reforming to acetic acid over copper-based catalysts. *Bull. Chem. Soc. Jpn.* **1991**, *64*, 2619–2623. [CrossRef]

3. Tu, Y.I.; Chen, Y.W.; Li, C. Characterization of unsupported copper-chromium catalysts for ethanol dehydrogenation. *J. Mol. Catal.* **1994**, *89*, 179–189. [CrossRef]

4. Wu, R.; Sun, K.; Chen, Y.; Zhang, M.; Wang, L. Ethanol dimerization to Ethyl acetate and hydrogen on the multifaceted copper catalysts. *Surface Sci.* **2021**, *703*, 121742. [CrossRef]

5. Inui, K.; Kurabayashi, T.; Sato, S. Direct synthesis of ethyl acetate from ethanol over Cu-Zn-Zr-Al-O catalyst. *Appl. Catal. A Gen.* **2002**, *237*, 53–61. [CrossRef]

6. Inui, K.; Kurabayashi, T.; Sato, S. Direct synthesis of ethyl acetate from ethanol carried out under pressure. *J. Catal.* **2002**, *212*, 207–215. [CrossRef]

7. Inui, K.; Kurabayashi, T.; Sato, S.; Ichikawa, N. Effective formation of ethyl acetate from ethanol over Cu-Zn-Zr-Al-O catalyst. *J. Mol. Catal. A Chem.* **2004**, *216*, 147–156. [CrossRef]

8. Colley, S.W.; Tabatabaei, J.; Waugh, K.C.; Wood, M.A. The detailed kinetics and mechanism of ethyl ethanoate synthesis over a Cu/Cr_2O_3 catalyst. *J. Catal.* **2005**, *236*, 21–33. [CrossRef]

9. Gaspar, A.B.; Barbosa, F.G.; Letichevsky, S.; Appel, L.G. The one-pot ethyl acetate syntheses: The role of the support in the oxidative and the dehydrogenative routes. *Appl. Catal. A Gen.* **2010**, *380*, 113–117. [CrossRef]

10. Semenov, I.P.; Men'shchikov, V.A.; Sycheva, O.I. Pilot tests of a catalyst for the production of ethyl acetate from ethanol. *Catal. Ind.* **2020**, *12*, 287–291. [CrossRef]

11. Usman, M.R. Hydrogen storage methods: Review and current status. *Renew. Sustain. Energy Rev.* **2022**, *167*, 112743. [CrossRef]

12. Phung, T.K. Copper-based catalysts for ethanol dehydrogenation and dehydrogenative coupling into hydrogen, acetaldehyde and ethyl acetate. *Int. J. Hydrogen Energy* **2022**, *47*, 42234–42249. [CrossRef]

13. Sekine, Y.; Higo, T. Recent trends on the dehydrogenation catalysis of liquid organic hydrogen carrier (LOHC): A review. *Top. Catal.* **2021**, *64*, 470–480. [CrossRef]

14. Rao, P.C.; Yoon, M. Potential liquid-organic hydrogen carrier (LOHC) systems: A review on recent progress. *Energies* **2020**, *13*, 6040. [CrossRef]
15. Niermann, M.; Beckendorff, A.; Kaltschmitt, M.; Bonhoff, K. Liquid organic hydrogen carrier (LOHC)—Assessment based on chemical and economic properties. *Int. J. Hydrogen Energy* **2019**, *44*, 6631–6654. [CrossRef]
16. Khamhaeng, P.; Laosiripojana, N.; Assabumrungrat, S.; Kim-Lohsoontorn, P. Techno-economic analysis of hydrogen production from dehydrogenation and steam reforming of ethanol for carbon dioxide conversion to methanol. *Int. J. Hydrogen Energy* **2021**, *46*, 30891–30902. [CrossRef]
17. Bueno, J. Copper-Based Catalysts, Process for Preparing Same and Use Thereof. WO2004/080589 A2, 12 March 2004.
18. Inui, K.K.; Takahashi, T.S.; Kurabayashi, T.I. Catalyst for Ester Production and Process for Producing Ester. U.S. Patent 2004/0242917 A1, 2 December 2004.
19. Inui, K.K.; Takahashi, T.S.; Kurabayashi, T.I. Catalyst for Ester Production and Process for Producing Ester. U.S. Patent 7,091,155 B2, 15 August 2006.
20. Colley, S.W.; Fawcett, C.R.; Sharif, M.; Tuck, M.W.M.; Watson, D.J.; Wood, M.A. Davy Process Tech., Republic of South Africa. Purification of Ethyl Acetate from Mixture Comprising Ethanol and Water by Pressure Swing Distillation. EP 1 117 629 B1, 15 December 2004.
21. Colley, S.W.; Fawcett, C.R.; Rathmell, C.; Tuck, M.W.M. Process for the Preparation of Ethyl Acetate. EP 1 117 631 B1, 17 November 2004.
22. Colley, S.W.; Fawcett, C.R.; Sharif, M.; Tuck, M.W.M.; Watson, D.J.; Wood, M.A. Process. U.S. Patent 7,553,397 B1, 30 June 2009.
23. Santacesaria, E.; Di Serio, M.; Tesser, R.; Carotenuto, G. Process for the Production of Ethyl Acetate from Ethanol. WO2011/104738A2, 1 September 2011.
24. Santacesaria, E.; Carotenuto, G.; Tesser, R.; Di Serio, M. Ethanol dehydrogenation to ethyl acetate by using copper and copper chromite catalysts. *Chem. Eng. J.* **2012**, *179*, 209–220. [CrossRef]
25. Carotenuto, G.; Tesser, R.; Di Serio, M.; Santacesaria, E. Bioethanol as feedstock for chemicals such as acetaldehyde, ethyl acetate and pure hydrogen. *Biomass Conv. Bioref.* **2013**, *3*, 55–67. [CrossRef]
26. Santacesaria, E.; Carotenuto, G.; Tesser, R.; Di Serio, M. Production of pure hydrogen by ethanol dehydrogenation. *Oil Gas Eur. Mag.* **2011**, *37*, 99–102.
27. Santacesaria, E.; Carotenuto, G.; Tesser, R.; Di Serio, M. Production of pure hydrogen by ethanol dehydrogenation. In Proceedings of the DGMK Conference, Berlin, Germany, 4–6 October 2010.
28. Carotenuto, G.; Tesser, R.; Di Serio, M.; Santacesaria, E. Kinetic study of ethanol dehydrogenation to ethyl acetate promoted by a copper/copper chromite catalyst. *Catal. Today* **2013**, *203*, 202–210. [CrossRef]
29. Luo, M.; Das, T.K.; Delibas, C.; Davis, B.H. Heterogeneous catalytic hydrogenation of ethyl acetate to produce ethanol. *Top. Catal.* **2014**, *57*, 757–761. [CrossRef]
30. Di, W.; Cheng, J.; Tian, S.; Li, J.; Chen, J.; Sun, Q. Synthesis and characterization of supported copper phyllosilicate catalysts for acetic ester hydrogenation to ethanol. *Appl. Catal. A Gen.* **2016**, *510*, 244–259. [CrossRef]
31. Zhu, Y.; Shi, L. Zn promoted Cu–Al catalyst for hydrogenation of ethyl acetate to alcohol. *J. Ind. Eng. Chem.* **2014**, *20*, 2341–2347. [CrossRef]
32. Zhu, Y.; Shi, X.W.L. Hydrogenation of ethyl acetate to ethanol over bimetallic Cu-Zn/SiO$_2$ catalysts prepared by means of coprecipitation. *Bull. Korean Chem. Soc.* **2014**, *35*, 141–146. [CrossRef]
33. Lu, Z.; Yin, H.; Wang, A.; Hu, J.; Xue, W.; Yin, H.; Liu, S. Hydrogenation of ethyl acetate to ethanol over Cu/ZnO/MO$_x$ (MO$_x$ = SiO$_2$, Al$_2$O$_3$, and ZrO$_2$) catalysts. *J. Ind. Eng. Chem.* **2016**, *37*, 208–215. [CrossRef]
34. Tran, B.L.; Johnson, S.I.; Brooks, K.P.; Autrey, S.T. Ethanol as a liquid organic hydrogen carrier for seasonal microgrid application: Catalysis, theory, and engineering feasibility. *ACS Sustain. Chem. Eng.* **2021**, *9*, 7130–7138. [CrossRef]
35. Mevawala, C.; Brooks, K.; Bowden, M.E.; Breunig, H.M.; Tran, B.L.; Gutiérrez, O.Y.; Autrey, T.; Müller, K. The ethanol–ethyl acetate system as a biogenic hydrogen carrier. *Energy Technol.* **2022**, *11*, 2200892. [CrossRef]
36. Stull, D.R.; Vestrum, E.F.; Sinke, G.C. *The Chemical Thermodynamics of Organic Compounds*, 1st ed.; John Wiley: New York, NY, USA, 1969.
37. Finger, P.H.; Osmari, T.A.; Costa, M.S.; Bueno, J.M.C.; Gallo, J.M.R. The role of the interface between Cu and metal oxides in the ethanol dehydrogenation. *Appl. Catal. A Gen.* **2020**, *589*, 117236. [CrossRef]
38. Pang, J.; Zheng, M.; Wang, C.; Liu, H.; Liu, X.; Sun, J.; Wang, Y.; Zhang, T. Hierarchical echinus-like Cu-MFI catalysts for ethanol dehydrogenation. *ACS Catal.* **2020**, *10*, 13624–13629. [CrossRef]
39. Freitas, I.C.; Damyanova, S.; Oliveira, D.C.; Marquez, C.M.P.; Bueno, J.M.C. Effect of Cu content on the surface and catalytic properties of Cu/ZrO2 catalyst for ethanol dehydrogenation. *J. Mol. Catal. A Chem.* **2014**, *381*, 26–37. [CrossRef]
40. Tu, Y.J.; Li, C.; Chen, Y.W. Effect of chromium promoter on copper catalysts in ethanol dehydrogenation. *J. Chem. Technol. Biotechnol.* **1994**, *59*, 141–147. [CrossRef]
41. Men'shchikov, V.A.; Gol'dshtein, L.K.; Semenov, I.P. Kinetics of ethanol dehydrogenation into ethyl acetate. *Kinet. Catal.* **2014**, *55*, 12–17. [CrossRef]
42. Carotenuto, G. Innovative Processes for the Production of Acetaldehyde, Ethyl Acetate and Pure Hydrogen by Ethanol. Ph.D. Thesis in Chemistry Macromolecular and Catalysis, University of Naples Federico II, Naples, Italy, November 2011.
43. Private Communication: Davy Process Technology Report; Ethyl Acetate Process. 13 April 2004.
44. Berg, L. Separation of Ethyl Acetate from Ethanol by Azeotropic Distillation. U.S. Patent 5,993,610 A, 30 November 1999.

45. Zhu, Z.; Ri, Y.; Jia, H.; Wang, Y.; Wang, Y. Process evaluation on the separation of ethyl acetate ethanol using extractive distillation with ionic liquid. *Sep. Purif. Technol.* **2017**, *181*, 44–52. [CrossRef]
46. Zhang, Q.; Liu, M.; Li, C.; Zeng, A. Design and control of extractive distillation process for separation of the minimum-boiling azeotrope ethyl-acetate and ethanol. *Chem. Eng. Res. Des.* **2018**, *136*, 57–70. [CrossRef]
47. Gadewar, S.B. Ethyl Acetate Production. U.S. Patent 2012/0035390A1, 9 February 2012.
48. Gadewar, S.B.; Vicente, B.C.; Norton, R.E.; Doherty, M.F. Ethyl Acetate Production. U.S. Patent 2013/0197266A1, 1 August 2013.
49. Gadewar, S.B. Ethyl Acetate Production. U.S. Patent 8,562,921B2, 22 October 2013.
50. Gadewar, S.B.; Stoimenov, P.K. Ethyl Acetate Production. U.S. Patent 2014/0012037A1, 9 January 2014.
51. Gadewar, S.B.; Vicente, B.C.; Norton, R.E.; Doherty, M.F. Ethyl Acetate Production. U.S. Patent 9,079,851B2, 14 July 2015.
52. Gadewar, S.B.; Stoimenov, P.K. Ethyl Acetate Production. U.S. Patent 9,447,018B2, 20 September 2016.
53. Senior Design Reports of the University of Pennsylvania: Ethanol to Ethyl Acetate. Available online: https://core.ac.uk/download/pdf/219382692.pdf (accessed on 27 January 2023).
54. Raich, B.A.; Foley, H.C. Ethanol dehydrogenation with a palladium membrane reactor: An alternative to Wacker chemistry. *Ind. Eng. Chem. Res.* **1998**, *37*, 3888–3895. [CrossRef]
55. Keuler, J.N.; Lorenzen, L. Pd-Ag membranes and their application for dehydrogenation of ethanol in a membrane reactor. In Proceedings of the 221st National Meeting ACS, San Diego, CA, USA, 1–5 April 2001.
56. Keuler, J.N.; Lorenzen, L. Comparing and modeling the dehydrogenation of ethanol in a plug-flow reactor and a Pd-Ag membrane reactor. *Ind. Eng. Chem. Res.* **2002**, *41*, 1960–1966. [CrossRef]
57. Lin, W.H.; Chang, H.F. A study of ethanol dehydrogenation reaction in a palladium membrane reactor. *Catal. Today* **2004**, *97*, 181–188. [CrossRef]
58. Sánchez, A.B.; Homs, N.; Miachon, S.; Dalmon, J.A.; Fierro, J.L.G.; de la Piscina, P.R. Direct transformation of ethanol into ethyl acetate through catalytic membranes containing Pd or Pd-Zn: Comparison with conventional supported catalysts. *Green Chem.* **2011**, *13*, 2569–2575. [CrossRef]
59. Zeng, G.; Chen, T.; He, L.; Pinnau, I.; Lai, Z.; Huang, K.W. A green approach to ethyl acetate: Quantitative conversion of ethanol through direct dehydrogenation in a Pd-Ag membrane reactor. *Chem. Eur. J.* **2012**, *18*, 15940–15943. [CrossRef]
60. Adkins, H.; Folkers, K. The catalytic hydrogenation of esters to alcohols. *J. Am. Chem. Soc.* **1931**, *53*, 1095–1097. [CrossRef]
61. Thakur, D.S.; Carrick, W. Copper Chromite Hydrogenation Catalysts for Production of Fatty Alcohols. WO 2012/074841, 7 June 2012.
62. Yuan, P.; Liu, Z.Y.; Zhang, W.; Sun, H.; Liu, S. Cu-Zn/Al$_2$O$_3$ catalyst for the hydrogenation of esters to alcohols. *Chin. J. Catal.* **2010**, *31*, 769–775. [CrossRef]
63. He, L.; Cheng, H.; Liang, G.; Yu, Y.; Zhao, F. Effect of structure of CuO/ZnO/Al$_2$O$_3$ composites on catalytic performance for hydrogenation of fatty acid ester. *Appl. Catal. A Gen.* **2013**, *452*, 88–93. [CrossRef]
64. Huang, H.; Cao, G.; Wang, S. An evaluation of alkylthiols and dialkyl disulfides on deactivation of Cu/Zn catalyst in hydrogenation of dodecyl methyl ester to dodecanol. *J. Ind. Eng. Chem.* **2014**, *20*, 988–993. [CrossRef]
65. Brands, D.S.; Poels, E.K.; Bliek, A. Ester hydrogenolysis over promoted Cu/SiO$_2$ catalysts. *Appl. Catal. A Gen.* **1999**, *184*, 279–289. [CrossRef]
66. Kim, S.M.; Lee, M.E.; Choi, J.W.; Suh, D.J.; Suh, Y.W. Role of ZnO in Cu/ZnO/Al$_2$O$_3$ catalyst for hydrogenolysis of butyl butyrate. *Catal. Commun.* **2011**, *12*, 1328–1332. [CrossRef]
67. Huang, X.; Ma, M.; Miao, S.; Zheng, Y.; Chen, M.; Shen, W. Hydrogenation of methyl acetate to ethanol over a highly stable Cu/SiO$_2$ catalyst: Reaction mechanism and structural evolution. *Appl. Catal. A Gen.* **2017**, *531*, 79–88. [CrossRef]
68. Zhao, Y.; Shan, B.; Wang, Y.; Zhou, J.; Wang, S.; Ma, X. An effective CuZn-SiO$_2$ bimetallic catalyst prepared by hydrolysis precipitation method for the hydrogenation of methyl acetate to ethanol. *Ind. Eng. Chem. Res.* **2018**, *57*, 4526–4534. [CrossRef]
69. Jiang, X.C.; Wang, Z.G.; Li, C.X. Intrinsic kinetics of ethyl acetate hydrogenolysis to ethanol over a Cu-Zn/Al$_2$O$_3$ catalyst. *J. Beijing Univ. Chem. Technol.* **2014**, *41*, 36–39.
70. Schittkowski, J.; Tölle, K.; Anke, S.; Stürmer, S.; Muhler, M. On the bifunctional nature of Cu/ZrO$_2$ catalysts applied in the hydrogenation of ethyl acetate. *J. Catal.* **2017**, *352*, 120–129. [CrossRef]
71. Natal Santiago, M.A.; Sánchez-Castillo, M.A.; Cortight, R.D.; Dumesic, J.A. Catalytic reduction of acetic acid, methyl acetate, and ethyl acetate over silica-supported copper. *J. Catal.* **2000**, *193*, 16–28. [CrossRef]
72. Niermann, M.; Drünert, S.; Kaltschmitt, M.; Bonhoff, K. Liquid organic hydrogen carriers (LOHCs)–techno-economic analysis of LOHCs in a defined process chain. *Energy Environ. Sci.* **2019**, *12*, 290–307. [CrossRef]
73. Gallucci, F.; Basile, A. Pd-Ag membrane reactor for steam reforming reactions: A comparison between different fuels. *Int. J. Hydrogen Energy* **2008**, *33*, 1671–1687. [CrossRef]
74. Rahimpour, M.R.; Samimi, F.; Babapoor, A.; Tohidian, T.; Mohebi, S. Palladium membranes applications in reaction systems for hydrogen separation and purification: A review. *Chem. Eng. Process.* **2017**, *121*, 24–49. [CrossRef]

Review

Research Progress on the Typical Variants of Simulated Moving Bed: From the Established Processes to the Advanced Technologies

Xiaotong Zhang [1], Juming Liu [1], Ajay K. Ray [2,*] and Yan Li [1,*]

[1] Chemical Engineering Institute, Inner Mongolia University of Technology, Inner Mongolia Autonomous Region, Hohhot 010000, China
[2] Department of Chemical and Biochemical Engineering, Western University, London, ON N6A 5B9, Canada
* Correspondence: aray@eng.uwo.ca (A.K.R.); yanli@imut.edu.cn (Y.L.)

Abstract: Simulated moving bed (SMB) chromatography is a highly efficient adsorption-based separation technology with various industrial applications. At present, its application has been successfully extended to the biochemical and pharmaceutical industrial sectors. SMB possesses the advantages of high product purity and yield, large feed treatment capacity, and simple process control due to the continuous operation mode and the efficient separation mechanism, particularly for difficult separation. Moreover, SMB performs well, particularly for multi-component separation or complicated systems' purification processes in which each component exhibits similar properties and low resolution. With the development of the economy and technology, SMB technology needs to be improved and optimized to enhance its performance and deal with more complex separation tasks. This paper summarizes the typical variants or modifications of the SMB process through three aspects: zone variant, gradient variant, and feed or operation variant. The corresponding modification principles, operating modes, advantages, limitations, and practical application areas of each variant were comprehensively investigated. Finally, the application prospect and development direction were summarized, which could provide valuable recommendations and guidance for future research in the SMB area.

Keywords: simulated moving bed; chromatographic separation; variants; three-zone SMB; operation mode

Citation: Zhang, X.; Liu, J.; Ray, A.K.; Li, Y. Research Progress on the Typical Variants of Simulated Moving Bed: From the Established Processes to the Advanced Technologies. *Processes* **2023**, *11*, 508. https://doi.org/10.3390/pr11020508

Academic Editors: Elio Santacesaria, Riccardo Tesser and Vincenzo Russo

Received: 12 December 2022
Revised: 3 February 2023
Accepted: 6 February 2023
Published: 8 February 2023

1. Introduction

The simulated moving bed (SMB) concept was proposed by Broughton et al. of UOP in 1961 for the separation of xylene isomers in the petrochemical field. It was then gradually applied to the sugar industry and chiral drug resolution areas [1–4]. The typical SMB system is developed based on the True Moving Bed (TMB), which involves several fixed-bed chromatographic columns. It divides into four zones by four inlet and outlet ports (feed, raffinate, desorbent, and extract). Different from the TMB process, the counter-current movement of the solid phase towards the fluid phase (as shown in Figure 1) is achieved by the simultaneous synchronous switch of four streams [5–14]. Therefore, the problems associated with the movement of the solid phase can be solved, such as particle attrition, bed voidage variation, unstable flow rate, and bed expansion.

An illustration of the SMB process is shown in Figure 2. The feed stream containing both strongly adsorbed (heavy) and weakly adsorbed (light) components enters the system between zones II and III. With the column switching, the heavy component moves backward into zone II with the solid phase, while the light component is desorbed by the eluent and moves forward into zone III with the liquid phase, thus achieving the separation purpose. Therefore, the II and III zones are normally called the separation zone, where the operating conditions are set such that the two components move counter currently. After that, the heavy component is desorbed from the solid phase in zone I, which makes the solid phase regenerate, so zone I is also called the solid phase regeneration zone, while the light component is adsorbed in zone IV, which regenerates the liquid phase, so zone IV is the liquid phase regeneration zone [15–17]. For efficient operation, the solid and liquid phase regeneration zones are typically operated in co-current mode by setting the operating parameters appropriately.

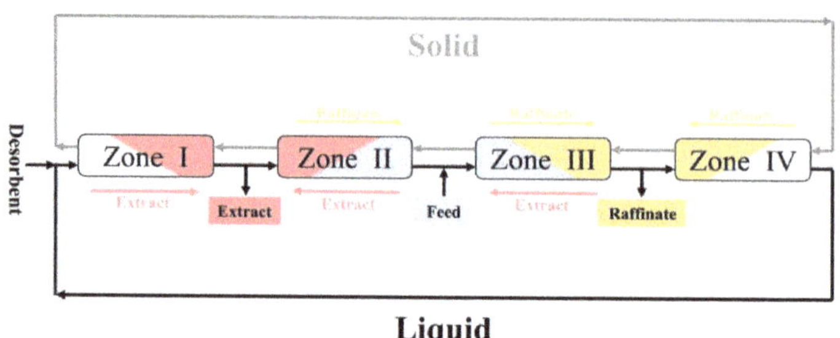

Figure 1. Schematic diagram of counter-current movement.

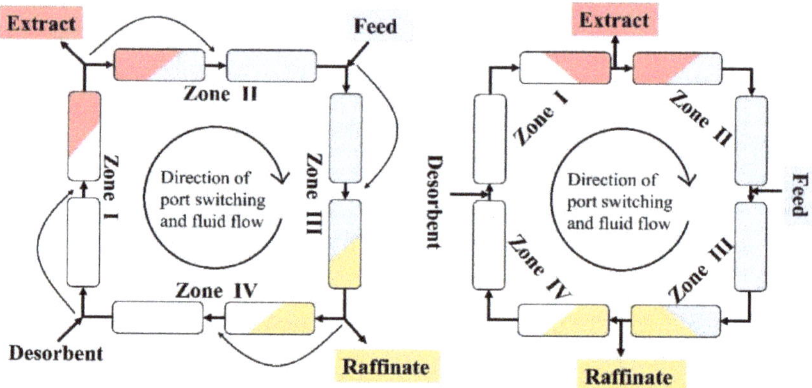

Figure 2. Schematic diagram of Simulated Moving Bed.

SMB technology has its specific advantages for binary systems and complicated systems whose components have similar properties and as such are difficult to separate by traditional methods. In recent years, with the development of the economy and the progress of science and technology, the industrial requirements for energy consumption, solvent consumption, separation efficiency, product specifications, and flexibility of process control become higher and more stringent. As a result, some new SMB modes have been successively proposed. For example, the Varicol system based on the non-synchronous switching of inlet/outlet ports proposed by Ludemann-Hombourger improves the performance of SMB within a cycle, making the operation more flexible and the requirement for the number of columns significantly reduced [18,19]. The SMB process with internal flow rate changes realizes the redistribution of each component through the change of flow rate. This technology can save solvent consumption, but the process control becomes complicated [20]. Gradient SMB systems include the introduction of concentration, temperature, and pressure gradient. The introduction of a gradient condition can realize the reallocation of each composition, improve the efficiency of separation, and reduce the solvent consumption; however, the implementation process is complicated, with a need to ensure the synchronicity of the switch and the gradient change [21–23]. In addition, the sequential simulated moving bed (SSMB) developed in recent years [24] divides a switching of the traditional SMB process into three steps, and each step presents a different operation mode. SSMB has shown excellent performance and great potential in terms of separation effect, process control, energy consumption and solvent consumption.

After years of research and development, diverse SMB-based variants have emerged with different operating modes and application areas [25,26]. This paper mainly focuses on an investigation of SMB variants and divides these into zone variant, gradient variant and feed or operation variant. The framework structure of this review is shown in Figure 3. According to the literature review and analysis, a detailed introduction and comparison were completed, and the separation mechanism, switching modes, advantages and disadvantages and applications of each SMB variant were summarized. Finally, this work can provide practical suggestions and references for SMB research works and industrial applications (intended for both the expert and novice), meanwhile putting forward the application prospect and future development direction.

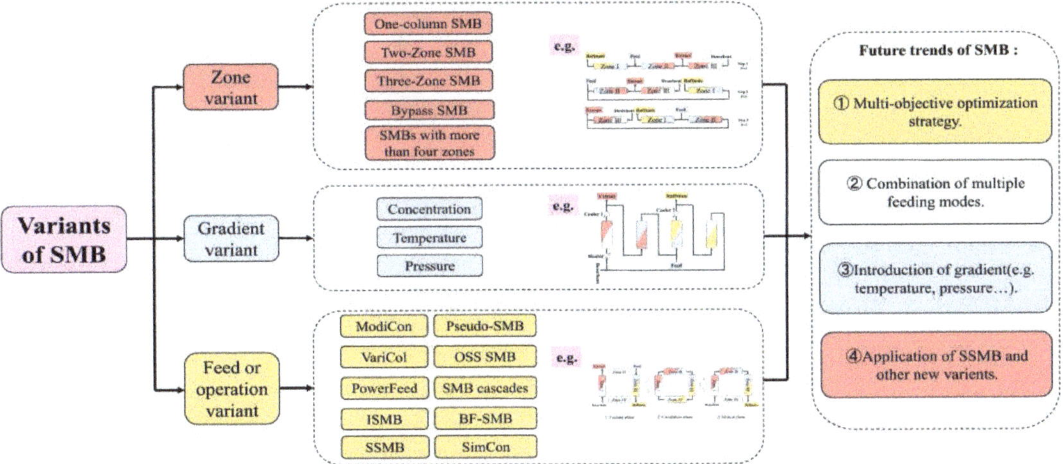

Figure 3. The framework structure of review.

2. Zone Variant

As the conventional four-zone SMB technology is relatively mature, researchers attempted to change the zone partition or structure to make the equipment simpler, meanwhile improving the separation performance (purity, productivity, etc.) or reducing operating costs [27–29]. The so-called zone variant is to reduce non-essential functional zones to combine or delete one or several areas of the four-zone SMB. Usually, the separation zones (zone II and zone III) are retained, and zone I (solid phase regeneration zone) or zone IV (liquid phase regeneration zone) are modified. However, the reduction of the regeneration zone means that the solid and liquid phases cannot be adequately regenerated and recycled, leading to problems such as increased desorbent consumption [30–33]. In the following sections, zone variant will be investigated from four aspects: one-column SMB, two-zone SMB, three-zone SMB and bypass SMB; furthermore, their advantages, disadvantages, and applications are analyzed, respectively.

2.1. One-Column SMB

One-column SMB was firstly developed by Wankat et al. [34,35] as a one-column chromatography with multiple tanks like a four-zone SMB cycle. As shown in Figure 4, the main principle is a four-step cycle. In the first step, the feed stream and the solution in tank 2 are fed into the column, the less retained component (raffinate) is collected at the output port, and meanwhile tank 1 is filled. In the second step, tank 1 is fed into the column and tank 4 is filled. Then, in the third step, fresh desorbent and the solution of tank 4 enter into the column, while the more retained component (extract) is collected and tank 3 is filled. In the final step, tank 3 and tank 2 work in the same way.

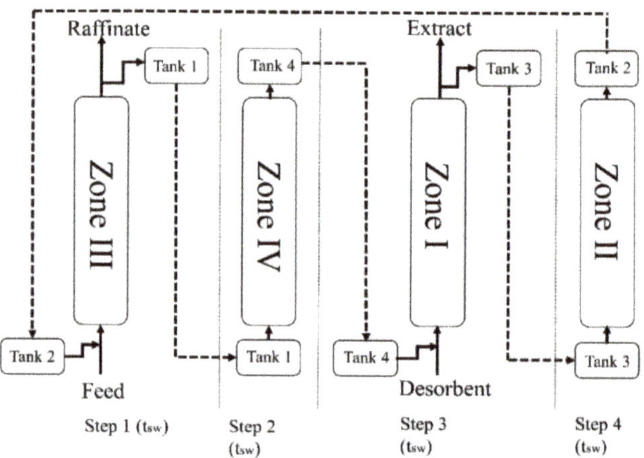

Figure 4. Schematic diagram of one-column SMB.

At present, there are few studies on this kind of SMB variant, and the one-column system exhibits obvious advantages and disadvantages. Specifically, it is cheap and flexible, and has the advantage if frequent desorbent changes are required. However, there is also a clear disadvantage that, due to the use of storage tanks, there is a mixing process during the whole operation and switching processes, which reduces the separation efficiency.

2.2. Two-Zone SMB

Two-zone SMB is proposed by Lee et al. [36]; its structure is shown in Figure 5. In this system, only the separation zones (zones II and III of the conventional SMB) are reserved, implying that the eluent will directly enter zone II, and the feed mixture will enter zone III. In the nth switching, the second half of the light component will move towards the end of zone III with the liquid phase and finally leave the column, while the heavy component moves backward with the solid phase and remains in zone II. At the $n + 1$st switching, the first column of the original zone II moves to the end of the zone III and becomes the last column of the zone III, so that the heavy component previously retained in zone II enters into zone III and leaves the column, while the first half of the adsorbed light component will immediately flow out from the exit, so that the two components can be collected separately.

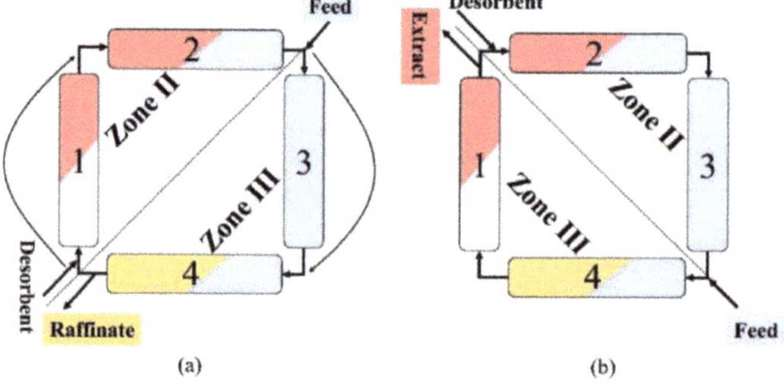

Figure 5. Schematic diagram of Two-zone SMB: (**a**) The end of the previous switch. (**b**) The beginning of the next switch.

Wankat et al. [37] designed another two-zone SMB using a two-step process, combining zones I and II of the conventional SMB into a new zone I, and zones III and IV into a new zone II. At first, the feed stream is introduced between zone I and zone II, while some desorbent circulates from zone I to zone II, and the remaining desorbent is sent to the tank. In the second step (no feed), the fresh desorbent and the desorbent in the tank are used to produce the product. The raffinate and extract products are collected from zone I and zone II, respectively. At the end of the second step, all the ports are switched and the whole operation process is repeated.

In conclusion, the two-zone SMB has the advantage of low cost and is more economical due to the isolation of two regeneration zones. In addition, relatively high purity can be achieved from this simplified equipment. For example, in Lee's work [36], compared to the conventional SMB, the two-zone SMB improved the purity and recovery of the fructose-rich product from 0.78% and 4.11% to 15.67% and 15.87%, respectively. As a result, the separation cost was reduced due to the low material consumption and simple column arrangement. Moreover, there still exist obvious disadvantages: (1) Although the port switch of two-zone SMB is similar to that of conventional four-zone SMB, it cannot achieve the countercurrent movement of the solid and liquid phases, so it is not available for continuous operation. (2) The purity and recovery of the two-zone SMB is lower than that of the four zone SMB, owing to the simplification. (3) The final purity cannot be easily increased by increasing the number of columns in each zone.

2.3. Three-Zone SMB

The three-zone SMB is the most studied mode among these variants; the schematic diagram is shown in Figure 6. In the typical three-step process, there is no desorbent cycle loop so the liquid phase regeneration zone is isolated. The desorbent enters from the end of zone III, and the binary mixture is fed between zones I and II. Finally, the heavy fraction (extract) exits between zones II and III, and the light fraction (raffinate) exits from the front end of zone I [38–41].

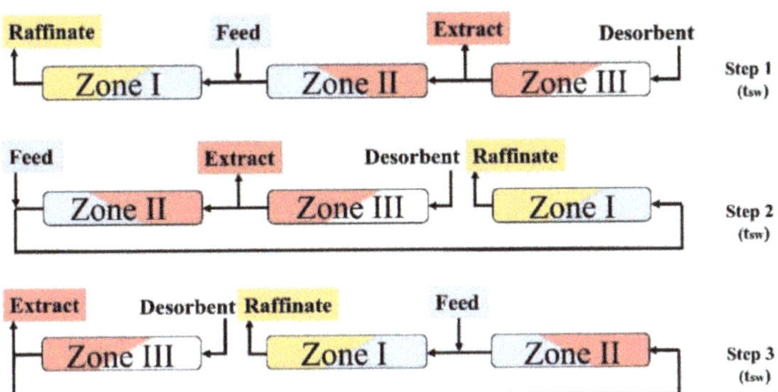

Figure 6. Standard three-zone SMB with three-step.

2.3.1. Three-Zone SMB without Zone IV

Wang et al. [42] designed an open-loop three-zone SMB in which zone IV is isolated. Then, this system is applied for the separation of Eicosapentaenoic acid (EPA) and docosahexaenoic acid (DHA) using a 1-1-2 column configuration, as shown in Figure 7. Zone I is completely independent and is used for elution. Zone II and zone III work as purification and adsorption sections, respectively. The mixture is separated in zone III, and the raffinate product EPA is obtained at the outlet (R port). The fluid phase is withdrawn in zone II to improve the separation of the components adsorbed in zone III. The eluent is withdrawn in zone I, and the extract product DHA is collected at the outlet (E port) to realize the regeneration of the column. Under the control of the automatic system, zones I to III move continuously along the fluid phase direction for one column length at each switching time.

The results showed that, in the 2-2-2 mode, the purity of both DHA and EPA reached 99% and the recovery was close to or at 100%. The solvent consumption was 1.11 L/g, which was significantly lower compared with 1.46 L/g in the 1-1-2 mode [42]. In consequence, this kind of variant could improve both the separation performance and the economical efficiency of SMB.

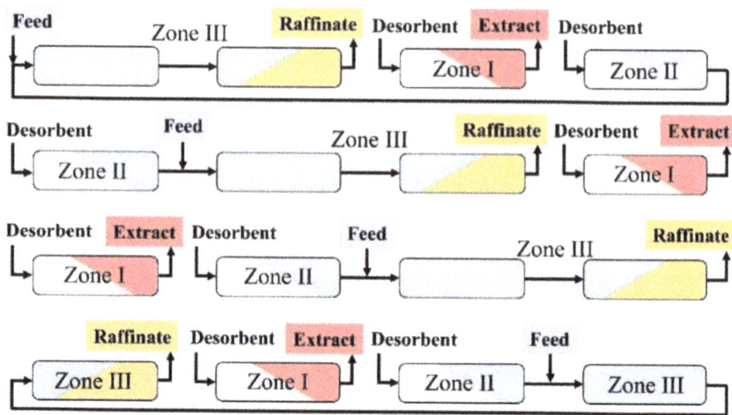

Figure 7. Schematic diagram of three-zone SMB (with 1-1-2 column distribution).

2.3.2. Three-Zone SMB without Zone I

Wei et al. [43] developed a three-zone SMB without zone I, which can simultaneously achieve the high purity and low desorbent consumption. For instance, zone I is isolated, and the desorbent directly enters zone II. During the switching time, due to the column switching, the part of the solution that retains more components partially enters zone IV and continues to move forward as the extract. The other part is at the first column in zone II, as the raffinate, and is converted from a discontinuous to continuous liquid stream under the action of the desorbent. The extractive residue and extractables were collected from zone III and zone IV, respectively.

2.3.3. Port Variant

Lee et al. [44] proposed a new three-zone SMB system called three-port SMB (TT-SMB). Actually, it is a combination of the above two three-zone operations. In the first step, the extract port is closed, then the desorbent and feed solution are injected into the inlet of zones I–IV and the feed node between zones II and III, respectively. When the extract port is closed, the solution from zone III is the raffinate. In the second step, the raffinate port is closed and the desorbent is supplied to the inlet of zone II. Since there is no raffinate port, the stream from zones I to IV is the extract solution. The results showed that the product purity was generally improved by 1–4%, the recovery was generally improved by 0.8–4.8%, and the productivity was increased by up to 13.8 g/L/h using TT-SMB compared to conventional SMB.

2.3.4. Other Variants

Wankat et al. [39] put forward two ways to improve the operation of the three-zone SMB, namely "partial withdrawal" and "partial feeding". The "partial withdrawal" mode is shown in Figure 8. In the first step, the obtained raffinate is recovered. In the second step, the desorbent is recycled during the switching time. In the third step, the raffinate is recovered and the cycle repeats. The "partial feed" mode is shown in Figure 9. Compared to the other three-zone SMB, this system only has a pulse feed in the second step during the switching time and no feed is introduced in the first and third steps. With the same feeding method, the three-zone SMB improved the recovery by up to 8.87% and the purity by up to 7.82% compared to the four-zone SMB.

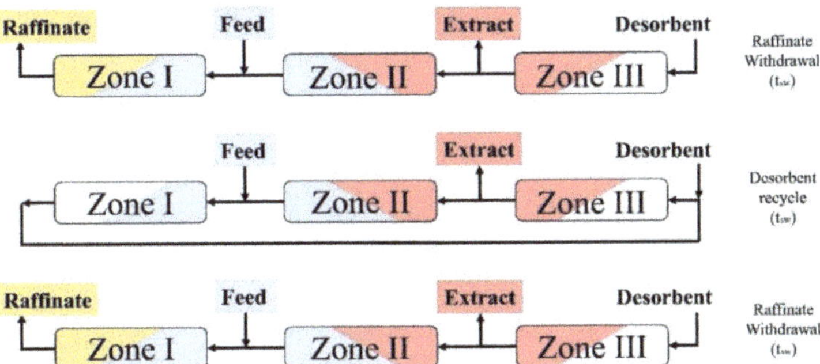

Figure 8. Schematic diagram of the three-zone SMB ("partial withdrawal" operation).

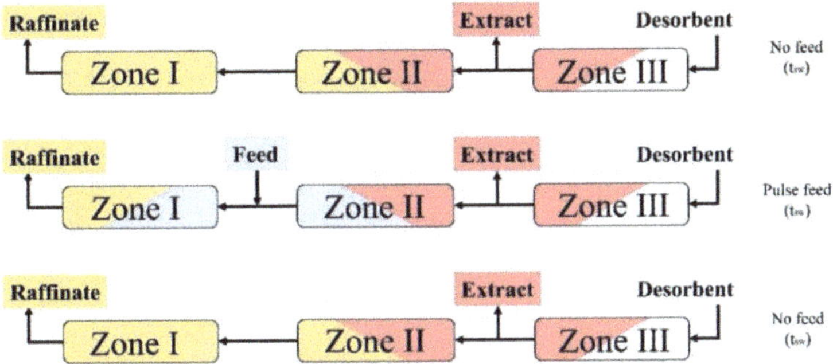

Figure 9. Schematic diagram of the three-zone SMB ("partial feed" operation).

In summary, compared to the conventional four-zone SMB, the separation performance in areas such as purity, recovery and eluent consumption of the three-zone SMB is inferior due to the open-loop structure and the diluted raffinate stream. However, this zone modification has several advantages: (1) Adsorbent consumption and the required equipment units such as valves and pumps are reduced. The productivity of the three-zone SMB becomes higher due to the saved amount of adsorbent. (2) The operation of three-zone SMB is relatively simple. (3) The productivity, desorbent efficiency, product purity and recovery can be improved by introducing the partial feed or partial withdrawal operation, which effectively overcome the drawbacks of the three-zone SMB.

2.3.5. Applications for Three-Zone SMB

The application of three-zone SMB in biological separation was investigated by Keßler et al. [45] and Kim et al. [46,47]. Keßler et al. studied the separation of IgG/lysozyme mixtures and the purification of dimeric BMP-2 from multicomponent mixtures by using the three-zone SMB with a concentration gradient. The results showed that solvent consumption was significantly reduced, and productivity was improved compared to the conventional SMB. In Kim's work, guanine and cytosine were successfully isolated from nucleotides with the three-zone SMB method. The final product purity of cytosine and guanine were achieved at 95% and 90%, respectively. Another work of Kim et al. used three-zone SMB to isolate immunoglobulin Y (IgY) from eggs, and the final IgY purity of 98% was obtained. The above research results show the excellent separation performance and development potential of three-zone SMB for bio-separation.

The application of three-zone SMB in enantiomeric drug separation was investigated by Cunha et al. [48] Two enantiomers (L-PZQ and D-PZQ) of praziquantel (PZQ) were successfully isolated using three-zone SMB. At least one enantiomer with high purity and productivity could be obtained. However, there are few research works focusing on the enantiomeric drug separation by using the three-zone SMB, indicating that the variant still has a great research potential in this area.

Yao et al. [49], Pangpromphan et al. [50,51] and Nam et al. [52] investigated the application of three-zone SMB in the food separation process. Yao et al. constructed the first asynchronous three-zone SMB for the separation of vanillin and syringaldehyde. They finally obtained relatively high product purity and effectively improved the feed flow rate. Pangpromphan et al. successfully separated Alpha-Tocopherol and Gamma-Oryzanol in rice bran oil using three-zone SMB. A mathematical model of adsorption kinetics was constructed and the corresponding operating conditions were optimized. Optimization results showed that the final purity of both products was quite high. The above examples reveal that three-zone SMB is a very effective technology for the separation of food ingredients, and meanwhile possesses high product purity and great modification potential.

2.4. Bypass SMB

Rajendran et al. [53] reported a new operation mode based on conventional four-zone SMB, as shown in Figure 10. The feed solution enters the system between zones II and III, and the extract and raffinate are recovered between zones I and II and between zones III and IV, respectively. After the separation and purification, the feed streams of the binary mixture are bypassed to the extract and raffinate streams with a certain volume. In this way, the desired product purity can be obtained by conjunctively purifying and mixing.

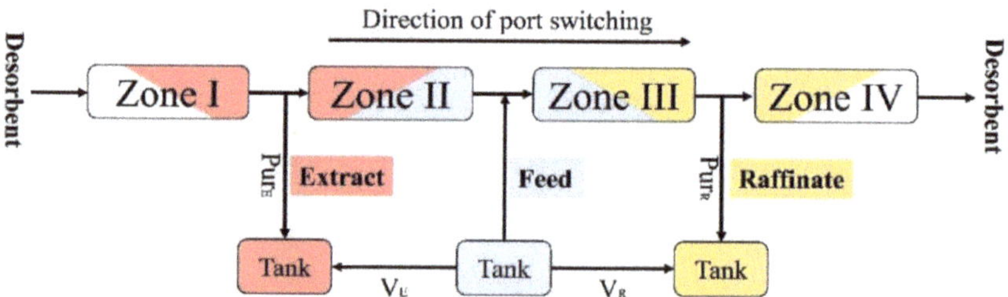

Figure 10. Schematic diagram of bypass SMB.

The advantages of bypass in SMB are high operating flexibility and good selectivity, which makes it suitable for cases where the purity of the target product is not strictly required. However, since this SMB mode is currently only targeted at producing specific products, its application range is narrow and the productivity is not significantly improved compared to the conventional SMB. Therefore, there are few research works and applications at present.

2.5. SMBs with More Than Four Zones

Generally, conventional four-zone SMBs only can handle the binary mixtures' separation task. To separate multi-mixtures, more zones need to be added to break through the limitations in terms of zone variants. The following is a brief description of the five-zone SMB and the nine-zone SMB.

A five-zone SMB is a closed loop with multiple chromatographic columns in series, generally equipped with two inlets (feed and desorbent ports) and three outlets (extractant 1, extractant 2, and extractive residue). The three inlets are assigned to low-affinity substance A, medium-affinity substance B and high-affinity substance C. Usually, the inlet is between zone III and zone IV, and low-affinity substance A is collected from the raffinate port (between zone IV and zone V), while high-affinity substance C and medium-affinity substance B are collected at the extract 1 port (between zone I and zone II) and extract 2 port (between zone II and zone III), respectively. For example, Mun [54] and Xie et al. [55] have successfully separated multiple components by designing and using a five-zone SMB with high yields and purity. The nine-zone SMB can be regarded as a five-zone SMB in parallel with the conventional four-zone SMB, and the whole system forms two closed loops with bypass stream. In Wooley et al.'s work [56], a nine-zone SMB was applied to extract two sugars from the biohydrolysis product with a purity close to 100% and a recovery of 88%.

3. Gradient Variant

The performance of the conventional four-zone SMB could be optimized by adjusting the adsorption behavior of each zone, which can be achieved by introducing a gradient parameter to change the operating conditions, such as temperature or solvent composition. Three kinds of gradient variants are frequently used: concentration gradient (or solvent gradient), temperature gradient and pressure gradient. Among them, the concentration gradient is most widely used and its operating conditions are relatively easier to achieve and fewer restrictions exist. The latter two variants are applicable to specific fluid phases and operating conditions.

3.1. Concentration Gradient

The separation in a conventional four-zone SMB is largely influenced by the adsorption affinity (or isotherm parameters) of the two components. To improve the performance of the SMB, the idea of distributing different isotherm parameters in different zones is applied by introducing different solvent intensities in the desorbent and feed, which resulted in different solvent intensities along the bed. Specifically, a concentration gradient is formed, and the elution ability of zones I–IV is gradually decreased. The elution intensity in zone II (between the extract port and the feed port) is greater than that in zone III (between the feed port and the raffinate port). The solute can, therefore, move forward in zone II and backward in zone III; thus, separation is achieved in these two zones. The solvent strength is commonly controlled by the concentration of the organic modifier in the fluid phase. The higher the concentration of the modifier, the lower the adsorption affinity (or isotherm parameter). Therefore, the concentration of the modifier in the desorbent should usually be set higher than the concentration in the feed material [57–60].

Wang et al. [60] used a concentration gradient SMB to separate paclitaxel and cephalosporin. The solvent composition, zone flow rate and switching time in the feed and desorbent were optimized using a non-dominated sequencing genetic algorithm with elite and jump genes (NSGA-II-JG) and rate model simulations. Compared to conventional SMB, optimal solvent gradient SMBs have substantially higher productivity and lower solvent consumption. Meanwhile, gradient SMBs can further improve productivity by eliminating limitations of flow rates in each zone.

Mun [61] applied the solvent gradient mode for the separation of phenylalanine and tryptophan. The amino acid separation process of SG-SMB was optimized to maximize the production efficiency under the constraints of pump capacity and purity. The inlet and outlet flow rates, switching times and local distribution of liquid phase along the chromatographic bed were optimized using genetic algorithms and rate model simulations. The results showed that the yield was increased, the desorbent consumption was reduced and the product concentration was increased compared to the conventional SMB. Specifically, in this case, the productivity of amino acid was increased up to 110%, and meanwhile the desorbent consumption was reduced up to 53%.

Compared to the conventional four-zone SMB, the concentration gradient SMB has higher productivity and lower desorbent consumption and, meanwhile, relatively high separation purity and efficiency could be obtained. Most operating designs employ an open-loop structure; the eluent flows out from the end of zone IV and is no longer circulated into zone I. However, in the concentration gradient mode, the mobile phase composition is not constant, which leads to cyclic steady-state characteristics when the inlet and outlet ports are switched periodically. Similarly, the internal adsorption equilibrium relationship of the solute also shows cyclic steady-state variations, which will reduce the stability of the system and thus make the process design more difficult.

3.2. Temperature Gradient

Conventional SMB units are operated under isothermal and isobaric conditions with constant adsorption intensities in all zones. Low adsorption intensities are favorable for zones I and II, while high adsorption intensities are favorable for zones III and IV. Therefore, it is desirable to introduce a gradient of adsorption intensity to improve the unit productivity and solvent consumption performance of the SMB. For the liquid fluid phase, the adsorption intensity can be effectively adjusted by changing the temperature [62–64].

Wankat et al. [62] combined the principles of SMB and thermal swing adsorption (TSA) and developed a traveling wave mode thermally assisted moving bed. A heat exchanger was used to control the fluid temperature, which resulted in a thermal wave passing through the column. As shown in Figure 11, the fluid is heated or cooled before it enters each adiabatic column, and the temperature within the column varies in each zone, thereby affecting the solute adsorption capacity and optimizing the performance of the SMB. Thermally assisted SMBs can be used to separate mixtures that are thermally stable and where the isotherm significantly shifts with temperature.

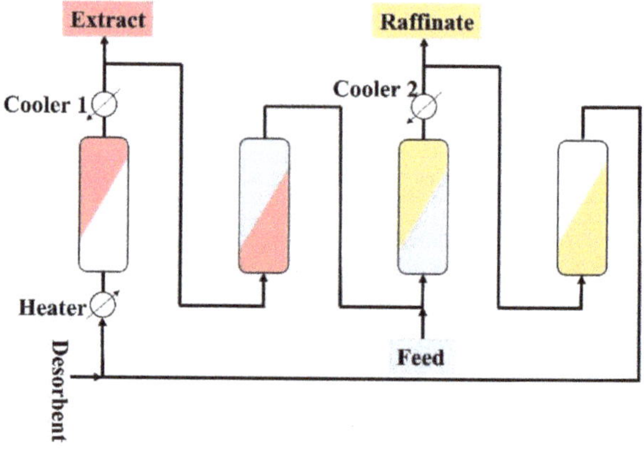

Figure 11. Schematic diagram of temperature gradient.

Yu et al. [65] investigated the feasibility of internal temperature gradient SMB for the enantiomeric binary separation process by introducing the temperature difference between feed and desorbent. The result demonstrated that a temperature gradient with a 20 K difference could significantly improve the productivity of the SMB device by 20%. In addition, increasing the flow rate ratio in zone IV could effectively reduce the solvent consumption.

In conclusion, the advantages of temperature gradient SMB are: (1) There are no restrictions on heat exchange rate, and the SMB and simulation system can be easily scaled up. (2) The adsorption intensity of each zone can be adjusted by changing the temperature to enhance the separation performance of the SMB. Nevertheless, some disadvantages are: (1) Although the traveling wave mode solves the serious heat transfer limitation problem, the external dead volume is increased by heat exchanger, and thus the hysteresis time is increased. (2) As the isotherm of separation mixture is certainly affected by temperature, its application scope is limited.

3.3. Pressure Gradient

The pressure gradient mode is, currently, mainly applied to supercritical fluid simulated moving beds (SF-SMB). Morbidelli et al. [66] designed a pressure gradient SMB, in which different pressure levels were applied in the four zones. Pressure control valves were, respectively, installed after each column and between the outlet and inlet valves, which could adjust different pressure values in the two adjacent zones. As a result, the Henry's constant of the components and their retention times were increased due to the decreased density of the supercritical fluid phase with decreasing pressure. Thus, the separation process could be optimized by introducing a decreased pressure gradient from zone I to zone IV to form a decreased gradient in the elution intensity of the fluid phase. When a pressure gradient is used, the product purity is increased by up to 2.3% and the productivity is increased by 0.29 g/kg compared to the isobaric mode.

Since the fluid phase of this variant uses supercritical fluids instead of conventional organic solvents, it has the obvious advantage of being green and environmentally friendly; meanwhile, the desorbent cost is lower than conventional SMB. However, it is not widely applicable due to the special fluid phase.

4. Feed or Operation Variant

Another modification method is to improve the performance of the conventional SMB by only changing the feed or operation mode, without altering the SMB configuration. Ten different variants are investigated below, which can be used either alone or in combination of two or more to improve the separation performance of SMBs in applications.

4.1. ModiCon

"ModiCon" feed mode, also called varying concentration feed model. Schramm et al. [67] proposed adjusting the feed concentration with a certain rule within the transition cycle appropriately. Then, the concentration spectral band changes its movement rate as it flows through the feed port and is usually applicable to nonlinear adsorption. Compared to conventional SMB, the ModiCon could regulate the feed concentration and increase productivity by about 50% and reduce solvent consumption by about 25% [64]. In the Langmuir adsorption model, the higher the concentration, the faster the concentration point moves. For the more retained component, when it passes through the inlet, its flow rate can be reduced by reducing the feed concentration, so that the less retained component with higher purity can be obtained [68,69].

When the ModiCon mode is used exclusively, the improvement of SMB performance is not obvious. Therefore, it usually needs to be combined with VariCol or other processes to improve the separation performance. After this combination, the separation efficiency can be improved and the solvent consumption can be reduced compared to the conventional SMB.

4.2. VariCol

VariCol, the asynchronous switching mode, is shown in Figure 12. In VariCol mode, only one inlet or outlet is switched within each switching time and each port is switched independently. This makes the length of the zone different for each stage; that is, the length of each zone is not fixed, and changes over time. According to the characteristics of the mixture and the spectral band distribution of the separation process, the column length can be adjusted timely to make the distribution reasonable and the separation more efficient [70–75].

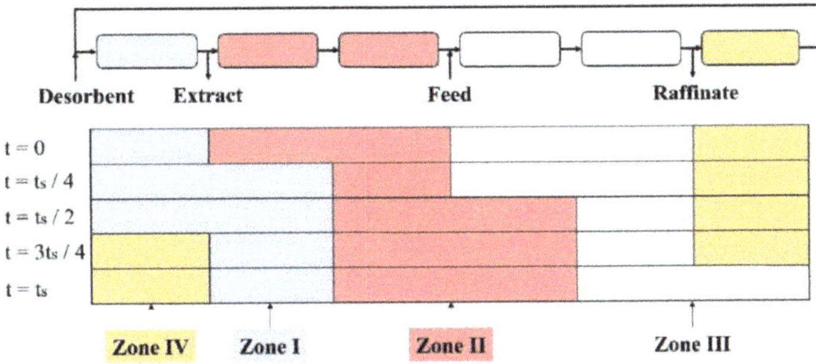

Figure 12. Schematic diagram of VariCol.

Supelano et al. [76] compared the separation performance of three SMB modes, ModiCon, VariCol, and ModiCon + VariCol, in terms of maximum throughput for a given product purity by using the resolution of guaiacol glycerol ether as an example. It was found that the separation performance was not significantly improved when only the ModiCon feed mode was applied, while when both ModiCon + VariCol feed modes were used in combination the throughput, number of columns, and the separation efficiency were obviously improved compared to the conventional SMB, due to the larger zone I and zone II.

Zhang et al. [72] conducted a systematic multi-objective optimization study of the SMB and VariCol processes for the chiral resolution of racemic pindolol using the NSGA-II-JG algorithm. The result showed that, under the condition of higher feed concentration and lower feed flow rate, a relatively high product purity can be obtained without consuming more desorption agent and, meanwhile, a better performance can be obtained by increasing the number of columns. When the VariCol process was used, the product recovery was generally improved by 0.15–0.56% and the product purity was generally improved by 0.1–0.52% compared to the conventional SMB [72]. It was finally proven that the VariCol process outperforms conventional SMB in terms of desorbent consumption with the same separation requirements.

Lin et al. [77] designed and optimized the operating conditions for the enantiomeric separation of aminoglutamine by SMB and VariCol methods using a mass transfer-diffusion model, while considering the intraparticle mass transfer resistance and axial dispersion effects. It was also verified that the separation performance of VariCol process was superior to the conventional SMB.

In summary, the VariCol process allows more flexibility in the chromatographic column use and breaks the limitations of constant zone length and constant solids flow rate. Compared to the conventional SMB, the VariCol mode can obtain higher productivity and lower desorbent consumption.

4.3. PowerFeed

PowerFeed is the separation process with variable feed flow rate. When the mixture solution flows through the feed port position, the feed flow rate is changed in each switching interval, which affects the components' movement rate in each zone by changing the concentration spectrum band. In this way, different components in a mixture are gradually separated [78–80].

The PowerFeed mode is similar to the ModiCon mode mentioned above, in which both methods achieve the separation purpose by changing the concentration of the spectral band. The ModiCon changes the feed concentration directly, whereas PowerFeed changes the concentration by adjusting the flow rate [81–83]. Similarly, PowerFeed generally needs to be used in conjunction with other modes to improve the separation performance.

The three feed variant modes mentioned above, ModiCon, VariCol and PowerFeed, have one thing in common; all of them improve the performance of SMB by increasing the degrees of freedom. All three types of feed modes can lead to improved separation efficiency and performance, and the performance can be further improved by effectively combining these three variants.

4.4. Intermittent Simulated Moving Bed (ISMB)

ISMB is fully known as intermittent SMB. In the ISMB process, as shown in Figure 13, each switching time t_s is divided into two sub steps with durations αt_s and $(1-\alpha)t_s$, respectively. In step 1, the ISMB also contains two inlets and two outlets, while zone IV is isolated. In step 2, all inlets and outlets are closed, and the liquid phase circulates along the column with the same flow rate in all four zones, thus redistributing the concentration profiles and adjusting the position of each component [84,85].

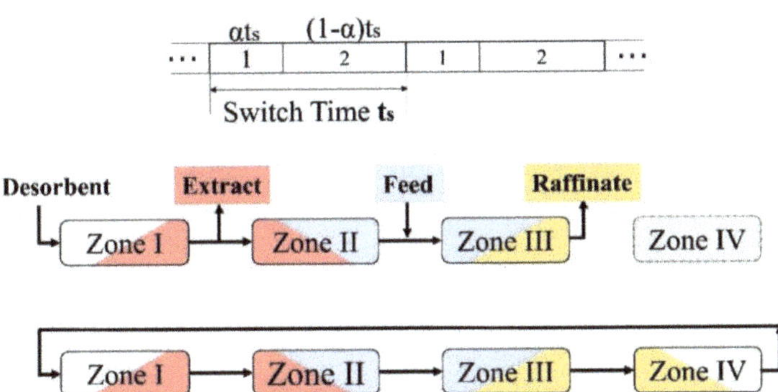

Figure 13. Schematic diagram of the ISMB process with port switching occurring at the end of Step 2 and the beginning of Step 1.

Mazzotti et al. [86–88] proposed a three-column ISMB as a new semi-continuous chromatographic process. Higher throughput can be achieved while using fewer columns due to the timely recycling of less retained components. The experiments were conducted for the binary separation process, and the ternary separation by using the three-column ISMB cascade chromatography was studied and designed. The final product purity of up to 97.8% was obtained with a productivity of 2.10 g/L/h and a solvent consumption of 12 g/L, which proved better than that of the conventional SMB process. The cascade operation could provide greater flexibility, better simulation accuracy, and improved purity and performance of the ISMB.

In conclusion, better concentration distribution could be obtained by reasonably designing the time interval in the ISMB process. In addition, complex separation tasks such as ternary separation can be conducted, which indicates the great application potential of ISMB. In summary, the main advantages of ISMB are: (1) High separation efficiency and performance can be obtained with a simple operation mode. (2) High operating flexibility. (3) The feasibility in multiple mixtures' separation process.

4.5. Sequential Simulated Moving Bed (SSMB)

SSMB is called sequential simulated moving bed, and its process is shown in Figure 14. SSMB divides the conventional SMB into three phases: "feeding", "circulation" and "elution". In the feeding phase, the feed solution and eluent, respectively, enter zones III and I simultaneously, while zones II and IV are isolated. In the second phase, all inlets and outlets are closed, and all the columns are connected into a closed loop. The liquid phase is circulated in this system and redistributes the concentration profiles. In the third "elution" phase, the eluent is passed into zones I to III, while zones IV is isolated. The extract products are collected in both the feed and elution phases, and the raffinate products are collected only in the feed phase [26,89].

For example, Li et al. [26,90] applied SSMB to the separation of glucose and fructose and compared the separation performance to the conventional SMB. The results revealed that the solvent consumption of SSMB was significantly less than that of SMB for the same purity and recovery requirements, which further proved the technical and economic superiority of the SSMB process [90]. In addition, the feasibility of SSMB for the separation and purification of xylo-oligosaccharides (XOS) under different constraints and objectives with multi-objective optimization was also investigated.

SSMB is not only an improvement of the conventional SMB, but also a modification of the ISMB. The main advantages are: (1) High utilization of the mobile phase and low water consumption can effectively reduce the cost, so the SSMB is suitable for industrial production. (2) When separating some specific mixtures, the separation performance of SSMB is significantly higher than that of SMB. (3) The back mixing problem existing in the separation process of SMB can be effectively solved by SSMB. (4) Ternary separation can be achieved by adding the inlet and outlet ports. In addition, SSMB also possesses some disadvantages: (1) The operation is more complicated and increases the control difficulty. (2) The utilization rate of the stationary phase is lower. (3) The flow rate ratio (m value) is influenced by various factors and is not constant during the switching time, which indicates that the SSMB cannot be directly designed by using the m value.

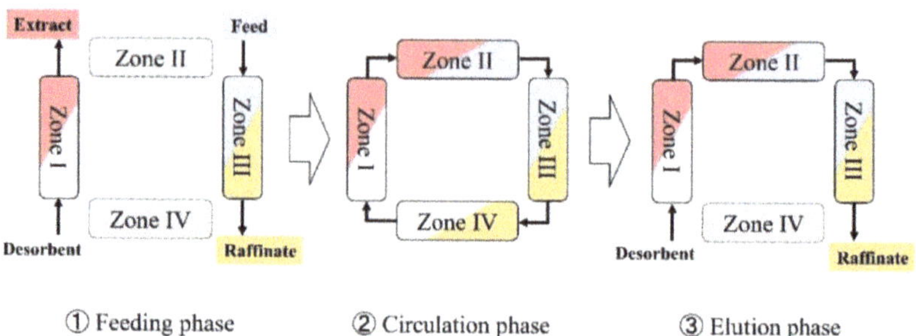

Figure 14. Schematic diagram of the SSMB process.

4.6. Pseudo-SMB

Pseudo-SMB is a new SMB technology mainly used to separate ternary mixtures. The process can be seen as a combination of the true moving bed and simulated moving bed. There are two main steps. The first step is similar to the TMB process, where the ternary mixture (A, B and C) is injected into the inlet and the desorbent (D1) is also injected, with the aim of separating component B, which has an intermediate affinity for the desorbent. The second step is similar to the SMB process, and closes the device with only the desorbent (D2) injected from the inlet in order to collect the components A and C, respectively. In the process of separating ternary mixtures, the pseudo-SMB is relatively easy to operate and has advantages for small-scale ternary separation [26,56,91–93].

4.7. Outlet Swing Stream (OSS) SMB

Gomes et al. [94] proposed an unusual SMB in which the flow rates of zones II and III were kept constant based on the conventional SMB, and the flow rates of each outlet (extract, raffinate and desorbent) were artificially manipulated to dynamically adjust the flow rates of zones I and IV for the separation operation. It can effectively improve the product purity and reduce the desorbent consumption [26].

4.8. Backfill-SMB (BF-SMB)

To improve the separation and chromatography performance of conventional SMB, Kim et al. [95] proposed a strategy called backfill-simulated moving bed (BF-SMB). A part of the product is refilled into the SMB from the feed node or intermediate node as a feed to simulate a TMB-like effect, enriching the main components near the product extraction node, thereby improving the separation of the SMB performance. This strategy can effectively improve product purity without compromising recovery and desorbent consumption [17,95].

4.9. SMB Cascades

Complex separation tasks such as the separation of ternary or more multiple mixtures can be handled by properly connecting multiple SMBs working in succession. This process is called SMB cascades, also known as tandem SMB. It is generally operated using two (or more) consecutive SMB units. A ternary (or multiple) mixture is fed from the inlet into the first SMB unit, and after separating a single material the remaining mixture is introduced into the second SMB unit for further separation. If the initial feed has only three mixtures, two SMB cascades will separate them all; if there are more than three, a third SMB unit will need to be passed through, and so on. Since each unit is independent of the others, it is possible to set independent parameters without unit interfering. After the cascade operation, the separation performance is greatly improved compared to the conventional SMB. However, it is important to note that the more SMBs are cascaded, the more diluted the sample becomes, and thus the productivity is reduced. Therefore, the idea of bypass SMB, as mentioned before, can be used for proper priming to ensure sample purity [26,96,97].

4.10. SimCon

Song et al. [98] proposed a novel SMB strategy called SimCon. Under the constraints of the maximum allowable pressure or flow rate, the feed flow and product flow are simultaneously controlled. This operation can make the flow and pressure fluctuations in the column as small as possible to improve the performance of the SMB. The SimCon operation consists of three steps. In the first step, when the desorbent is injected, only the raffinate port is opened. In the second step, all inlet and outlet ports are opened, which is consistent with the conventional SMB process and is called an intermediate step. In the final step, only the raffinate port is closed, and the other ports are opened [17]. Experimental data have confirmed that, compared with conventional SMB, SimCon operation can increase the product purity by 3.2%, the recovery rate by 3.1%, the productivity by 0.9 g/L/h, and the desorbent consumption by 0.04 L/g [98]. The separation performance and process cost can be effectively optimized by the SimCon strategy.

In conclusion, changing the different feeding or operating modes can effectively improve the separation efficiency, reduce the solvent consumption, or increase the product purity; a brief description of three main variants is listed in Table 1. Through these changes, while improving the performance of the equipment, SMB can also handle more challenging separation tasks, improve flexibility, or make the operation simple.

Table 1. A summary and comparison of three SMB variants.

	Modification Mechanism	Switching Mode	Advantage	Disadvantage	Classification
Zone variant	Non-essential functional zones are reduced by combing or deleting one or several columns.	Three-zone SMB possesses a unique three-step switching mode; the others' is similar to the conventional SMB.	The economical efficiency is improved due to the simplified device and the better separation performance.	The solid and liquid phases cannot be adequately regenerated and recycled.	One-column SMB, two-zone SMB, three-zone SMB, bypass SMB, and SMBs with more than four zones.
Gradient variant	The adsorption behavior of each zone is adjusted by introducing a gradient parameter.	Similar to conventional SMB switching mode, basically.	Higher productivity, purity, and lower desorbent consumption can be achieved.	Poor stability, high design difficulty, may only be used under limited conditions.	Concentration gradient, temperature gradient and pressure gradient.
Feed or operation variant	The feed or operation mode is changed without altering the SMB configuration.	For VariCol mode, only one inlet or outlet is switched within each switching time, and each port is switched independently. SSMB has unique three step switching mode. Others' are similar to that of conventional SMB.	High utilization of mobile phase and low water consumption. Complicated and multiple systems can by separated by this variant.	The operation and optimization works become more complex due to the internal instability.	ModiCon, VariCol, PowerFeed, ISMB, SSMB, Pseudo-SMB, OSS, BF-SMB, SMB cascades, SimCon.

5. Conclusions

In this paper, three significant types of SMB variants were introduced and analyzed. First, modifications of the conventional SMB process based on zone structure changes were reviewed. In most zone variants, the separation performance and process economy could be improved by simplifying the operating zone construction. Secondly, gradient variants of the SMB process were investigated, in which SMB's performance was effectively enhanced by introducing concentration, temperature, or pressure gradients with a result of altering the adsorption behavior of each zone. Finally, the SMB variants with different feed or operation modes were researched. This revealed that the separation performance could be adjusted by using ModiCon, VariCol and PowerFeed modes, alone or in combination, or choosing new SMB technologies such as ISMB, SSMB, Pseudo-SMB, OSS, BF-SMB, SMB cascades, and SimCon. According to the literature review and analysis results, it can be concluded that: (1) The use of new SMB technology or the combination of ModiCon, VariCol, and PowerFeed modes have a promising application in the future. (2) Multi-component separation by using new SMB technology will also be an important and challenging research direction. (3) The combination of the new SMB methods and multi-objective optimization (MOO) strategy can effectively improve the separation performance of SMB, which will simplify the process design and provide valuable guidance for practical industrial applications.

Author Contributions: Conceptualization, X.Z. and Y.L.; validation, Y.L., J.L. and A.K.R.; investigation, X.Z. and Y.L.; writing—original draft preparation, X.Z.; writing—review and editing, Y.L. and A.K.R.; supervision, Y.L., J.L. and A.K.R.; funding acquisition, Y.L. and A.K.R. All authors have read and agreed to the published version of the manuscript.

Funding: This research was funded by [NSERC Canada], grant number [RP438568]; National Natural Science Foundation of China, grant number [22268031]; Natural Science Foundation of the Inner Mongolia Autonomous Region, grant number [2021BS02003]; Basic Research Funding for Universities Directly Under Inner Mongolia Autonomous Region, grant number [JY20220212].

Data Availability Statement: Not applicable.

Acknowledgments: The authors express their great thanks for the support from the NSERC Canada (Grant No. RP438568), National Natural Science Foundation of China (Grant No. 22268031), the Natural Science Foundation of the Inner Mongolia Autonomous Region (Grant No. 2021BS02003), and the Basic Research Funding for Universities Directly Under Inner Mongolia Autonomous Region (Grant No. JY20220212).

Conflicts of Interest: The authors declare no conflict of interest.

References

1. Ruthven, D.M.; Ching, C.B. Counter-current and simulated counter-current adsorption separation processes. *Chem. Eng. Sci.* **1989**, *44*, 1011–1038. [CrossRef]
2. Deveant, R.M.; Jonas, R.; Schulte, M.; Keil, A.; Charton, F. Enantiomer Separation of a Novel Ca-Sensitizing Drug by simulated moving bed (SMB)-chromatography. *J. Für Prakt. Chem./Chem.-Ztg.* **1997**, *339*, 315–321. [CrossRef]
3. Francotte, E.; Richert, P.; Mazzotti, M.; Morbidelli, M. Simulated moving bed chromatographic resolution of a chiral antitussive. *J. Chromatogr. A* **1998**, *796*, 239–248. [CrossRef]
4. Juza, M.; Mazzotti, M.; Morbidelli, M. Simulated moving-bed chromatography and its application to chirotechnology. *Trends Biotechnol.* **2000**, *18*, 108–118. [CrossRef]
5. Nagy, M.Z.M.; Hanák, L.; Argyelán, J.; Szánya, T.; Ravasz, B.; Aranyi, A. Temesvári Separation of Organic Compounds by Simulated Moving Bed Preparative Liquid Chromatography. *Chromatographia* **2004**, *60*, S181–S187. [CrossRef]
6. Houwing, J.; Billiet, H.A.H.; van der Wielen, L.A.M. Optimization of azeotropic protein separations in gradient and isocratic ion-exchange simulated moving bed chromatography. *J. Chromatogr. A* **2002**, *944*, 189–201. [CrossRef] [PubMed]
7. Antos, D.; Seidel-Morgenstern, A. Two-step solvent gradients in simulated moving bed chromatography. *J. Chromatogr. A* **2002**, *944*, 77–91. [CrossRef]
8. Szanya, T.; Argyelan, J.; Kovats, S.; Hanak, L. Separation of steroid compounds by overloaded preparative chromatography with precipitation in the fluid phase. *J. Chromatogr. A* **2001**, *908*, 265–272. [CrossRef] [PubMed]
9. Pynnonen, B. Simulated moving bed processing: Escape from the high-cost box. *J. Chromatogr. A* **1998**, *827*, 143–160. [CrossRef]
10. Zhong, G.; Guiochon, G. Steady-state analysis of simulated moving-bed chromatography using the linear, ideal model. *Chem. Eng. Sci.* **1998**, *53*, 1121–1130. [CrossRef]

11. Pais, L.S.; Loureiro, J.; Rodrigues, A.E. Separation of 1,1′-bi-2-naphthol enantiomers by continuous chromatography in simulated moving bed. *Chem. Eng. Sci.* **1997**, *52*, 245–257. [CrossRef]

12. Wu, D.J.; Xie, Y.; Ma, Z.; Wang, N.H.L. Design of Simulated Moving Bed Chromatography for Amino Acid Separations. *Ind. Eng. Chem. Res.* **1998**, *37*, 4023–4035. [CrossRef]

13. Zhang, Y.; Hidajat, K.; Ray, A.K. Modified reactive SMB for production of high concentrated fructose syrup by isomerization of glucose to fructose. *Biochem. Eng. J.* **2007**, *35*, 341–351. [CrossRef]

14. Yu, W.; Hidajat, K.; Ray, A.K. Optimal operation of reactive simulated moving bed and Varicol systems. *J. Chem. Technol. Biotechnol.* **2003**, *78*, 287–293. [CrossRef]

15. Rajendran, A.; Paredes, G.; Mazzotti, M. Simulated moving bed chromatography for the separation of enantiomers. *J. Chromatogr. A* **2009**, *1216*, 709–738. [CrossRef]

16. Kim, Y.; Cho, S.; Jang, K.; Lee, J.; Kim, M.; Moon, I. Effect of radial distribution of injected flow on simulated moving bed performance. *J. Chromatogr. A* **2022**, *1662*, 462703. [CrossRef]

17. Kim, K.-M.; Lee, J.W.; Kim, S.; Santos da Silva, F.V.; Seidel-Morgenstern, A.; Lee, C.-H. Advanced Operating Strategies to Extend the Applications of Simulated Moving Bed Chromatography. *Chem. Eng. Technol.* **2017**, *40*, 2163–2178. [CrossRef]

18. Toumi, A.; Engell, S.; Ludemann-Hombourger, O.; Nicoud, R.M.; Bailly, M. Optimization of simulated moving bed and Varicol processes. *J. Chromatogr. A* **2003**, *1006*, 15–31. [CrossRef] [PubMed]

19. Calderon Supelano, R.; Barreto, A.G., Jr.; Andrade Neto, A.S.; Secchi, A.R. One-step optimization strategy in the simulated moving bed process with asynchronous movement of ports: A VariCol case study. *J. Chromatogr. A* **2020**, *1634*, 461672. [CrossRef]

20. Zang, Y.; Wankat, P.C. Variable Flow Rate Operation for Simulated Moving Bed Separation Systems: Simulation and Optimization. *Ind. Eng. Chem. Res.* **2003**, *42*, 4840–4848. [CrossRef]

21. Wang, J.; Tian, Y.; Li, Y.; Xu, J.; Yu, W.; Ray, A.K. Multi-objective optimization of non-isothermal simulated moving bed reactor: Methyl acetate synthesis. *Chem. Eng. J.* **2020**, *395*, 125041. [CrossRef]

22. Wei, F.; Shi, L.; Wang, Q.; Zhao, Y. Fast and accurate separation of the paclitaxel from yew extracum by a pseudo simulated moving bed with solvent gradient. *J. Chromatogr. A* **2018**, *1564*, 120–127. [CrossRef] [PubMed]

23. Cristancho, C.A.M.; Seidel-Morgenstern, A. Purification of single-chain antibody fragments exploiting pH-gradients in simulated moving bed chromatography. *J. Chromatogr. A* **2016**, *1434*, 29–38. [CrossRef]

24. Li, Y.; Xu, J.; Yu, W.; Ray, A.K. Multi-objective optimization of sequential simulated moving bed for the purification of xylo-oligosaccharides. *Chem. Eng. Sci.* **2020**, *211*, 115279. [CrossRef]

25. Aniceto, J.P.S.; Silva, C.M. Simulated moving bed strategies and designs: From established systems to the latest developments. *Sep. Purif. Rev.* **2015**, *44*, 41–73. [CrossRef]

26. Faria, R.P.V.; Rodrigues, A.E. Instrumental aspects of simulated moving bed chromatography. *J. Chromatog-Raphy A* **2015**, *1421*, 82–102. [CrossRef]

27. Lee, C.G.; Jo, C.Y.; Song, Y.J.; Mun, S. Continuous-mode separation of fucose and 2,3-butanediol using a three-zone simulated moving bed process and its performance improvement by using partial extract-collection, partial extract-recycle, and partial desorbent-port closing. *J. Chromatogr. A* **2018**, *1579*, 49–59. [CrossRef]

28. Ching, C.B.; Chu, K.H.; Hidajat, K.; Uddin, M.S. Comparative study of flow schemes for a simulated countercurrent adsorption separation process. *AIChE J.* **1992**, *38*, 1744–1750. [CrossRef]

29. Hur, J.S.; Wankat, P.C. Two-Zone SMB/Chromatography for Center-Cut Separation from Ternary Mixtures: Linear Isotherm Systems. *Ind. Eng. Chem. Res.* **2006**, *45*, 1426–1433. [CrossRef]

30. Mun, S. Relationship between desorbent usage and the recovery of a target product in three-zone simulated moving bed processes designed under the conditions of positive and negative flow-rate-ratios of liquid to solid phases. *J. Chromatogr. A* **2019**, *1603*, 388–395. [CrossRef]

31. Ching, C.B.; Ruthven, D.M. Analysis of the performance of a simulated counter-current chromatographic system for fructose-glucose separation. *Can. J. Chem. Eng.* **1984**, *62*, 398–403. [CrossRef]

32. Barker, P.E.; Knoechelmann, A.; Ganetsos, G. Simulated counter-current moving column chromatography used in the continuous separation of carbohydrate mixtures. *Chromatographia* **1990**, *29*, 161–166. [CrossRef]

33. Hashimoto, K.; Yamada, M.; Adachi, S.; Shirai, Y. A simulated moving-bed adsorber with three zones for continuous separation of L-phenylalanine and NaCl. *J. Chem. Eng. Jpn.* **1989**, *22*, 432–434. [CrossRef]

34. Abunasser, N.; Wankat, P.C.; Kim, Y.S.; Koo, Y.M. One-column chromatograph with recycle analogous to a four-zone simulated moving bed. *Ind. Eng. Chem. Res.* **2003**, *42*, 5268–5279. [CrossRef]

35. Abunasser, N.; Wankat, P.C. One-column chromatograph with recycle analogous to simulated moving bed adsorbers: Analysis and applications. *Ind. Eng. Chem. Res.* **2004**, *43*, 5291–5299. [CrossRef]

36. Lee, K. Two-Section Simulated Moving-Bed Process. *Sep. Sci. Technol.* **2000**, *35*, 519–534. [CrossRef]

37. Jin, W.; Wankat, P.C. Two-Zone SMB Process for Binary Separation. *Ind. Eng. Chem. Res.* **2005**, *44*, 1565–1575. [CrossRef]

38. Nam, H.-G.; Mun, S. Optimal design and experimental validation of a three-zone simulated moving bed process based on the Amberchrom-CG161C adsorbent for continuous removal of acetic acid from biomass hydrolyzate. *Process Biochem.* **2012**, *47*, 725–734. [CrossRef]

39. Zang, Y.; Wankat, P.C. Three-Zone Simulated Moving Bed with Partial Feed and Selective Withdrawal. *Ind. Eng. Chem. Res.* **2002**, *41*, 5283–5289. [CrossRef]

40. Peng, B.; Wang, S. Separation of p-xylene and m-xylene by simulated moving bed chromatography with MIL-53(Fe) as stationary phase. *J. Chromatogr. A* **2022**, *1673*, 463091. [CrossRef]
41. Kung, H.-C.; Liang, K.-Y.; Mutuku, J.K.; Huang, B.-W.; Chang-Chien, G.-P. Separation and purification of caulerpin from algal *Caulerpa racemosa* by simulated moving bed chromatography. *Food Bioprod. Process.* **2021**, *130*, 14–22. [CrossRef]
42. Wei, B.; Wang, S. Separation of eicosapentaenoic acid and docosahexaenoic acid by three-zone simulated moving bed chromatography. *J. Chromatogr. A* **2020**, *1625*, 461326. [CrossRef] [PubMed]
43. Shen, B.; Chen, M.; Jiang, H.; Zhao, Y.; Wei, F. Modeling Study on a Three-Zone Simulated Moving Bed without Zone I. *Sep. Sci. Technol.* **2011**, *46*, 695–701. [CrossRef]
44. Kim, K.M.; Song, J.Y.; Lee, C.H. Three-port operation in three-zone simulated moving bed chromatography. *J. Chromatogr. A* **2014**, *1340*, 79–89. [CrossRef]
45. Keβler, L.C.; Gueorguieva, L.; Rinas, U.; Seidel-Morgenstern, A. Step gradients in 3-zone simulated moving bed chromatography. Application to the purification of antibodies and bone morphogenetic protein-2. *J. Chromatogr. A* **2007**, *1176*, 69–78. [CrossRef]
46. Song, S.-M.; Park, M.-B.; Kim, I.H. Three-zone simulated moving-bed (SMB) for separation of cytosine and guanine. *Korean J. Chem. Eng.* **2012**, *29*, 952–958. [CrossRef]
47. Song, S.-M.; Kim, I.H. A three-zone simulated moving-bed for separation of immunoglobulin Y. *Korean J. Chem. Eng.* **2013**, *30*, 1527–1532. [CrossRef]
48. Cunha, F.C.; Secchi, A.R.; de Souza, M.B., Jr.; Barreto, A.G., Jr. Separation of praziquantel enantiomers using simulated moving bed chromatographic unit with performance designed for semipreparative applications. *Chirality* **2019**, *31*, 583–591. [CrossRef]
49. Yao, C.; Chen, J.; Lu, Y.; Tang, S.; Fan, E. Construction of an asynchronous three-zone simulated-moving-bed chromatography and its application for the separation of vanillin and syringaldehyde. *Chem. Eng. J.* **2018**, *331*, 644–651. [CrossRef]
50. Tangpromphan, P.; Duangsrisai, S.; Jaree, A. Development of separation method for Alpha-Tocopherol and Gamma-Oryzanol extracted from rice bran oil using Three-Zone simulated moving bed process. *Sep. Purif. Technol.* **2021**, *272*, 118930. [CrossRef]
51. Tangpromphan, P.; Budman, H.; Jaree, A. A simplified strategy to reduce the desorbent consumption and equipment installed in a three-zone simulated moving bed process for the separation of glucose and fructose. *Chem. Eng. Process.-Process Intensif.* **2018**, *126*, 23–37. [CrossRef]
52. Nam, H.-G.; Park, C.; Jo, S.-H.; Suh, Y.-W.; Mun, S. Continuous separation of succinic acid and lactic acid by using a three-zone simulated moving bed process packed with Amberchrom-CG300C. *Process Biochem.* **2012**, *47*, 2418–2426. [CrossRef]
53. Maruyama, R.T.; Karnal, P.; Sainio, T.; Rajendran, A. Design of bypass-simulated moving bed chromatography for reduced purity requirements. *Chem. Eng. Sci.* **2019**, *205*, 401–413. [CrossRef]
54. Xie, Y.; Chin, C.Y.; Phelps, D.S.C.; Lee, C.H.; Lee, K.B.; Mun, S.; Wang, N.H.L. A Five-Zone Simulated Moving Bed for the Isolation of Six Sugars from Biomass Hydrolyzate. *Ind. Eng. Chem. Res.* **2005**, *44*, 9904–9920. [CrossRef]
55. Wooley, R.; Ma, Z.; Wang, N.H.L. A Nine-Zone Simulating Moving Bed for the Recovery of Glucose and Xylose from Biomass Hydrolyzate. *Ind. Eng. Chem. Res.* **1998**, *37*, 3699–3709. [CrossRef]
56. Jiang, C.; Huang, F.; Wei, F. A pseudo three-zone simulated moving bed with solvent gradient for quaternary separations. *J. Chromatogr. A* **2014**, *1334*, 87–91. [CrossRef]
57. Antos, D.; Seidel-Morgenstern, A. Application of gradients in the simulated moving bed process. *Chem. Eng. Sci.* **2001**, *56*, 6667–6682. [CrossRef]
58. Abel, S.; Mazzotti, M.; Morbidelli, M. Solvent gradient operation of simulated moving beds. *J. Chromatogr. A* **2002**, *944*, 23–39. [CrossRef]
59. Abel, S.; Mazzotti, M.; Morbidelli, M. Solvent gradient operation of simulated moving beds. 2. Langmuir isotherms. *J. Chromatogr. A* **2004**, *1026*, 47–55. [CrossRef]
60. Mun, S.; Wang, N.-H.L. Optimization of productivity in solvent gradient simulated moving bed for paclitaxel purification. *Process Biochem.* **2008**, *43*, 1407–1418. [CrossRef]
61. Mun, S. Optimal Design of Solvent Gradient Simulated Moving Bed Chromatography for Amino Acid Separation. *J. Liq. Chromatogr. Relat. Technol.* **2011**, *34*, 1518–1535. [CrossRef]
62. Kim, J.K.; Abunasser, N.; Wankat, P.C.; Stawarz, A.; Koo, Y.-M. Thermally Assisted Simulated Moving Bed Systems. *Adsorption* **2005**, *11*, 579–584. [CrossRef]
63. Jin, W.; Wankat, P.C. Thermal Operation of Four-Zone Simulated Moving Beds. *Ind. Eng. Chem. Res.* **2007**, *46*, 7208–7220. [CrossRef]
64. Migliorini, C.; Wendlinger, M.; Mazzotti, M.; Morbidelli, M. Temperature Gradient Operation of a Simulated Moving Bed Unit. *Ind. Eng. Chem. Res.* **2001**, *40*, 2606–2617. [CrossRef]
65. Jiang, X.; Zhu, L.; Yu, B.; Su, Q.; Xu, J.; Yu, W. Analyses of simulated moving bed with internal temperature gradients for binary separation of ketoprofen enantiomers using multi-objective optimization: Linear equilibria. *J. Chromatogr. A* **2018**, *1531*, 131–142. [CrossRef]
66. Mazzotti, M.; Storti, G.; Morbidelli, M. Supercritical fluid simulated moving bed chromatography. *J. Chromatogr. A* **1997**, *786*, 309–320. [CrossRef]
67. Schramm, H.; Kaspereit, M.; Kienle, A.; Seidel-Morgenstern, A. Improving Simulated Moving Bed Processes by Cyclic Modulation of the Feed Concentration. *Chem. Eng. Technol.* **2002**, *25*, 1151–1155. [CrossRef]

68. Schramm, H.; Kienle, A.; Kaspereit, M.; Seidel-Morgenstern, A. Improved operation of simulated moving bed processes through cyclic modulation of feed flow and feed concentration. *Chem. Eng. Sci.* **2003**, *58*, 5217–5227. [CrossRef]

69. Lübke, R.; Seidel-Morgenstern, A.; Tobiska, L. Numerical method for accelerated calculation of cyclic steady state of ModiCon–SMB-processes. *Comput. Chem. Eng.* **2007**, *31*, 258–267. [CrossRef]

70. Ludemann-Hombourger, O.; Nicoud, R.M.; Bailly, M. The "VARICOL" Process: A New Multicolumn Continuous Chromatographic Process. *Sep. Sci. Technol.* **2000**, *35*, 1829–1862. [CrossRef]

71. Zhang, Z.; Hidajat, K.; Ray, A.K.; Morbidelli, M. Multiobjective optimization of SMB and varicol process for chiral separation. *AIChE J.* **2002**, *48*, 2800–2816. [CrossRef]

72. Zhang, Y.; Hidajat, K.; Ray, A.K. Multi-objective optimization of simulated moving bed and Varicol processes for enantioseparation of racemic pindolol. *Sep. Purif. Technol.* **2009**, *65*, 311–321. [CrossRef]

73. Zhang, Y.; Hidajat, K.; Ray, A.K. Enantio-separation of racemic pindolol on α1-acid glycoprotein chiral stationary phase by SMB and Varicol. *Chem. Eng. Sci.* **2007**, *62*, 1364–1375. [CrossRef]

74. Zhang, Z.; Mazzotti, M.; Morbidelli, M. Multiobjective optimization of simulated moving bed and Varicol processes using a genetic algorithm. *J. Chromatogr. A* **2003**, *989*, 95–108. [CrossRef]

75. Matos, J.; Faria, R.P.V.; Loureiro, J.M.; Ribeiro, A.M.; Nogueira, I.B.R. Design and Optimization for Simulated Moving Bed: Varicol Approach. *IFAC-Pap.* **2021**, *54*, 542–547. [CrossRef]

76. Calderon Supelano, R.; Barreto, A.G., Jr.; Secchi, A.R. Optimal performance comparison of the simulated moving bed process variants based on the modulation of the length of zones and the feed concentration. *J. Chromatogr. A* **2021**, *1651*, 462280. [CrossRef]

77. Lin, X.; Gong, R.; Li, J.; Li, P.; Yu, J.; Rodrigues, A.E. Enantioseparation of racemic aminoglutethimide using asynchronous simulated moving bed chromatography. *J. Chromatogr. A* **2016**, *1467*, 347–355. [CrossRef]

78. Kim, K.; Kim, J.I.; Park, H.; Koo, Y.M.; Lee, K.S. Bi-level optimizing control of a simulated moving bed process with nonlinear adsorption isotherms. *J. Chromatogr. A* **2011**, *1218*, 6843–6847. [CrossRef]

79. Vignesh, S.V.; Hariprasad, K.; Athawale, P.; Bhartiya, S. An optimization-driven novel operation of simulated moving bed chromatographic separation. *IFAC-Pap.* **2016**, *49*, 165–170. [CrossRef]

80. Zhang, Z.; Mazzotti, M.; Morbidelli, M. Continuous chromatographic processes with a small number of columns: Comparison of simulated moving bed with Varicol, PowerFeed, and ModiCon. *Korean J. Chem. Eng.* **2004**, *21*, 454–464. [CrossRef]

81. Kloppenburg, E.; Gilles, E.D. A New Concept for Operating Simulated Moving-Bed Processes. *Chem. Eng. Technol.* **1999**, *22*, 813–817. [CrossRef]

82. Zhang, Z.; Mazzotti, M.; Morbidelli, M. PowerFeed operation of simulated moving bed units: Changing flow-rates during the switching interval. *J. Chromatogr. A* **2003**, *1006*, 87–99. [CrossRef] [PubMed]

83. Kawajiri, Y.; Biegler, L.T. Optimization strategies for simulated moving bed and PowerFeed processes. *AIChE J.* **2006**, *52*, 1343–1350. [CrossRef]

84. Song, M.; Cui, L.; Kuang, H.; Zhou, J.; Yang, P.; Zhuang, W.; Chen, Y.; Liu, D.; Zhu, C.; Chen, X.; et al. Model-based design of an intermittent simulated moving bed process for recovering lactic acid from ternary mixture. *J. Chromatogr. A* **2018**, *1562*, 47–58. [CrossRef] [PubMed]

85. Katsuo, S.; Mazzotti, M. Intermittent simulated moving bed chromatography: 2. Separation of Troger's base enantiomers. *J. Chromatogr. A* **2010**, *1217*, 3067–3075. [CrossRef]

86. Jermann, S.; Mazzotti, M. Three column intermittent simulated moving bed chromatography: 1. Process description and comparative assessment. *J. Chromatogr. A* **2014**, *1361*, 125–138. [CrossRef]

87. Jermann, S.; Meijssen, M.; Mazzotti, M. Three column intermittent simulated moving bed chromatography: 3. Cascade operation for center-cut separations. *J. Chromatogr. A* **2015**, *1378*, 37–49. [CrossRef]

88. Jermann, S.; Alberti, A.; Mazzotti, M. Three-column intermittent simulated moving bed chromatography: 2. Experimental implementation for the separation of Troger's Base. *J. Chromatogr. A* **2014**, *1364*, 107–116. [CrossRef]

89. Karlsson, S.; Pettersson, F.; Skrifvars, H.; Westerlund, T. Optimizing the Operation of a Sequential-Simulated Moving-Bed Separation Process Using MINLP. *Comput. Aided Chem. Eng.* **2000**, *8*, 463–468. [CrossRef]

90. Li, Y.; Ding, Z.; Wang, J.; Xu, J.; Yu, W.; Ray, A.K. A comparison between simulated moving bed and sequential simulated moving bed system based on multi-objective optimization. *Chem. Eng. Sci.* **2020**, *219*, 115562. [CrossRef]

91. Wei, F.; Shen, B.; Chen, M.; Zhao, Y. Study on a pseudo-simulated moving bed with solvent gradient for ternary separations. *J. Chromatogr. A* **2012**, *1225*, 99–106. [CrossRef]

92. Kurup, A.S.; Hidajat, K.; Ray, A.K. Optimal operation of a Pseudo-SMB process for ternary separation under non-ideal conditions. *Sep. Purif. Technol.* **2006**, *51*, 387–403. [CrossRef]

93. Borges, D.A.; Silva, E.A.; Rodrigues, A.E. Design of chromatographic multicomponent separation by a pseudo-simulated moving bed. *AIChE J.* **2006**, *52*, 3794–3812. [CrossRef]

94. Sá Gomes, P.; Rodrigues, A.E. Outlet Streams Swing (OSS) and MultiFeed Operation of Simulated Moving Beds. *Sep. Sci. Technol.* **2007**, *42*, 223–252. [CrossRef]

95. Kim, K.M.; Lee, C.H. Backfill-simulated moving bed operation for improving the separation performance of simulated moving bed chromatography. *J. Chromatogr. A* **2013**, *1311*, 79–89. [CrossRef]

96. Kim, J.K.; Wankat, P.C. Designs of simulated-moving-bed cascades for quaternary separations. *Ind. Eng. Chem. Res.* **2004**, *43*, 1071–1080. [CrossRef]

97. Wankat, P.C. Simulated moving bed cascades for ternary separations. *Ind. Eng. Chem. Res.* **2001**, *40*, 6185–6193. [CrossRef]
98. Song, J.Y.; Kim, K.M.; Lee, C.H. High-performance strategy of a simulated moving bed chromatography by simultaneous control of product and feed streams under maximum allowable pressure drop. *J. Chromatogr. A* **2016**, *1471*, 102–117. [CrossRef]

MDPI

St. Alban-Anlage 66

4052 Basel

Switzerland

Tel. +41 61 683 77 34

Fax +41 61 302 89 18

www.mdpi.com

Processes Editorial Office

E-mail: processes@mdpi.com

www.mdpi.com/journal/processes